S0-ALG-475

COMPUTER NUMERICAL CONTROL
OPERATION AND PROGRAMMING

SECOND EDITION

KELLY CURRAN
JON STENERSON

FOX VALLEY TECHNICAL COLLEGE

Prentice
Hall

Upper Saddle River, New Jersey
Columbus, Ohio

Library of Congress Cataloging-in-Publication Data

Computer numerical control : operation and programming / Kelly Curran,
    Jon Stenerson.
    p. cm.
    Includes index.
    Previous edition has main entry under Stenerson, Jon.
    ISBN 0-13-011980-6
    1. Machine-tools—Numerical control. 2. Machine-tools—Numerical
    control—Programming. I. Curran, Kelly. II. Stenerson, Jon. III. Title.
    TJ1189 .C87 2001
    621.9'023'0285—dc21
                                                            00-036763
                                                                 CIP

Vice President and Publisher: Dave Garza
Editor in Chief: Stephen Helba
Executive Editor: Ed Francis
Production Editor: Christine M. Buckendahl
Production Coordination: Janet Bixler, Custom Editorial Productions, Inc.
Design Coordinator: Karrie Converse-Jones
Cover Designer: Linda Fares
Cover Photo: FPG
Production Manager: Brian Fox
Marketing Manager: Jamie Van Voorhis

This book was set in Stone Serif by Custom Editorial Productions, Inc., and was printed and
bound by R. R. Donnelley & Sons Company. The cover was printed by Phoenix Color Corp.

**Copyright © 2001, 1997 by Prentice-Hall, Inc., Upper Saddle River, New Jersey
07458.** All rights reserved. Printed in the United States of America. This publication is
protected by copyright and permission should be obtained from the publisher prior to any
prohibited reproduction, storage in a retrieval system, or transmission in any form or by any
means, electronic, mechanical, photocopying, recording, or likewise. For information
regarding permission(s), write to: Rights and Permissions Department.

10   9   8   7   6   5   4   3   2   1
ISBN: 0-13-011980-6

To my wife, Sheryl, and my daughters, Megan and Melissa, for their patience, support, and inspiration.

Kelly Curran

To Jane.

Jon Stenerson

Contents

···

Preface

Greater world competition has forced manufacturers to stay current with technology. Over the past 20 years computer numerical control has revolutionized the way manufacturers produce their products. To compete in the global market, most manufacturers have seen the need to use computers to efficiently produce quality products. Computer numerical control (CNC), computer-aided manufacturing (CAM), and computer-integrated manufacturing (CIM) have gained a foothold in industry and have revolutionized the way manufacturers do business.

The ever-increasing popularity of CNC machines has created a need for people who are knowledgeable about CNC. There is a huge demand for people who are capable of programming, setting up, and operating CNC machine tools. They will continue to be in demand as long as they stay current with technology.

We decided to write this text when we were unable to find a practical, easy-to-understand book with enough examples and programming assignments. We saw gaps in the books that were available. We believe that students must have a firm, practical understanding of carbide tooling to use the capability of CNC machines. Wire-EDM technology and programming have been ignored in most texts. We also think it is essential to provide a basic, practical understanding of statistical quality control. We devoted two chapters to statistical quality control. Chapter 16 is devoted to an easy-to-understand examination of ISO 9000 because quality systems such as this have become prevalent in industry. There are many other differences that we hope you will appreciate as you use the book.

The information in this textbook is based on our years of experience teaching CNC courses to students at Fox Valley Technical College in Appleton, Wisconsin, as well as industrial CNC courses for local business and industry, university and technical school students, and on-the-job trainees. We sincerely hope that our logical, easy-to-understand approach will enable readers to accomplish more than would otherwise be possible.

Kelly Curran

Jon Stenerson

Acknowledgments

This text would not have been possible without the help we received from corporations, companies, and individuals. We would especially like to thank the following companies for their assistance.

Bridgeport Machines, Inc.

Fanuc LTD.

Giddings & Lewis, Inc.

Koike Aronson, Inc.

Kennametal, Inc.

Laser Lab, International, Inc.

Mori Seiki Co., Ltd.

Yamazaki Mazak Corp.

We would also like to especially thank Brian Meyer for his assistance with Chapter 12, *Electrical Discharge Machining (EDM)*.

Chapter 1

INTRODUCTION TO COMPUTER NUMERICAL CONTROL

INTRODUCTION

The trend toward automation of production equipment is putting great demands on people. Since the early 1970s manufacturers have worked to increase productivity, quality, process capability, reliability, and flexibility. In the early 1970s American manufacturers began to realize that they had to be competitive with foreign manufacturers if they were to survive. This meant using technologies that could improve quality and productivity.

OBJECTIVES

Upon completion of this chapter, the reader will be able to:

- *Describe the evolution of computer numerical control (CNC).*
- *Identify the different types of numerical control machines and their parts.*
- *Explain the advantage of CNC over NC.*
- *Describe the difference between point-to-point control and continuous path control.*
- *Describe a closed-loop system.*
- *Name three ways to load a program into a machine control.*
- *Describe the different axis coordinate systems.*
- *Identify positions on a Cartesian coordinate grid using absolute and incremental programming methods.*

HISTORY OF NUMERICAL CONTROL

Anyone working in the machine tool field cannot ignore the influence of the computer in manufacturing. The capabilities that these machine tools have given to the industry have forced managers and owners of companies to update their thinking to stay competitive. The inherent accuracy and repeatability of

1

these machine tools have helped quality process tools such as statistical process control gain a foothold in machine shops.

EVOLUTION OF THE NC/CNC MACHINE

Numerical control is nothing new. As early as 1808 weaving machines used metal cards with holes punched in them to control the pattern of the cloth being produced. Each needle on the machine was controlled by the presence or absence of a hole on the punched cards. The cards were the program for the machine. If the cards were changed, the pattern changed.

The player piano is also an example of numerical control. The player piano uses a roll of paper with holes punched in it. The presence or absence of a hole determined if that note was played. Air was used to sense whether a hole was present.

The invention of the computer was one of the turning points in numerical control. In 1943 the first computer, called ENIAC (Electronic Numerical Integrator and Computer) was built.

The ENIAC computer was very large. It occupied more than 1500 square feet and used approximately 18,000 vacuum tubes to do its calculations. The heat generated by the vacuum tubes was a constant problem. The computer could operate only a few minutes without a tube failing. In addition, the computer weighed many tons and was very difficult to program. ENIAC was programmed through the use of thousands of switches. The $15 calculator available today is much more powerful than this early attempt.

The real turning point in computer technology was the invention of the transistor in 1948. The transistor was the replacement for the vacuum tube. It was very small, cheap, dependable, used very little power, and generated very little heat: the perfect replacement for the vacuum tube. The transistor did not see much industrial use until the 1960s.

INTEGRATED CIRCUITRY

In 1959 a new technology emerged: integrated circuits (ICs). Integrated circuits were actually control circuits on a chip. When manufacturers discovered how to miniaturize circuits, it helped reduce the size and improve the dependability of electronic control even more than the transistor had. Large-scale integrated circuits first were produced in 1965.

In 1974 the microprocessor was invented. This made the microcomputer, and thus small applications, possible. Great strides in the manufacture of memory for computers helped make computers more powerful and affordable.

The original conception of numerically controlled machine tools occurred in the 1950s as a method of producing airfoils of great accuracy for the government.

These complex parts were made by manual machining methods and inspected by comparing them to templates. The templates also had to be manufactured by manual methods, which was very time consuming and inaccurate.

However, in a shop in Traverse City, Michigan, a man named John Parsons was working on a method to improve the production of inspection templates for helicopter rotor blades. Parsons started as a tool room apprentice and had no college degree. Parsons' method involved calculating the coordinate points along the airfoil surface. By calculating a large number of intermediate points and then manually moving the machine tool to each of these points, the accuracy of the templates was improved. Parsons came up with the idea of using punched cards for the many calculations. The data could then be used to position the machine tool. Parsons submitted a proposal to the Air Force to develop a machine to produce these templates and received a development contract in 1948. His first attempts at automatic position control used punch-card tabulating machines to calculate the positions along the airfoil curve and an ordinary manual milling machine to position the tool to the tabulated positions. He had two operators, one to move each axis of the machine. This method produced airfoils tens of times more accurate than the preceding method, but was still a very time-consuming process.

In 1949, the Air Force awarded Parsons a contract to produce a control system that could move the axis of a machine to calculated points automatically. The Massachusetts Institute of Technology (MIT) was subcontracted by Parsons to develop a motor that could control the axis of the machines. The servo motor was born.

Parsons envisioned the following system. A computer would calculate the path that the tool should follow and store that information on punched cards. A reader at the machine would then read the cards. The machine control would take the data from the reader and control the motors attached to each axis.

In 1951 MIT was awarded the prime contract to develop the machine control. The first machine produced by Parsons and MIT was demonstrated in 1952. Called a Cincinnati Hydrotel, it was a three-axis vertical spindle milling machine. The machine control used vacuum tubes.

One of the first attempts at making programming easier for people was called APT (Automatically Programmed Tool) Symbolic Language. APT, invented in 1954, used English-like symbolic language to produce a program that the machine tool could understand. Remember, a machine needs the geometry of the part and machining instructions such as speeds, feeds, and coolant to operate. APT made it easier for people to write these programs, which were then translated to a program that the machine could understand.

In 1955 the Air Force awarded $35 million in contracts to manufacture numerical control machines. The first numerically controlled machine tools were very bulky. The machine control was vacuum-tube operated and needed a separate computer to generate its binary tape codes. (Binary coding systems use

1s and 0s.) Programming complex parts took highly specialized people. Developments and refinements continued, and by the early 1960s numerical control machines became much more common in industry. As the acceptance of numerical control machines grew, they became easier to use and more powerful.

Up until about 1976 these machines were called NC (numerical control) machines. In 1976 CNC (computer numerical control) machines were produced. These machine controls used microprocessors to give them additional capability. They also featured additional memory. The NCs typically read one short program step (block) at a time and executed it; however, CNC machines could store whole programs.

Improvements in computer technology in the late 1970s and 1980s brought the cost of numerical control machines down to a level where most manufacturing companies cannot afford to be without them.

TYPES OF NUMERICAL CONTROL MACHINES

With the refinement of the computer and numerical systems, the applications of this technology have become more varied. The most common of these machine tools are the turning and machining centers; together, these two are more than half of the numerical control machines on the market. Other types of machine tools that have been married to the computer include ram and wire feed-type electrical discharge machines (EDMs); cylindrical, reciprocating, and creep-feed grinders; cut-off saws; flame cutting, laser cutting, plasma cutting, and water-jet cutting machines; and coordinate measuring machines.

One of the newer types of machines, the mill-turn center, incorporates live tooling in the turret of the turning center. This machine can turn complex parts and mill grooves and slots on cylindrical parts. Although each machine works a little bit differently, the basic method of programming remains the same. The standards and codes developed in the early years of numerical control still apply today.

CNC PLASMA MACHINES

The CNC plasma cutting machine uses numerical control to position a plasma cutting torch (see Figure 1–1). Plasma is created by passing a gas through an electric arc. The gas is ionized by the arc, and an extremely high-temperature cutting arc is produced, which allows it to cut ferrous and non-ferrous metals. CNC plasma cutting machines are extremely fast and will cut smooth, accurate parts from plate stock.

CNC SPRING FORMING MACHINES

Computer numerical control spring forming machines create springs by coiling flat or round spring steel into complex shapes through the use of dies (see Figure

FIGURE 1–1 *Computer numerical control plasma and oxy-fuel cutting machines integrate numerical control with flame cutting. (Courtesy of Koike Aronson, Inc.)*

1–2). The dies are positioned by numerical control and can be programmed for different shape and size springs. They can automatically produce many thousands of springs each hour.

CNC LASER CUTTING MACHINES

Computer numerical control laser cutting machines are very similar to plasma cutters. Laser cutting machines use coherent light as a cutting tool (see Figure 1–3). Laser cutting machines can cut plate stock into intricate shapes.

VERTICAL MACHINING CENTERS

Vertical machining centers are vertical milling machines that use numerical control positioning and automatic tool changers to produce complex machine parts in one setup (see Figure 1–4).

NSF-2U

FIGURE **1–2** *Computer numerical control spring forming machines are capable of making many different styles of springs with great accuracy and repeatability. (Courtesy of Asahi-Seiki Manufacturing Co., Ltd.)*

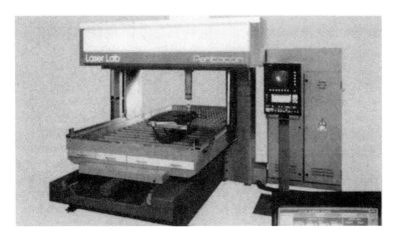

FIGURE **1–3** *Computer numerical control laser cutting machines couple the speed and accuracy of numerical control with the versatility of laser technology. (Courtesy of Laser Lab International, Inc.)*

FIGURE 1–4 *Computer numerical control vertical machining centers use automatic tool changing systems to enhance multiple machining processes. (Courtesy of Bridgeport Machines, Inc.)*

HORIZONTAL MACHINING CENTERS

Horizontal machining centers are horizontal milling machines that are numerically controlled (see Figure 1–5). Horizontal machining centers are equipped with automatic tool changers along with a variety of other features to increase their versatility and production capabilities.

VARIABLE AXES MACHINING CENTER

The variable axes machining center is capable of machining in six different directions simultaneously (see Figure 1–6). The VARIAX may be an example of what complex, multi-axis machines will look like in the future. The variable axes capability allows the VARIAX to move and position the cutting tool in literally any direction. This contouring ability makes the variable axes machine capable of machining complex parts in one setup.

CNC PRESS BRAKES

Computer-controlled press brakes are used to bend and form sheet metal into parts. For example, the steel control cabinet for a CNC machine is probably formed on a CNC press brake. These machines are programmed to control the down stroke of the ram, which helps control the angle of the bend. Interchangeable dies on the bottom also control the angle and programmable back

FIGURE 1–5 *Computer numerical control horizontal machining centers use automatic tool changers and their physical size to become the workhorses of the machine shop. (Courtesy of Giddings & Lewis, Inc.)*

stops control how the work is positioned before the bend. They are programmed in step-like fashion. The operator puts the sheet metal into the press and makes a bend. The operator then repositions the work, and another bend is made. The control prompts each step. In the case of a control cabinet, the seams are then welded together, perhaps by a robot welder.

CNC PUNCH PRESS

In the past, a die had to be made for stamped parts. This die was good for only one purpose and required a long leadtime. Dies were exceptionally costly for large parts. If changes were made, the die had to be reworked at great expense of money and time.

CNC punch presses are programmable. Many punches of different shapes and sizes are built into the machine. A CNC program controls the machine, which actually moves the sheet metal to the punches. The punches are used to punch holes, slots, or other internal part shapes or features and to cut the outside shape of the part, small or large. The only leadtime required is to

FIGURE 1–6 *The variable axes machining center (VARIAX)
is a prototype of the newest technology in machine tools.
(Courtesy of Giddings & Lewis, Inc.)*

create the program. If a change is desired, the program is changed. This has drastically changed the metal stamping business. Programming software can optimize the job so the maximum number of parts can be made from the piece of material with minimum waste, called *part nesting.*

POINT-TO-POINT VS. CONTINUOUS PATH

Many of the early machines were point-to-point machines. Point-to-point emphasizes product over process. For example, if you decide to go rent a video, the path you take to get to the video store is very unimportant. The important thing is that you get to the store.

In some machining operations, we do not care about the path; we only care about the destination. For example, consider the drilling of holes. The important thing is that the holes end up in the correct location. We don't care how the machine got from the first hole to the second, and so on (see Figure 1–7).

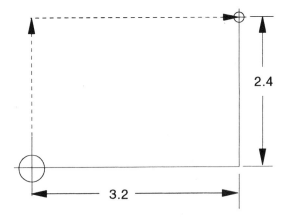

FIGURE 1–7 *Point-to-point machines were really only good for operations such as drilling holes where the path to get to the holes was unimportant.*

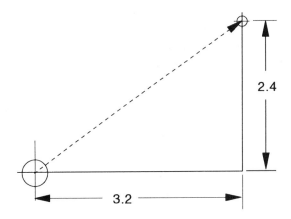

FIGURE 1–8 *Continuous path machines can control both axes individually to create a controlled path to the next point.*

Many of the early numerical control drilling machines were point-to-point. On the other hand, imagine that you enter a marathon. It is now very important that you follow the prescribed path to the finish line or you will be disqualified. This would equate to operations such as milling the outside of a piece we are manufacturing, when the tool must follow an exact path (see Figure 1–8), called *continuous path control.* Obviously it is more difficult to control an exact path. The machines of today use the continuous path method.

MACHINE TOOL AXES

Numerical control machines can also be classified on the number of axes, or directions of motion that they are capable of. Machining centers usually have two-and-a-half, three, four, or five axes with three and four axes being the most common. The four and five axes machines incorporate a rotary table of some sort to be capable of continuous motion in all axes simultaneously.

Lathes and turning centers generally have between two and four axes. The standard configuration consists of a two-axis lathe with one axis parallel to the spindle and one axis perpendicular to the spindle. Both axes are fully controllable to allow turning of chamfers and radii with standard tooling. With the onset of the mill-turn, the two-axis lathe becomes a three-axis machine, with the third axis being the rotation of the spindle. Besides controlling spindle speed, the programmer can control the position of the spindle radially, similar to a rotary table. When combined with a powered milling or drilling attachment (live tooling), the three-axis turning machine can now do secondary operations such as slotting and off-center drilling.

A four-axis turning center has a rear or second turret that can be programmed independently of the master turret.

COMPONENTS OF CNC MACHINES

The numerical control machine can be divided into three basic areas. The first area is the control, which processes the commands from the input media. The second area is the drive mechanisms, and the last area is the machine itself.

NC/CNC CONTROLS

Machine controls are divided into two types: numerical control (NC) and computer numerical control (CNC). NC controls must read the program each time a part is run; they have no means of storing or editing existing programs. CNC controls can store and allow editing of loaded programs. All machines built today are CNC machines.

Machine controls have changed greatly with the age of the computer. Controls today are "soft wired," which allows greater flexibility in changing and upgrading the computer control. In fact, many of the machine's operating characteristics can be changed by the operator so that the machine operates the way he/she wants it to. Parameter tables in CNC machines today allow each machine to be personalized to the needs of the job to be run. The modern CNC machine tool is software driven. Simply put, computer controls are programmed instead of hard wired.

The latest advancement in CNC controls has been the marriage of the CNC control and the microcomputer. CNC controls have always been based on microprocessors. Microcomputers have made so many advances in speed and memory that they are now being used as the processor for CNC control. Different manufacturers have taken different approaches to this technology. Some have used a microcomputer for all of the CNC control functions. Others have added the microcomputer to the CNC controller. The microcomputer has several advantages in CNC applications. Memory has become very cheap for microcomputers. This allows PC-based CNC machines to store incredible amounts of data and programs. It allows the addition of inexpensive computer hardware such as additional memory, CD-ROM, disk drives, and network cards. The ability to add an inexpensive network card and add CNC machines to your regular computer network is a big advantage. This eliminates the need to make a serial cable that is specific to the particular machine. It also allows any CNC machine to share programs with any other CNC machine or any other computer on the network (see Figure 1–9). You can even share programs with other sites over the Internet.

The PC-based controller also allows the manufacturer to attach to the CNC machine over the Internet for troubleshooting of programs and for maintenance. The operator can even surf the Internet while machining. A PC-based

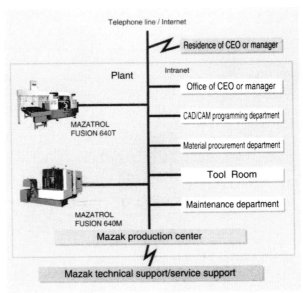

Mazatrol Fusion 640 communication

FIGURE 1–9 *A PC-based mill control and a PC-based turning center control connected to a plant computer network. (Courtesy Yamazaki Mazak Corporation.)*

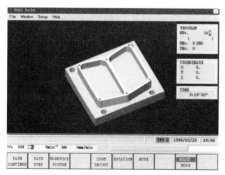

Tool path solid model screen

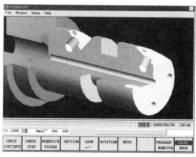

Tool path solid model screen

FIGURE 1–10 *A PC-based mill control and a PC-based turning center control. These are Mazak Fusion 640 CNC controls. (Courtesy Yamazaki Mazak Corporation.)*

controller may even access to machining conditions so that a record can exist of how best to run the job. Figure 1–10 shows PC-based mill and turning center controller displays. The bottom of the figure shows the tool-path graphics that can be displayed. The top screens show an example of some of the information that is available about machining conditions. The display can show the current machine load status, cutting conditions, spindle load, and so on. The control can "learn" from high-efficiency cutting conditions that have occurred in the past. The operator can then automatically select these conditions for subsequent machining.

PC-based controllers also allow the user to add programming software directly to the machine. For example, one could install Gibbs software on a PC-based controller and use it to develop programs for the machine. Thus, all PC-based CNC machines on a plant floor could be programmed using the same programming system.

CRT DISPLAYS

CRT or cathode-ray-tube displays were added to CNC machines to allow the operator easy access to information. On the screen of the machine tool, the operator can see the program, tool and cutter offsets, machine positions, variables, alarms, error messages, spindle RPM, and horsepower usage. Machine tool builders also offer a tool-path simulation. The control reads the program and simulates the tool path on the screen. The tool path simulation can be used to prove the program before the program is run to eliminate programming errors that could damage tools, parts, machine, or operator. Some companies also use the simulation to approximate the machining time so that they can accurately bid on jobs.

LCDS

LCD or liquid-crystal-displays are now widely used as display units on newer CNCs. CRT displays are very deep in relation to their screen size. Their depth requires the display cabinet to be very deep. LCDs are very shallow. This allows display to be mounted directly on the front, even on the doors of the machine. These LCD displays use active matrix technology to improve the visibility and clarity of the display. This is the same technology that is used on portable computers.

DRIVE MOTORS

Most of the early machine drives were hydraulic. The hydraulic pump would provide oil under pressure to servo valves. The servo valves would direct oil to the hydraulic motors, which were attached to the axis drive screws. Newer machine tools have electric servo motors that turn ball screws, which in turn drive the different axes of the machine tool.

STEPPING MOTORS AND OPEN-LOOP SYSTEMS

The stepping motor is an electric motor that rotates a set amount every time the motor receives an electronic pulse from the master control unit (MCU). The stepping motor's rotary motion is converted to the linear motion of the machine table through the use of lead screws. Note that there is no feedback and no sensors to check if the machine actually made the move. This type of feedback or error checking is known as an "open-loop system" (see Figure 1–11).

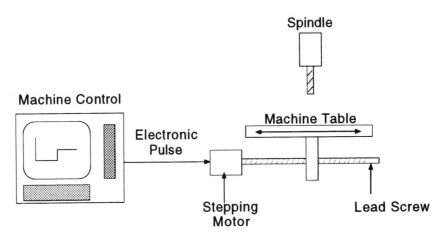

FIGURE **1–11** *Open-loop system configuration.*

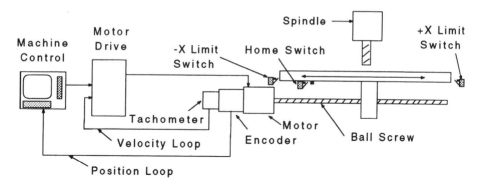

FIGURE **1–12** *Closed-loop system configuration.*

SERVO MOTORS AND CLOSED-LOOP SYSTEMS

Servo motors permit untended operation and closed-loop control makes sure that the machine actually does what the program told it to do (see Figure 1–12). The increasing capability and speed of computers make it possible to continuously monitor the machine's position and velocity while it is operating. For example, if we tell the machine to move 10 inches at a feedrate of 5 inches per minute, the computer will constantly monitor the axis to be sure it is properly executing the move. If the table were to run into an obstruction, the computer would know instantly that it should be moving but it is not. The computer would then stop the motion and signal an error condition. This usually occurs before serious damage is done to the machine.

The advantages of the servo motor include motor and feedback mechanism in one housing, increased travel and spindle speeds, and increased accuracy and repeatability.

Velocity Loop

The velocity loop consists of a motor drive, the motor, and a tachometer. The drive is a device that gets a velocity command from the CNC control. For example, if our program told the machine to run the spindle clockwise at 800 RPM, the CNC control might output +3.5 volts to the motor drive. The motor drive amplifies this voltage to a higher voltage and current for the motor to operate. As the motor begins to turn, the tachometer turns too.

The tachometer is really a generator. The faster it turns, the more voltage it outputs. It is mechanically attached to the motor shaft so that it turns at the same speed as the motor. The voltage from the tachometer is called *feedback*. It is fed back to the drive so the drive can correct itself.

The drive has a circuit called a *comparator* (or *summing junction*). This circuit compares the command from the CNC control and the voltage from the tachometer (feedback on the actual speed) and outputs the difference between the two, called the *error*. The feedback from the tachometer is connected so that it will be the opposite polarity of the command from the CNC control. The drive outputs the difference (or error) of the two and assures that the drive maintains almost perfect velocity. For example, assume we begin some heavy cuts in tool steel, which would tend to slow the motor down. If the motor slows down, the tachometer slows down and outputs less voltage.

The drive then sums the command with the smaller feedback signal, and there is a larger error. The drive then outputs more voltage (current) and the motor speeds up and runs at the correct speed. This all happens instantaneously, so the speed really never varies.

If the load becomes too great, such as when the table runs into a wall, the drive is protected by a current limit. The current limit stops the drive, probably before any damage was done. The drive would then have to be reset.

Position Loop

The position loop is closed by a device called an *encoder*. An encoder is a disk with three rings around it (see Figure 1–13). Each of these rings consists of many slits. A typical encoder might have 1000 slits in the first ring, 1000 in the second, and only one in the third. There are three light receivers on the other side of the disk, one for each ring. A light is shone through each of these rings and the encoder passes light to the receiver when the slits line up with the light and receiver. This creates pulses, which are sent back to the CNC control. The CNC control knows that there are 1000 pulses per revolution, so by counting the pulses it knows the encoder's position exactly. The slits in the first ring are not in line with the second ring. Where there is a slit in the first

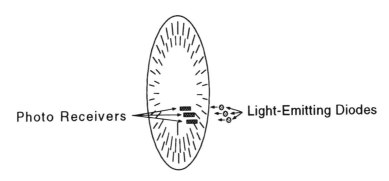

FIGURE **1–13** *A typical encoder. Note the light-emitting diodes and the light receivers.*

ring there is none in the second. The CNC control knows when the first ring pulse begins and ends. It also knows when the second ring's pulse begins and ends. This means that the CNC control can actually increase the resolution of the encoder by four times. Instead of one revolution of the motor being broken into 1000 pieces, it can be broken into 4000 pieces. This is very accurate. The CNC control knows which direction the machine is moving by determining which pulse occurs first, the first ring pulse or the second ring pulse.

Homing

When a machine is initially started, it has no idea where each axis is located. It needs to be "homed." Homing moves each axis to a known position where each axis is initialized. It is crucial that a machine home at the exact same position each time so that programs and fixture locations can be accurate and consistent.

The procedure is very simple. The operator moves each axis to a safe location and then hits a home key for each axis. Safe location means that there are no fixtures, parts, or other tooling that the machine will run into as it homes.

To see what happens when one axis is homed, study Figure 1–12, especially the position loop and the table switches. Note that there are three switches that the table can contact. There is a –X limit switch, a +X limit switch, and a home switch. The limit switches provide overtravel protection, which disables the motor drive if the table ever moves far enough to contact these switches. If we contact the –X limit switch, for example, the motor drive will be disabled for any further movement in the –X direction, and the control will go into an error condition. The operator will have to reset the error and move the axis in the +X direction until the table is off the –X limit switch. On many machines this will require the operator to hold an override key down while moving the table off the limit switch. The same is true if the +X switch is hit by the table. The operator would have to reset the error and move the table in the –X direction.

The home switch is used to roughly position the table during the homing routine. The operator moves the table to a safe position and hits the home key for the X axis (in this example). The table begins to move slowly toward the home switch. When the table contacts the switch, a signal is sent to the CNC control, which senses that the switch has closed.

The CNC control then reverses the motor and very slowly turns it. The CNC control monitors the encoder pulses until it sees the home pulse. When the CNC control sees the home pulse from the encoder, it initializes its position. It now knows exactly where this axis is. The home switch assures that the table is close to position and that we are in the right revolution of the encoder. The encoder home pulse establishes the home position exactly. Each axis homes in the same manner.

Note that most machines also have software axis limits stored in computer memory. For example, there may be a limit of –12 inches for the X axis. If the X axis ever gets to a –12 inch position the CNC control inhibits the –X axis and turns on an error code. The software limits are set up to be hit before the actual limit switches are hit, providing extra protection.

Ball Screws

The rotary motion generated by the drive motors is converted to linear motion by recirculating ball screws. The ball lead screw uses rolling motion rather than the sliding motion of a normal lead screw. Sliding motion is used on conventional Acme lead screws. Acme lead screws work on a friction and backlash principle; ball screws do not (see Figure 1–14). Acme lead screws were used on conventional machine tools.

The balls, located inside of the ball screw nut, contact the hardened and ground lead screw and recirculate in and out of the thread (see Figure 1–15). The contact points of the ball and screw are directly opposing one another and virtually eliminate backlash. The contact points are also very small, so very little friction is generated between them.

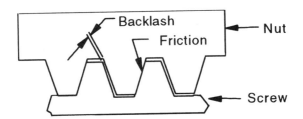

FIGURE 1–14 *Friction and backlash are two disadvantages of the conventional Acme lead screw.*

FIGURE 1–15 *Recirculating ball screws have virtually no backlash. (Courtesy of Giddings & Lewis Inc.)*

Other advantages of the ball lead screw over the Acme lead screw are

1. *less wear*
2. *high speed capability*
3. *precise position and repeatability*
4. *longer life*

Backlash Compensation

Note that although ball screws greatly reduce backlash, there is no such thing as zero backlash. Machine tool manufacturers have incorporated an electronic pitch error or backlash compensator into most CNC machines. The backlash compensator corrects errors detected by lasers at the time the machine is assembled. The amount of compensation is loaded into storage area within the control by the machine tool representative. As the machine moves, the control adjusts the position of the machine according to the stored data. This number can be changed in memory as the machine wears.

THE CNC MACHINE

The machine tool itself has evolved along with the computer control and drive systems. In the beginning, the majority of numerical control machines were just conventional machine tools with a control added on to the manual machine. This is called a *retrofit*. Modern CNC machine tools have been completely redesigned for numerical control machining and bear little resemblance to their conventional counterparts. Requirements of new machine tools include rigidity, rapid mechanical response, low inertia of moving parts, and low friction along the slideways.

Tool Changing

Tool changers are provided on machining centers and turning centers to allow automatic changing of cutting tools without operator intervention. Machining centers can have anywhere from 16 to 100 tooling stations. Lathe or turning centers typically have 8–12 tool turrets, which are indexed automatically. Quickness of tool changing is an important factor on machines used in production environments. Some machines have bi-directional turrets that can select the quickest route to the desired tool.

Programming and Input Media

Numerical control programs may be produced in a variety of ways. One method involves writing a program down on paper and then typing it on a tape preparation system (tape punch) or on an editor. The editor may be on

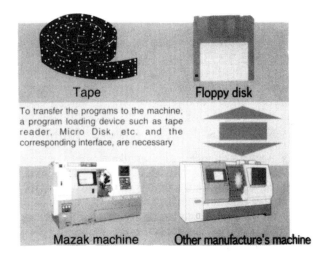

FIGURE **1–16** *Program loading and storage options.*
(Courtesy of Yamazaki Mazak Corp.)

the computer control itself or on a remote computer (personal computer). In the case of a remote editor or computer, the program is loaded into the control by means of a punched tape, floppy disk, diskette, or wires connected to the control, called *direct numerical control* or *distributed numerical control* (DNC) (see Figure 1–16). The newer PC-based CNC machines can connect directly to the company computer network to share programs. You will probably never see a tape punch nor a punched tape. The first NC machines read the programs from punched tape. Punched tape has disappeared from machine shops.

The second method involves a computer-aided parts programming system (CAPP), which allows the programmer to describe the part and the tool paths using a special language or screen prompts. The computer then post-processes the tool paths into numerical control language, which is loaded into the control via one of the methods described.

The third method of programming is to generate the program using the conversational or symbolic programming systems built into some machine controls. In most systems the computer program appears as machine language (i.e., G-code and M-code programs). This is why programmers and operators need a strong background in G-code language.

WHY CNC?

Consider a typical job (see Figure 1–17). A good operator first studies the blueprint and plans the correct machining methods and sequence. Next, the operator gathers the tools necessary to do the job. In this case there is a fixture

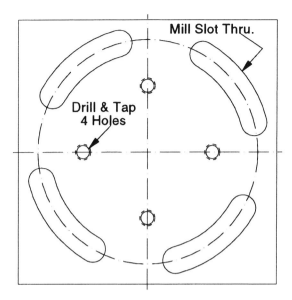

FIGURE 1–17 *This typical job shop part could be done in one setup using CNC machine tools.*

for milling and a fixture for drilling. The operator also gets an endmill and a drill. Our operator then finds an available milling machine and properly locates and fastens the milling fixture onto the table. After loading the endmill into the spindle and adjusting the speeds and feeds of the machine, the operator turns on the coolant and spindle. The operator then very carefully moves the table handles to move the table and spindle to the proper location. The machine is then engaged and milling begins.

When the milling is complete, the operator chooses a drill press. The operator carefully mounts the drilling fixture on the table and loads the drill. Proper speeds and feeds are set, and the coolant and spindle are turned on. The operator then guides the drill into the fixture to correctly locate the holes to be machined.

Contrast how this job is done on the CNC machine. The operator is given the job. The operator downloads the part program into the CNC machining center and locates the first piece in the vise. After using an edge finder to locate the piece, the operator runs the program. The machine turns on the coolant, loads the proper tools, sets the proper speeds and feeds, and machines the part while the operator watches.

We do not need drill fixtures to guide the drills to assure accurate location; the machine drills the holes precisely. In many cases the operator can just clamp the part in the vise on the machine. This helps reduce the cost of producing a part because we do not have to design and build, or buy, a fixture. It helps reduce leadtime because we can make the part right away; we do not have to wait until a fixture is built. Leadtime is the time between when we receive an order and when we can make and ship it.

If our customer wants a change in a dimension of the part or a hole moved, we can do it immediately with a CNC machine. We would not have to rework the fixture. Setup time is drastically reduced with a CNC machine. Remember that the machine is making money only while we are actually machining. We are not making money while the operator loads tools, lines up and clamps fixtures on the table, and so on.

CNC machines drastically reduce the nonmachining time. The CNC can make very rapid positioning moves between machining operations. Many CNC machines have tool changers and tool carousels that can hold many tools. Most of the common tools will already be available in the CNC so the operator doesn't have to load tools. The CNC can change a tool in seconds, while an operator might take minutes.

Remember that the machine is making money only while actual machining is taking place. CNC helps drastically reduce nonmachining time, and is appropriate for small- and large-lot production.

CNC machines have been very beneficial in tool and die shops. Skilled craftsmen can use the machines to produce complex, one-of-a-kind parts that would be very time consuming if done manually. CNC machines are very effective for large lot sizes too. Some would argue that if huge numbers of parts are repetitively made it is more cost effective to produce the parts using a special purpose machine or fixtures (hard tooling). This is not true if changes occur in the design of the parts. CNC will allow more rapid change.

AXES AND COORDINATE SYSTEMS

To fully understand numerical control programming you must understand axes and coordinates. Think of a part that you would have to make. You could describe it to someone else by its geometry. For example, the part is a 4-inch by 6-inch rectangle. All parts can be described in this fashion. Any point on a machined part, such as a hole to be drilled, can be described in terms of its position. The system that allows us to do this, called the *Cartesian coordinate* or *rectangular coordinate system,* was developed by a French mathematician, René Descartes (see Figure 1–18).

THE CARTESIAN COORDINATE SYSTEM

Consider just one axis first. Imagine a line with zero marked in the center (see Figure 1–19). Now imagine that to the right of zero we have marked every inch with a positive number representing how far the mark is to the right of zero.

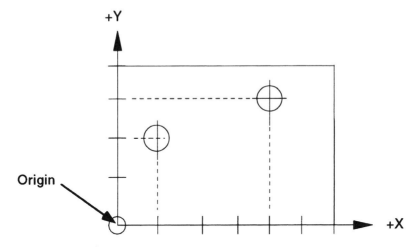

FIGURE 1–18 *An example of coordinate positioning.*

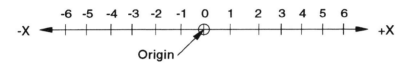

FIGURE 1–19 *Single-axis coordinate line.*

Mark the inches to the left of zero beginning with –1, –2 ,–3 and so on. This line is called the X axis. We could describe a particular position on our line by giving its position in inches from zero. This would be called a position's *coordinate*.

Let's add a perpendicular line (axis) that crosses our first line (X axis) at zero. The horizontal line is the X axis and the vertical line is the Y axis. The point where the lines cross is the zero point, usually called the *origin*. Points are described by their distance along the axis and by their direction from the origin by a plus (+) or minus (–) sign (see Figure 1–20).

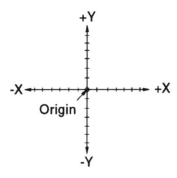

FIGURE 1–20 *Dual-axis coordinate grid.*

Quadrants

The axes divide the work envelope into four sections called *quadrants* (see Figure 1–21). The quadrants are numbered counterclockwise, starting from the upper right.

> *Points in the upper right, quadrant 1, have positive X (+X) and positive Y (+Y) values.*
>
> *Points in quadrant 2 have negative X (–X) and positive Y (+Y) values.*
>
> *Points in quadrant 3 have negative X (–X) and negative Y (–Y) values.*
>
> *Points in quadrant 4 have positive X (+X) and negative Y (–Y) values.*

To locate a point such as (X3.0, Y3.0) in the two-axis system, start at the zero point and count to the right (+ move) three units on the X axis and up (+ move) three units on the Y axis.

Figure 1–22 shows a point in the Cartesian coordinate system. The point's coordinates are identified as X3, Y3. Note that only one point would match these coordinates. Figure 1–23 shows another Cartesian coordinate system with

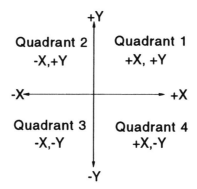

FIGURE **1–21** *The four quadrants of the Cartesian coordinate system.*

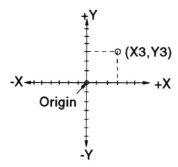

FIGURE **1–22** *Locating a position using the Cartesian coordinate system.*

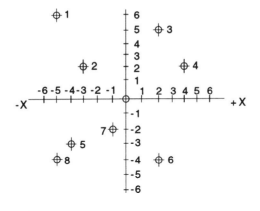

	X	Y			X	Y
Point 1	-5	6		Point 5	-4	-3
Point 2	-3	2		Point 6	2	-4
Point 3	2	5		Point 7	-1	-2
Point 4	4	2		Point 8	-5	-4

FIGURE 1–23 *Cartesian coordinate system and the XY coordinates for eight points.*

eight points identified with their coordinates. Add one more axis (see Figure 1–24) to represent depth. If we are going to drill a hole, we would describe its location by its X and Y coordinates. We would use a Z value to represent the depth of the hole. The Z axis is added perpendicular to the X and Y axes.

Think about a vertical milling machine. The X axis is the table movement right and left as you face the machine. The Y axis is the table movement toward and away from you. The Z axis is the spindle movement up and down. A move toward the work is a negative Z (–Z) move. A move up in this axis would be a positive Z (+Z) move.

Milling machines use all three axes, as seen in Figure 1–25. The X axis usually has the longest travel. On a common vertical milling machine, the X axis moves to the operator's left and right. The Y axis moves toward and away from the operator. The Y axis usually has the shortest travel. The Z axis always denotes movement parallel to the spindle axis, the up and down movement. Toward the work is a negative Z move.

Lathes or turning centers typically use only the X and Z axes. The Z denotes movement parallel to the spindle axis and controls the lengths of parts or shoulders. The X axis is perpendicular to the spindle and controls the diameters of the parts (see Figure 1–26).

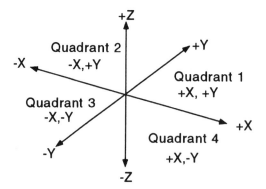

FIGURE 1–24 *Three-axes coordinate grid.*

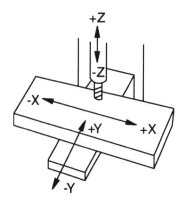

FIGURE 1–25 *Typical milling machine configuration illustrating the X, Y, and Z axes.*

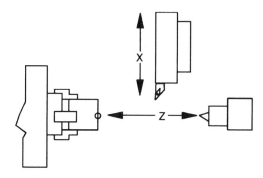

FIGURE 1–26 *Typical lathe configuration illustrating the X and Z axes.*

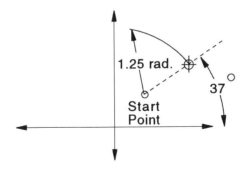

FIGURE 1–27 *Polar coordinates are described from the present position, not from the absolute axes origin (X0,Y0).*

POLAR COORDINATES

Within the coordinate systems, it is also possible to describe the position of points by stating angles and distances along the angles. The direction of the angular line is viewed from the X axis datum line. A positive angular dimension runs counterclockwise from your present position (A+45). A negative angular dimension would run clockwise from your present position (A–45). The desired position therefore is a point on the angular line, the desired radial distance from your present position (A+37, 1.250) (see Figure 1–27).

ABSOLUTE AND INCREMENTAL PROGRAMMING

Absolute and incremental programming coordinates specify the relative tool moving position with respect to the program zero or to the tool moving distance from its current position. Computer programs can be written in absolute programming format, incremental programming format, or may use both formats in the same program.

Absolute programming specifies a position or an end point from the workpiece coordinate zero (datum). It is an absolute position (see Figure 1–28).

Incremental programming specifies the movement or distances from the point where you are currently located (see Figure 1–29). Remember, a move to the right or up from this position is always a positive move (+); a move to the left or down is always a negative move (–). With an incremental move we are specifying how far and in what direction we want the machine to move.

Absolute positioning systems have a major advantage over incremental positioning. If the programmer makes a mistake when using absolute positioning, the mistake is isolated to the one location.

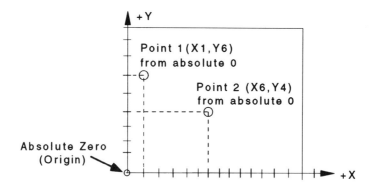

FIGURE 1–28 *Absolute coordinate positions are always located from the program zero.*

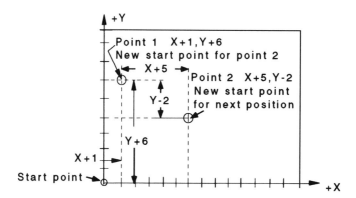

FIGURE 1–29 *In incremental coordinate positioning, your present position becomes the program start position. From the start point, point 1 is one position to the right on the X axis (X+1) and six positions up on the Y axis (Y+6). When we move to point 2, point 1 becomes our start point. Point 2 is 5 positions to the right of our present position on the X axis (X+5) and 2 positions down on the Y axis (Y–2). Remember, in incremental programming, moves down or moves to the left are negative moves.*

When the programmer makes a positioning error using incremental positioning, all future positions are affected. Most CNC machines allow the programmer to mix absolute and incremental programming. There will be times when using both systems in one program will make your programs easier to write.

CHAPTER QUESTIONS

1. Describe the difference between numerical control and computer numerical control.

2. What developments in the electronic industry helped make computers more powerful and affordable?

3. Name the two basic types of positioning.

4. Make a sketch of the four Cartesian quadrants and identify the signs of each quadrant.

5. Name the two most common computer numerical control machines and state the basic axes associated with each.

6. Describe the incremental positioning mode.

7. Name one disadvantage of the incremental positioning mode.

8. Define the zero or origin point.

9. State the biggest advantage that ball screws have over Acme screws.

10. Identify the positions marked on the Cartesian coordinate grid below. Use absolute positioning.

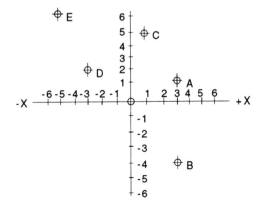

	X	Y
Point A		
Point B		
Point C		
Point D		
Point E		

11. Using the Cartesian coordinate system, write down the X and Y values for the incremental moves.

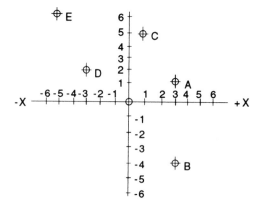

	X	Y
Origin to Point A		
Point A to Point B		
Point B to Point C		
Point C to Point D		
Point D to Point E		

Chapter 2

··

FUNDAMENTALS OF PROGRAMMING

INTRODUCTION

Numerical control programming is the process of taking the information that you would normally get from a part print to do manual machining and converting it to a language that a machine will understand. This book focuses on the machine programming language called word address programming. Word address programming is the most common machine language in use today.

OBJECTIVES

Upon completion of this chapter, the reader will be able to:

- *Identify and explain program words.*
- *Explain the parts of a program.*
- *Arrange and explain blocks of information.*
- *Describe preparatory and miscellaneous functions.*
- *Describe G92 and G54 workpiece coordinate settings.*
- *Write simple programs using word address format.*
- *Explain how to program an arc.*
- *Describe the purpose of tool height and tool diameter offsets.*

WORD ADDRESS PROGRAMMING

The word address format for numerical control programming precisely controls machine movement and function through the use of short sentence-like commands. These commands consist of addresses, words, and characters. Let's consider a simple operation that could be performed on a manual milling machine and convert it to a word address command: Turn the spindle on in a clockwise direction at a spindle speed (RPM) of 600. An operator would adjust the gearing on the machine to choose an RPM of 600 and then turn the spindle on in a clockwise direction.

The command to do this on a CNC machine would be: N0010 M03 S600;.

The N0010 is a line number. The M03 tells the spindle to start in a clockwise direction. The S600 tells the spindle how fast to turn. The semicolon tells the control that it is done with this command block and that it should move down to the next command. A full command, or block of information, is made up of addresses and characters that carry out a command.

LETTER ADDRESS COMMANDS

The command block controls the machine tool through the use of letter address commands. Following are abbreviated descriptions of the most common letter address commands.

N is the letter address for the line number or sequence number of the block.

G codes are preparatory functions, which set up the mode in which the rest of the operation(s) are to be executed.

F is the feed rate of the controlled axis.

S is a spindle speed setting.

T is the letter address for a tool call.

M is a miscellaneous function. These machine functions include coolant on/off, spindle forward, and many others.

H and D are examples of auxiliary letter address codes used for tool offset storage.

PART PROGRAMMING

A part program is simply a series of command blocks that execute motions and machine functions to manufacture a part. Let's take a look at a simple program that cuts around the outside of a 3-inch by 4-inch block with a 1/2-inch diameter endmill (see Figure 2–1). Note that the coordinate dimensions reflect the position of the center of the spindle. This means that we have to compensate for the size of the cutter.

N0010 G00 X-1.00 Y-1.00; (point 1)

N0020 G01 X-.25 Y-.25 F10.0; (point 2)

N0030 G01 Y3.25; (point 3)

N0040 G01 X4.25; (point 4)

N0050 G01 Y-.25; (point 5)

N0060 G01 X-.50; (point 6)

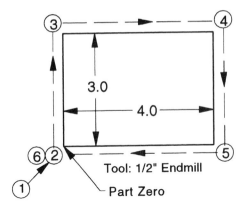

FIGURE 2–1 *Profile mill a 3-inch by 4-inch block.*

N0010 G00 X-1.00 Y-1.00;

Line number 10 rapid positions the tool to a position just off the lower left-hand corner of the part (X-1, Y-1).

N0020 G01 X-.25 Y-.25 F10.0;

Line number 20 feeds the tool to a position that is one tool radius value to the left of the side of the part (X-.250). The tool will feed at 10 inches per minute. It is now in alignment with the left-hand corner of the part. This gets us ready to cut the left side of the part.

N0030 G01 Y3.25;

Line number 30 cuts the left side of the part and positions the spindle center past the top of the part by the tool radius (Y3.25). This positions the edge of the tool for the cut across the top of the part.

N0040 G01 X4.25;

Line 40 cuts the top of the part. The feed rate is still 10 inches per minute, because we have not changed it since line number 20. The spindle center is now positioned .25 to the right of the part. This gets us ready to cut the right-hand side of the part.

N0050 G01 Y-.25;

Line number 50 positions the tool at point 5. The right-hand side of the part is now complete. The tool center is also positioned one tool radius below the bottom of the part, ready to cut.

N0060 G01 X-.50;

Line number 60 cuts the bottom of the part to size and moves the tool completely off the part (X-.50).

Now that you know how simple command blocks look, let's take a more in-depth look at the individual parts of a word address part program.

PART DATUM LOCATION

To program a part, you need to determine where the workpiece zero (or part datum) should be located. The part datum is a feature of the part from which the majority of the dimensions of the part are located. Because all of the dimensions of the part in Figure 2–1 come from the lower left-hand corner, this was the logical choice for the workpiece zero point. Good programmers will choose a part feature that is easy to access with an edge finder and one that will involve the least amount of dimensional calculations. This approach will avoid potential errors in the machining of the part.

SEQUENCE NUMBERS (NXXXX)

Sequence numbers are a means of identifying blocks of information within the program. In most cases sequence numbers are not required because the machine will execute blocks of information in the order in which it reads them, although line or sequence numbers can be very helpful in identifying problems. The machine controller can be commanded to find blocks of information by their line numbers. In addition, line or sequence numbers are needed in the use of some canned cycles, which will be covered later in this book.

PREPARATORY FUNCTIONS (G-CODES)

Preparatory functions are used to set the control for various machine movements such as linear interpolation (G01) and rapid traverse (G00). Linear interpolation means that the cutter moves on a controlled linear path (line). A "G" followed by a two-digit number determines the machining mode in that block or line.

G-codes or preparatory functions fall into two categories: modal or non-modal. Non-modal or "one-shot" G-codes are those command codes that are only active in the block in which they are specified. Modal G-codes are those command codes that will remain active until another G-code in the same group overrides it. For example, if you had five command lines that were all linear feed moves, you would only have to put a G01 in the first line. The other four lines would be controlled by the previous G01 code.

It is important to note that the G-codes shown in Figure 2–2 are commonly used machining center G-codes, but some of them may be slightly different from those used on your machine tool. Consult the manual for your machine to be sure.

G00	Rapid traverse	Modal
G01	Linear positioning at a feedrate	Modal
G02	Circular interpolation clockwise	Modal
G03	Circular interpolation counter-clockwise	Modal
G28	Zero or home return	Non-modal
G40	Tool diameter compensation cancel	Modal
G41	Tool diameter compensation left	Modal
G42	Tool diameter compensation right	Modal
G43	Tool height offset	Modal
G49	Tool height offset cancel	Modal
G54	Workpiece coordinate preset	
G70	Inch Programming	Modal
G80	Canned cycle cancel	Modal
G81	Canned cycle drill	Modal
G83	Canned peck cycle drill	Modal
G84	Canned tapping cycle	Modal
G85	Canned boring cycle	Modal
G90	Absolute coordinate positioning	Modal
G91	Incremental positioning	Modal
G92	Workpiece coordinate preset	
G98	Canned cycle initial point return	Modal
G99	Canned cycle R point return	Modal

FIGURE 2–2 *List of commonly used machining center G-codes.*

SPINDLE CONTROL FUNCTIONS (S)

Spindle speeds are controlled with an "S" followed by up to four digits. When programming the machining center the spindle speed is programmed in revolutions per minute (RPM). A spindle speed of 600 RPM would be programmed S600. Spindle speeds may also be programmed in surface feet per minute (SFPM) through the use of a G96 preparatory code. Surface feet per minute is the cutting speed of the material you would be working with. Most turning center controls will typically program in SFPM. This allows the spindle speed to automatically change as the diameter of the workpiece changes, maintaining a constant surface speed. The proper surface speed for cutting tool materials and workpiece materials can usually be found in reference texts or in charts provided by cutting tool manufacturers. For example, a cutting speed for mild steel and a carbide tool might be 400 SFPM. A spindle speed of 400 SFPM would be programmed G96 S400. The spindle is turned on using a miscellaneous (M) code of either M03 or M04. An M03 will turn the spindle on in a clockwise direction, while an M04 will turn the spindle on in a counterclockwise direction. A M05 turns the spindle off.

MISCELLANEOUS FUNCTIONS (M-CODES)

Miscellaneous functions or M-codes perform miscellaneous machine functions such as tool changes, coolant control, and spindle operations. An M-code is a two- or three-digit numerical value preceded by a letter address, "M." M-codes, similar to G-codes, can be modal or non-modal.

Figure 2–3 lists commonly used machining center miscellaneous functions (M-codes).

TOOL CALLS

The tool call block is fairly straightforward, although the machining center differs slightly from the turning center. The tool call always starts with a "T"

M00	Program stop	Non-modal
M01	Optional stop	Non-modal
M02	End of program	Non-modal
M03	Spindle start clockwise	Modal
M04	Spindle start counter clockwise	Modal
M05	Spindle stop	Modal
M06	Tool change	Non-modal
M07	Mist coolant on	Modal
M08	Flood coolant on	Modal
M09	Coolant off	Modal
M30	End of program & reset to the top	Non-modal
M40	Spindle low range	Modal
M41	Spindle high range	Modal
M98	Subprogram call	Modal
M99	End subprogram & return to main program	Modal

FIGURE **2–3** *Miscellaneous functions.*

and then the tool number (T02). On a machining center you have to command a tool change with a miscellaneous code of M06. You then tell the control which tool to change to. A typical tool change block for a machining center is N0010 M06 T02;. This line of code would tell the CNC controller to change to tool 2.

On a turning center the tool call also starts with a "T" and the tool number (T02), but then you add the tool offset number. T0202 would be the tool call for tool number 02 with an offset of number 02. It is written 02 because you typically have more than 10 tools and 10 offsets available to you, for example, T1212 (tool 12, offset 12). The offset lets the operator correct any errors in the size of the part. It is not necessary to use an M06 on the turning center to do a tool change; in fact, an M06 on the turning center usually unclamps the chuck.

CONTROLLED AXES WORDS (X, Y, Z)

The movable axes that can be controlled on a machine are known as controlled axes. On a milling machine we have three axes: X, Y, and Z. Each controlled axis is specified by a letter address (X, Y, or Z), which may be preceded by a direction sign (+). Chapter 1 covered the Cartesian coordinate system, which laid the groundwork on how the axes of the machines are set up as well as the direction of travel. A simple command block to rapid position the tool of the milling machine to 1 inch above the workpiece zero could be done by this line: N0010 G00 Z1.00;. The G00 means a rapid move, and the Z1.0 means 1 inch above the part zero.

MOTION BLOCKS

Tool or table motion can be controlled in three ways: rapid positioning, linear feed, and circular feed.

Rapid Traverse Positioning (G00)

A rapid positioning block consists of a preparatory G-code (G00) and the coordinate to which you want to move. A rapid move to a location of X10, Y5, and Z1 would be programmed: G00 X10.0 Y5.0 Z1.0;. This block would command the machine to move at a rapid traverse rate to this position, moving all of the axes simultaneously. The rapid traverse rate for each machine is different, but it normally ranges from 100 to 600 or more inches per minute. The rapid traverse rate can usually be overridden using the rapid traverse override switch located on the control. This means that an operator can choose to reduce the rapid rate from 0 to 100 percent. This is often done when trying a new program.

Linear Feed Mode (G01)

A G01 linear interpolation code moves the tool to a commanded position in a straight line at a specific feed rate. The feed rate is the speed at which the machine axes move. Linear feed blocks are normally cutting blocks. The rate at which the metal is removed is controlled through an F or feed rate code. Machining centers, similar to manual milling machines, use feed rates in inches per minute or IPM. Turning centers are typically programmed in inches per revolution of the spindle or IPR. To make a straight-line cutting motion on a machining center, the block of information would look like this: G01 X10.00 F10.00;. The tool would move to an X-axis position of 10.00 inches at a feed rate of 10 inches per minute. Remember, straight-line moves can also be angular. CNC machine controls are capable of making simultaneous axes moves (see Figure 2–4).

The G00, G01, and F codes are all modal. Modal commands are active unless changed by another preparatory code. If you were programming a series of straight-line moves, you would have to put the G01 and the feed rate in the first line only. The lines that follow would be controlled by the previous G01 and feed rate. If you wanted to change to a rapid positioning mode, you would use a G00 at the beginning of that line.

The example shown in Figure 2–5 incorporates the programming procedures we have discussed to this point. The machining procedure is to mill around the outside of the part, .500 inches deep, with a .50 inch endmill at a feed rate of 5 inches per minute. The first step is to set up the control by doing our preliminary procedures. The second step is the tool call. The third step sets the WPC or workpiece zero point. In our fourth step we need to start the spindle and set the RPM. Next we rapid position close to the part and start our linear

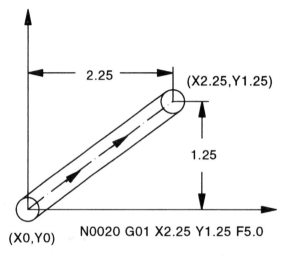

FIGURE 2–4 *G01 linear interpolation example.*

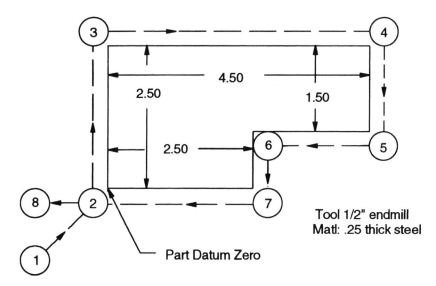

FIGURE 2–5 *Simple contour programming example.*

cutting moves. After we have cut the profile of the part, we need to return to the home position and end the program.

N0010 G70 G90; (Inch programming, absolute programming)

N0020 M06 T02; (Tool change, tool #2)

N0030 G54 X-10.250 Y-8.750 Z-7.525; (Workpiece zero setting)

N0040 M03 S800; (Spindle start clockwise, 800 RPM)

N0050 G00 X-1.00 Y-1.00; (Rapid to position #1, just off the lower left corner of the part)

N0060 G00 Z.100; (Rapid down to .100 clearance above the part)

N0070 G01 Z-.50 F5.0; (Feed down to depth at 5 inches per minute)

N0080 G01 X-.25 Y-.25; (Feed to position 2)

N0090 G01 Y2.750; (Feed to position 3)

N0100 G01 X4.75; (Feed to position 4)

N0110 G01 Y.75; (Feed to position 5)

N0120 G01 X2.75; (Feed to position 6)

N0130 G01 Y-.25; (Feed to position 7)

N0140 G01 X-1.00; (Feed to position 8)

N0150 G28; (Return all axes to home position)

N0160 M06 T0; (Tool change, puts the tool away and empties the spindle)

N0170 M30; (Rewinds the program, resets the control, and ends the program)

PROGRAMMING PROCEDURES

Whether you are programming a machining center or a turning center, all CNC programs have the same general format:

1. *startup or preliminary procedures*
2. *tool call*
3. *workpiece location block*
4. *spindle speed control*
5. *tool motion blocks*
6. *home return*
7. *program end procedures*

STARTUP OR PRELIMINARY PROCEDURES

Every machine shop has its own procedures for startup blocks. In our previous example the only thing we did in the startup block was to tell the control we were programming in the absolute positioning mode (G90) and in inches (G70). Some startup procedures contain a great deal more information, such as telling the control we are programming in the X and Y axes (G17) or canceling any tool compensation that may be active (G40). As you gain more experience programming you will probably develop a set of preparatory codes you will always use at the start of your programs.

TOOL CHANGE AND TOOL CALL BLOCK

An M06 calls for a tool change, and the T06 tells the control which pocket the tool is in.

WORKPIECE COORDINATE (WPC) SETTING

The machine must know where the part is on the table. This is called workpiece coordinate setting. Two important factors deal with workpiece coordinate setting: where the part datum (part zero) is located and where the part datum is located on the machine table. The WPC tells the machine the position of the part datum (see Figure 2–6).

The workpiece datum point (part zero) may be located at the corner or any other part of the workpiece, but we have to tell the control where this point is on the machine table. The technique for locating the workpiece datum varies for each machine tool. Some controls use a button to set the zero point. The setup person or operator uses the jog buttons to position the spindle center over the part datum and then press the zero set or zero shift button to set the

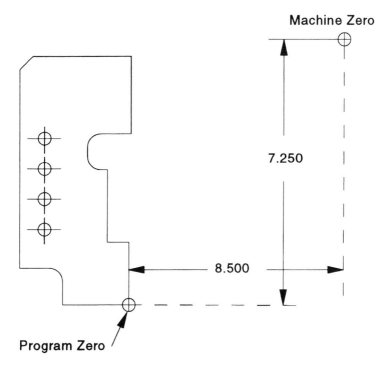

FIGURE 2–6 *The workpiece coordinate setting is used to reference the machine to the part location.*

coordinate system to zero. On other types of controls the WPC is set with a G-code. Systems of this type use either a G54-G59 or a G92 followed by X, Y, and Z dimensions.

The G54-G59 workpiece coordinate is the absolute coordinate position of the part datum (see Figure 2–7). These are not available on all machines. Six are available, and all serve the same function. This allows the programmer to have six different workpiece coordinates established on a machine. This would be very beneficial for repetitive jobs that could be located at the same position on the machine table. For example, a job that is run once each week might use the G59. The G59 would be used to establish the location for that particular job.

To locate this position, the operator would position the center of the spindle directly over the part datum using an edge finder or probe and then take note of the machine position (see Figure 2–8). The coordinates of this position would be placed in the G54 line. A typical workpiece coordinate setting of this type would be written: N0010 G54 X-8.500 Y-7.250 Z-15.765;.

The G92 workpiece coordinate is the incremental distance from the workpiece datum to the center of the spindle (see Figure 2–9). In effect, it tells the

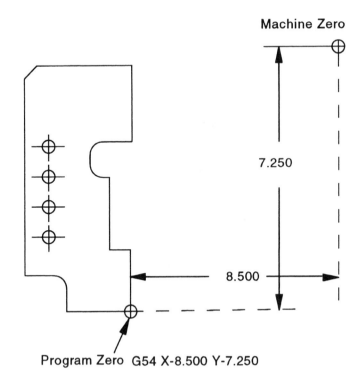

FIGURE 2–7 *The G54 workpiece coordinate setting is an absolute coordinate position from the machine home position. In this case the position is written with negative values because the part is located to the left and down on the X and Y axes. For clarity the Z was omitted.*

machine where the spindle is in relation to the workpiece. The spindle must be in that position when you start to run the program.

When the G92 is called, the center of the spindle has to be in the pre-programmed position. If it is not, the control will start machining at the wrong position. If, for example, the center of the spindle at the home position is 10 inches to the right on the X axis, 5 inches back on the Y axis, and 8 inches above the part on the Z axis, the G92 would be written: G92 X10.00 Y5.00 Z8.00;. If the center of the spindle were located any other distance away from the part datum, when the G92 was called, the tool would try to cut the part in the wrong location. This is why using a G92 can be very dangerous! The G54 type of WPC setting is a lot safer than a G92. No matter where you are when the G54 is called, the control knows exactly where your part is located because it is an absolute position, not an incremental distance.

It is important to remember that a G54 or G92 will not move the machine tool to this point; it merely tells the control where the part (G54) or spindle (G92) is.

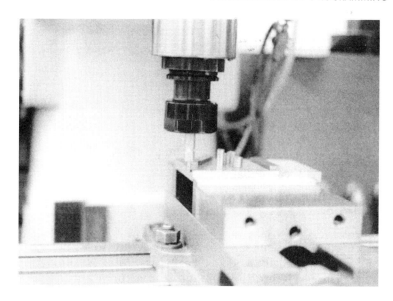

FIGURE 2–8 *Locating the corner of the part using an edge finder. An edge finder or probe can be used to precisely position the center of the spindle over the part datum (zero).*

Tool Movement	Absolute Commands	Incremental Commands
Point 1 to point 2	X-.25 Y3.25	X0.0 Y3.5
Point 2 to point 3	X4.25 Y3.25	X4.5 Y0.0
Point 3 to point 4	X4.25 Y-.25	X0.0 Y-3.5
Point 4 to point 1	X-.25 Y-.25	X-4.50 Y0.0

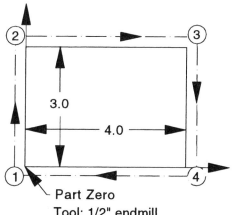

Part Zero
Tool: 1/2" endmill

FIGURE 2–9 *Incremental positioning.*

SPINDLE START BLOCK

The spindle is controlled through the use of two codes. The M03 tells the spindle to start in a clockwise direction, and the S1000 tells it how fast to turn.

TOOL MOTION BLOCKS

The tool motion blocks are the body of the program. The tool is positioned and the cutting takes place in these blocks.

HOME RETURN

The tool needs to be returned to home whenever a tool change takes place. Some machine controls use a G28 command to return to home; other controls return to home automatically when a tool change (M06) is commanded. When a tool change or home return is commanded it is important to know how the tool gets there. Does the Z axis move straight up to a clearance position first, or do all of the axes move simultaneously toward the home position? If they all move simultaneously, you need to be careful that the tool will not run into the fixture or part as it rapids to home position.

PROGRAM END BLOCKS

There are a number of different ways to end the program. Some controls require that you turn off the coolant and spindle with individual miscellaneous function codes. Other controls will end the program, rewind the program, and turn off all miscellaneous functions with an M30 code.

INCREMENTAL POSITIONING

Now that we have established the basics of absolute coordinate positioning we can look at another type of positioning, incremental positioning. Absolute positioning is when all of the coordinates of the part program are related to an absolute zero point. Incremental positioning defines the coordinates of the part in relationship to the present position.

Incremental positioning is also known as point-to-point positioning. The point where you are presently is the datum for the next coordinate position (see Figure 2–9). Incremental positions are the direction (+ or –) and the distance to the next point.

This type of programming can be used to program the whole part or just certain sections of the program. Incremental positioning is programmed with a G91 preparatory code and can be very useful when programming a series of holes that are incrementally located on the part print. Figure 2–10 would be a typical application for incremental programming. If you wished to switch back to absolute programming at any point in the program, you use a G90 preparatory code.

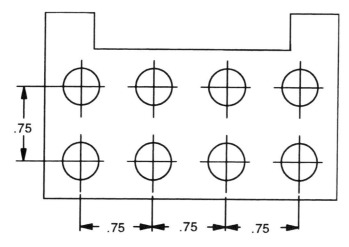

FIGURE 2-10 *Incremental programming example.*

CIRCULAR INTERPOLATION

Up to this point we have discussed only straight-line moves. If a computer numerical control machine was capable only of straight-line moves it would be very limited. One of the most important features of a CNC machine is the ability to do circular cutting motions. CNC machines are capable of cutting any arc of any specified radius value. Arc or radius cutting is known as *circular interpolation*. Circular interpolation is carried out with a G02 or G03 code.

To cut an arc, the programmer needs to follow a very specific procedure. When we start cutting an arc, the tool is already positioned at the start point of the arc. First, we need to tell the direction of the arc. Is it a clockwise (G02) or counterclockwise (G03) arc? The second piece of information the control needs is the end point of the arc. The last piece of information is the location of the arc center or, if you are using the radius method of circular interpolation, the radius value of the arc (see Figure 2–11).

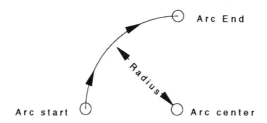

FIGURE 2-11 *The critical pieces of information needed to cut an arc are the arc start point, arc direction, arc end point, and arc centerpoint location.*

ARC START POINT

The arc start point is the coordinate location where the arc starts. The tool is moved to the arc start point in the line prior to the arc generation line. Simply stated, the start point of the arc is the point where you are when you want to generate an arc.

ARC DIRECTION (G02, G03)

Circular interpolation can be carried out in two directions, clockwise and counterclockwise. There are two G-codes that specify arc direction (see Figure 2–12). The G02 code is used for circular interpolation in a clockwise direction. The G03 code is used for circular interpolation in a counterclockwise direction. Both G02 and G03 codes are modal and are controlled by a feed rate (F) code, just like a G01.

ARC END POINT

As stated earlier, the computer control requires that the tool be positioned at the start point of the arc prior to a G02 or G03 command. The current tool position becomes the arc starting point. The arc end point is the coordinate position for the end point of the arc. The arc start point and arc end point set up the tool path, which is generated according to the arc center position (see Figure 2–13).

ARC CENTERPOINTS

To generate an arc path, the controller has to know where the center of the arc is. There are two methods of specifying arc centerpoints: the coordinate arc centerpoint method and the radius method. When using the coordinate arc center method, a particular problem arises. How do we describe the position of the arc center? If we use the traditional X, Y, Z coordinate position words to describe the end point of the arc, how will the controller discriminate between the end

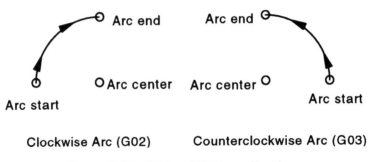

FIGURE **2–12** *G02 and G03 arc direction.*

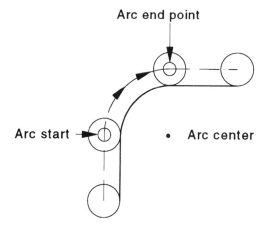

FIGURE 2–13 *Arc end-point location.*

point coordinates and the arc center coordinates? We use different letters to describe the same axes. Secondary axes addresses are used to designate arc centerpoints. The secondary axes addresses for the axes are

I = X axis coordinate of an arc centerpoint

J = Y axis coordinate of an arc centerpoint

K = Z axis coordinate of an arc centerpoint

Because we will be cutting an arc in only two axes directions, only two of three secondary addresses will be used to generate an arc. When cutting arcs in the X/Y axes, the I/J letter addresses will be used. If we were cutting an arc on a turning center, the X/Z axes would be the primary axes, and the I/K letter addresses would be used to describe the arc centerpoint.

The type of controller you are using dictates how these secondary axes are located. With most controllers, such as the Fanuc controller, the arc centerpoint position is described as the incremental distance from the arc start point to the arc center (see Figure 2–14).

On some other types of controls the arc centerpoint position is described as the absolute location of the arc centerpoint from the workpiece zero point (see Figure 2–15).

Modern CNC controllers have become so advanced that they are capable of calculating the centerpoint of the arc by merely stating the arc size and the end point of the arc. Because these controls are still in their infancy, we will concentrate on the incremental method of arc center locating. As we practice generating some arcs using this method, keep in mind that we are locating the arc center using the incremental method (direction and distance) (Figure 2–16). If the arc centerpoint is located down or to the left of the start point, a negative sign (–) must precede the coordinate dimension.

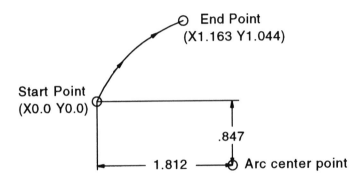

G02 X1.163 Y1.044 I.812 J-.847

FIGURE 2-14 *The centerpoint of the arc is the distance from the start point position to the arc center position.*

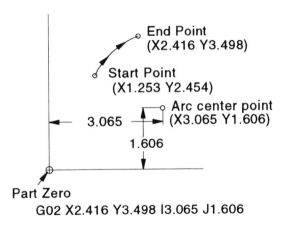

Part Zero
G02 X2.416 Y3.498 I3.065 J1.606

FIGURE 2-15 *When using absolute centerpoint positioning, the centerpoint is the coordinate position from the part zero.*

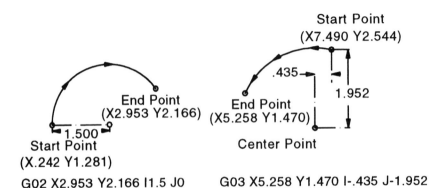

G02 X2.953 Y2.166 I1.5 J0   G03 X5.258 Y1.470 I-.435 J-1.952

FIGURE 2-16 *Circular interpolation example. The incremental arc method was used.*

COMPREHENSIVE PROGRAMMING EXERCISE

Figure 2–17 shows a base plate that involves linear cutting and arc cutting. To simplify the programming of the part, we have eliminated the holding device and are programming the center of the spindle.

N0010 G90 G70; (Absolute programming, inch programming)

N0020 M06 T3; (Tool change, tool #3, .500 dia. endmill)

N0030 G54 X-8.500 Y-6.75; (Workpiece coordinate setting)

N0040 M03 S800; (Spindle start forward, 800 RPM)

N0050 G00 X6.00 Y-.25; (Rapid traverse to position #1)

N0060 G00 Z.100; (Rapid traverse down to .100 above the part)

N0070 G01 Z-.500 F6.0; (Feed tool down to part depth)

N0080 G01 X1.00; (Feed tool to position #2)

N0090 G02 X-.25 Y1.00 I0.0 J1.25; (Clockwise arc to position #3)

N0100 G01 Y2.00; (Linear feed to position #4)

N0110 G02 X1.00 Y3.25 I1.25 J0.0; (Clockwise arc to position #5)

N0120 G01 X4.00; (Linear feed to position #6)

N0130 G02 X5.25 Y2.00 I0.0 J-1.25; (Clockwise arc to position #7)

N0140 G01 Y1.00; (Linear feed to position #8)

N0150 G02 X4.00 Y-.25 I-1.25 J0.0; (Clockwise arc to position #9)

N0160 G01 Y-1.00; (Position tool off of the part to position #10)

N0170 G28; (Return tool and all axes to home position)

N0180 M05; (Spindle off)

N0190 M06 T4; (Tool change to tool #4, .500 dia. slotting endmill)

N0200 M03 S750; (Spindle start forward, 750 RPM)

N0210 G00 X1.00 Y1.00; (Rapid traverse to position #11)

N0220 G00 Z.100; (Rapid traverse down to .100 above the part)

N0230 G01 Z-.525 F2.0; (Feed tool down through the part at 2 inches per minute feed rate)

N0240 G01 Y2.00 F5.0; (Cut the slot to position #12 at 5 IPM)

N0250 G01 Z.100; (Feeds the tool up to a clearance of .100 above the part)

N0260 G00 X4.00; (Rapid traverse to position #13)

N0270 G01 Z-.525 F2.0; (Feed tool down through the part at 2 IPM)

N0280 G01 Y1.00 F5.0; (Cut the slot to position #14 at 5 IPM)

N0290 G01 Z.100; (Feeds the tool up to a clearance of .100 above the part)

N0300 G28; (Return the tool and all axes to home position)

N0310 M06 T0; (Tool change to tool 0, spindle empties)

N0320 M30; (Rewind program, end program)

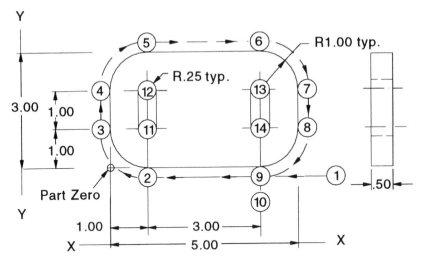

FIGURE 2–17 *Base plate.*

TOOL LENGTH OFFSET

Up to this point we have not dealt with tool length offsets, which make it possible for the controller to adjust to different tool lengths. Every tool is going to be a different length, but modern machine controls can deal with this quite easily. CNC controllers have a special area within the control to store tool length offsets. The tool length offset is the distance from the tool tip at home position to the workpiece Z zero position (see Figure 2–18). This distance is stored in a table that the programmer can access using a G-code or tool code. On a machining center that has a Fanuc control, a G43 code is used. The letter address G43 code is accompanied by an "H" auxiliary letter and a two-digit number. The G43 tells the control to compensate the Z axis, while the H and the number tell the control which offset to call out of the tool offset storage table. The tool length offset typically needs to be accompanied by a Z axis move to activate it.

A typical tool length offset block would look like this:

N0010 G43 H10;

The line number is 10. The G43 calls for a tool length offset, and the H10 is the number of the offset, which is found in register 10 of the tool length offset file. It is always a good idea to somehow correspond the tool length offset register number to the tool number. For example, if you are using tool number 10 (T10), try to correspond the height offset by using height offset number 10 (H10).

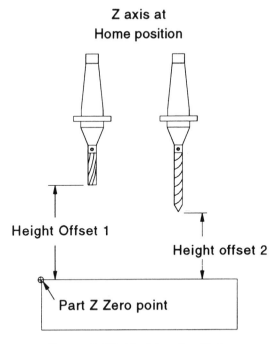

FIGURE **2–18** *Tool length offsets.*

On some other types of machining center controls, such as Mazatrol, the height offset is called up with the tool number. If the program calls for tool number 10, the control automatically accesses the tool file and offsets the tool according to the tool length registered in the tool file under the tool number 10.

Because machine controls vary, it is a good idea to find out how your control deals with variations in tool lengths.

TOOL DIAMETER OFFSETS

Tools differ in length and to compensate for that difference we use height offsets. Tools also differ in diameter and to compensate for this we use tool diameter offsets. Tool diameter offsets are also used to control the size of milled features.

In Figure 2–19 we had to compensate for the 1/2-inch diameter tool by offsetting the tool path by the radius of the tool. The control can offset the path of the tool so we can program the part just as it appears on the part print. This saves us from having to mathematically calculate the cutter path (see Figure 2–20). The diameter offset also allows the programmer to use the same program for any size cutter. Without diameter offsets the programmer would have to state the precise size of the tool to be used and program the center of the spindle accordingly.

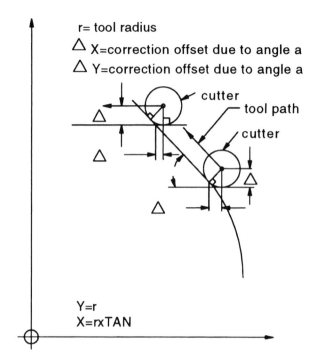

r= tool radius

△ X=correction offset due to angle a

△ Y=correction offset due to angle a

cutter

tool path

cutter

Y=r
X=rxTAN

FIGURE **2–19** *Tool diameter compensation allows us to program the part, not the tool path. The mathematical calculations that are needed to program a part profile with angles and radii, without the aid of cutter diameter compensation, can be very involved.*

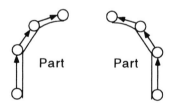

Part   Part

FIGURE **2–20** *Left and right tool compensation.*

With cutter compensation capabilities, the cutter size can be ignored and the part profile can be programmed. The radius of the cutting tool to be used is entered into the offset file, and when the offset is called the tool path will automatically be offset by the tool radius. If, after the part has been inspected, it is found to be too big or too small, the offset can be changed and the part can be re-machined to proper size.

Cutter compensation can be to the right or to the left of the part profile. To determine which offset you need, imagine yourself walking behind the cutting tool. Do you want the tool to be on the left of the programmed path or to the right (see Figure 2–19)?

Compensation direction is controlled by a G-code. When compensation to the left is desired, a G41 is used. When compensation to the right is desired, a G42 is used. When using the cutter compensation codes, you need to tell the controller which offset to use from the offset table. The offset identification is a number that is placed after the direction code. A typical cutter compensation line would look like this: N0030 G41 D12;.

To use cutter compensation, the programmer will have to make a machine move (ramp on). This additional move must occur before cutting begins. This move allows the control to evaluate its present position and make the necessary adjustment from centerline positioning to cutter periphery positioning. This move must be greater than the radius value of the tool.

To cancel the cutter compensation and return to cutter centerline programming, the programmer must make a linear move (ramp off) to invoke a cutter compensation cancellation (G40). This is an additional move after the cut is complete. Figure 2–21 shows a typical programming example that uses tool length and tool diameter compensations.

N0010 G70 G90; (Inch programming, absolute programming)

N0020 M06 T02; (Tool change, tool #2)

N0030 G54 X-10.250 Y-8.750 Z-7.525; (Workpiece zero setting)

N0040 M03 S800; (Spindle start clockwise, 800 RPM)

N0050 G00 X-1.00 Y-1.00; (Rapid to position #1, just off the lower left corner of the part)

N0060 G43 Z.100 H01; (Rapid down to .100 clearance above the part, invoke tool height offset)

N0070 G01 Z-.50 F5.0; (Feed down to depth at 5 inches per minute)

N0080 G41 D2 X0.0 Y0.0; (Ramp on and invoke cutter compensation stored as D2, feed to position 2)

N0090 G01 Y3.00; (Feed to position 3)

N0100 G01 X4.00; (Feed to position 4)

N0110 G01 Y-0.25; (Feed to position 5)

N0120 G40 Y-1.00; (Ramp off and cancel cutter compensation, feed to position 6)

N0130 G00 Z0.100; (Rapid to .100 clearance above the part)

N0150 G28; (Return all axes to home position)

N0160 M06 T0; (Tool change puts the tool away and empties the spindle)

N0170 M30; (Rewinds the program, resets the control, and ends the program)

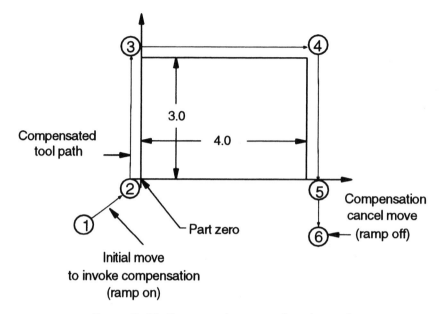

FIGURE 2–21 *Programming example using tool length and cutter diameter compensation.*

It is very important that the programmer's tooling intentions are communicated to the setup person or operator. This is usually done using the part manuscript and setup sheets. These will be examined in following chapters.

CHAPTER QUESTIONS

1. What is the most common CNC language in use today?
2. What primary role do preparatory functions serve?
3. Name three functions that miscellaneous codes control.
4. Name two considerations that must be taken into account when selecting a part datum location.
5. What does modal mean?
6. Describe the difference between G92 and G54 workpiece coordinate settings.
7. What is incremental positioning?
8. Describe the most common method of locating the arc centerpoint.
9. What purpose do tool length offsets serve?
10. What must be done to invoke a tool diameter compensation?
11. Complete the blocks for the part shown in Figure 2–22. Use a .50 endmill and cut to a depth of .25 around the part.

   N0010 G__ G__ ; (Inch programming, absolute programming)

   N0020 M__ T02; (Tool change, tool #2)

   N0030 G54 X_____ Y_____; (Workpiece zero setting)

   N0040 M__ S800; (Spindle start clockwise, 800 RPM)

   N0050 G__ X-1.00 Y-1.00; (Rapid to position #1)

   N0060 G__ Z___; (Rapid down to .100 clearance above the part)

   N0070 G__ Z___ __ 5.0; (Feed down to depth at 5 inches per minute)

   N0080 G01 X___ Y____; (Feed to position #2, offsetting for the tool radius)

   N0090 G01 X___ Y____; (Feed to position #3)

   N0100 G01 X___ Y____; (Feed to position #4)

   N0110 G01 X___ Y____; (Feed to position #5)

   N0120 G01 X___ Y____; (Feed to position #6)

   N0130 G01 X___ Y____; (Feed to position #7)

   N0140 G01 X___ Y____; (Feed to position #8)

   N0150 G01 X___ Y____; (Feed to position #9)

   N0160 G01 X___ Y____; (Feed to position #10, 1 inch to the left of the part)

   N0170 G__; (Return all axes to home position)

   N0180 M__ T__; (Tool change, tool 0)

   N0190 M__; (End program, rewind program)

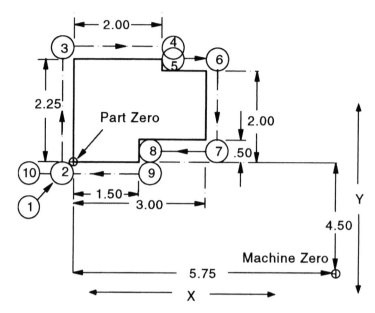

FIGURE **2–22** *Use with question 11.*

12. Program the part shown in Figure 2–23. Use a .5 endmill to machine the outside shape of the part and machine to a depth of .375. Assume it is tool number 5. Make sure you use offsets.

13. Program the part shown in Figure 2–24. Use a .5 endmill to machine the outside shape of the part and machine to a depth of .375. Assume it is tool number 5. Make sure you use offsets.

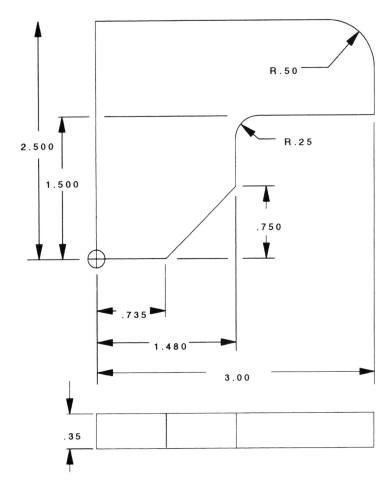

Figure 2–23 *Use with question 12.*

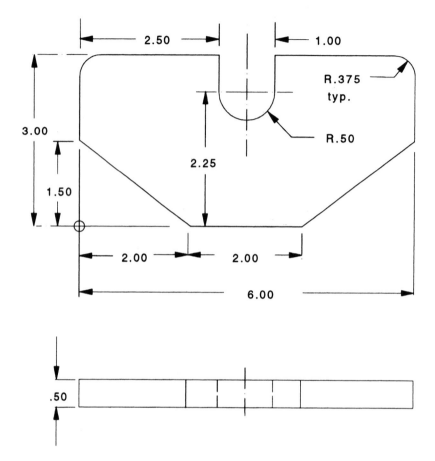

FIGURE 2–24 *Use with question 13.*

Chapter 3

BASIC TRIGONOMETRY

INTRODUCTION

This chapter will review simple right-angle trigonometry. Occasionally there may be dimensions missing on a blueprint that the programmer must have to complete the program. In these cases, simple trigonometry can be used to find the missing dimensions.

OBJECTIVES

Upon completion of this chapter, the reader will be able to:

- *Use the Pythagorean Theorem to calculate missing lengths.*
- *Find the sine, cosine, and tangent, using a table or calculator.*
- *Calculate missing lengths using functions such as sine, cosine, tangent, and so on.*
- *Calculate angles using functions such as sine, cosine, tangent, and so on.*

PYTHAGOREAN THEOREM

Some basic math is sometimes required to calculate dimensions needed for a cutter path. The math that is used for this purpose is called *trigonometry*. It involves the use of knowledge about triangles to solve for unknown dimensions. The calculations a programmer would normally have to make can be done with basic knowledge of right-angle trigonometry (see Figure 3–1). The sum of the angles of any triangle is always equal to 180 degrees. In a right triangle, one of the angles is always 90 degrees. In fact, a 90-degree angle is called a right angle. There are three sides and three angles in a right triangle. If we know the values of any two of these (excluding the 90-degree angle), we can calculate the unknown values.

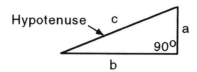

FIGURE **3–1** *This figure shows a typical right triangle. Note that one angle is 90 degrees. The sum of the angles of a triangle is always 180 degrees, so the sum of the other two angles of a right triangle must always equal 90. The longest side is the hypotenuse.*

Note that the longest side in a right triangle is called the *hypotenuse.* The hypotenuse is always the side opposite the 90-degree angle. There is a unique way to calculate the length of one side if the other two are known (see Figure 3–2). Suppose the length of side a is 4 inches, side b is 3 inches, and side c (hypotenuse) is 5 inches.

If you use the Pythagorean Theorem you see that if we take the square of side a, we get 16. If we take the square of side b, we get 9. If we add 16 and 9, we get 25. The square root of 25 is 5. This is the length of the hypotenuse. In another example, if we take the square of the hypotenuse, we get 25. If we take the square of side b, we get 9. If we then subtract 9 from 25, we get 16. The square root of 16 is 4, the length of side a. This formula is very helpful.

In Figure 3–3, the length of the hypotenuse is unknown. We do know the lengths of two sides: side a is 2.5 inches, and side b is 3.122 inches. If we

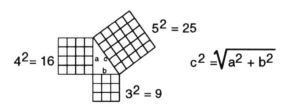

FIGURE **3–2** *The Pythagorean Theorem is used to solve for an unknown side if two sides are known.*

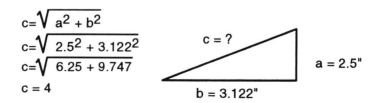

FIGURE **3–3** *The lengths of side a and side b are known. By squaring each, adding them, and taking the square root, we get the length of the hypotenuse.*

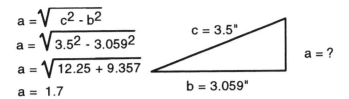

FIGURE **3–4** *We know the length of one side and the hypotenuse. By squaring the hypotenuse, subtracting the square of the side from it, and taking the square root, we find the length of the unknown side.*

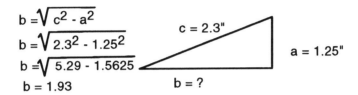

FIGURE **3–5** *The lengths of the hypotenuse and side a are known. By squaring the hypotenuse and subtracting the square of side a from it, and then taking the square root, we find the length of side b.*

square the length of side a (2.5 * 2.5), square the length of side b (3.122 * 3.122), and then add them (6.25 + 9.747) and take the square root, we find that c equals 4 inches.

In Figure 3–4, we know the length of the hypotenuse (3.5 inches) and the length of side b (3.059). If we square the hypotenuse, and subtract the square of side a, and take the square root, we find that the length of side a is 1.7 inches.

Examine Figure 3–5. In this case we know the length of the hypotenuse (2.3 inches) and the length of side a (1.25 inches). We need to calculate the length of side b. In this case we will square the hypotenuse (2.3 * 2.3) and then square side a (1.25 * 1.25). We will subtract the square of side a from the square of the hypotenuse and take the square root. The answer is 1.93 inches.

SINE, COSINE, AND TANGENT

Sometimes two sides are unknown. We might know only one side and an angle, for example, but we can still calculate the lengths. Three main formulas express the relationship between the sides and angles. These are called the *sine*, *cosine*, and *tangent* (see Figure 3–6). The formulas are easy to use. The sine is equal to the opposite side divided by the hypotenuse. The cosine is equal to

$$\frac{\text{Opposite}}{\text{Hypotenuse}} = \text{Sine}$$

$$\frac{\text{Adjacent}}{\text{Hypotenuse}} = \text{Cosine}$$

$$\frac{\text{Opposite}}{\text{Adjacent}} = \text{Tangent}$$

FIGURE 3–6 *Three common formulas that can be used to find the angles or sides of a right triangle if two of the three values are known.*

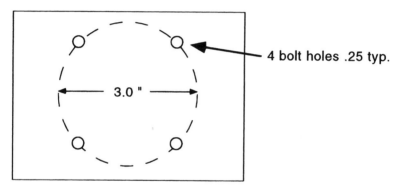

FIGURE 3–7 *A part with four holes through it on a 3.0-inch bolt circle.*

the side adjacent divided by the hypotenuse. The tangent is equal to the side opposite divided by the side adjacent. The sine, cosine, and tangent are variables that have a specific value for a specific angle. You can look them up in a table, or many calculators will calculate them. For example, Figure 3–7 shows a part that has four holes through it. We need to calculate the X and Y position for each hole. Assume the center of the part is X0, Y0.

If we calculate the values for one hole, we can use them to find the other positions because the holes are symmetric. Note that four holes are on a 3-inch bolt circle. We then know that the radius of the circle is 1.500 inches. We can then draw a right triangle to find the hole location (see Figure 3–8). We know the length of the hypotenuse is 1.500 inches. We also know that the angle is 45 degrees because the holes are spaced 90 degrees apart. The angle is 45 degrees from horizontal.

We can use one of the formulas in Figure 3–6 to calculate our unknown values. We will use the sine formula to calculate the length of the opposite side. The hypotenuse is always the longest side. The opposite and adjacent sides are dependent on which angle we know. The side that touches the

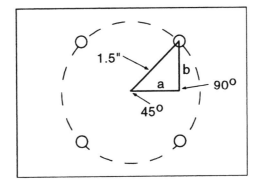

FIGURE **3–8** *This figure shows how we can draw a right triangle to find the position of the upper right hole. We know that the hypotenuse is 1.5 inches. We also know that a 45-degree angle is formed between the horizontal and the hypotenuse.*

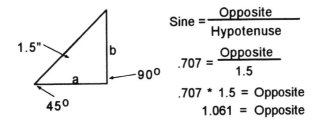

FIGURE **3–9** *The sine formula is used to calculate the length of side b. The sine of 45 degrees was found to be .707 with a calculator. The length of b is 1.061 inches.*

known angle becomes the side adjacent. The other side is the opposite side (side opposite the known angle). We know the hypotenuse (1.5) and the angle of 45 degrees. We can look up the sine of 45 degrees in a table or use a calculator with trigonometric functions. The sine of 45 degrees is .707. Figure 3–9 shows how the length of the opposite side was calculated to be 1.061 inches.

Next we will calculate the length of the adjacent side (see Figure 3–10). We know the length of the opposite side (1.061) and the angle of 45 degrees. Often several formulas can be used to find the length. Figure 3–17 can be used to find the formulas to calculate the value of an unknown side or angle. Figure 3–10 uses the tangent formula. The tangent for a 45-degree angle is equal to 1. This can be found in a table or with a calculator. The length is 1.061. Both sides are the same length because this was a 45 degree right triangle. In a 45-degree right triangle there are two 45-degree angles. This means that the two sides must be equal.

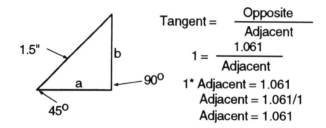

FIGURE **3–10** *Calculating the length of the adjacent side (a).*

Next, calculate the length of the hypotenuse for practice and to prove that the formulas work (see Figure 3–11). In this case, the cosine formula was used. The length of the adjacent side is 1.061 and the cosine of 45 degrees is .707. Figure 3–12 shows an example of this calculation to find hole locations. (Note that the sine of 45 degrees [.707] is equal to the cosine of 45 degrees [.707]. This is a unique case and only true for 45 degrees.) Using the formula, we calculate the length to be 1.500 inches.

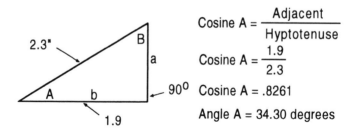

FIGURE **3–11** *Using the cosine formula to calculate the length of the hypotenuse. We already know that it is 2.3 inches, but it was calculated to prove the formula*

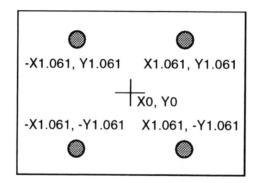

FIGURE **3–12** *Hole locations.*

Three other formulas express the relationships of right triangles in different ways (see Figure 3–13). They can be used in the same manner.

We can also calculate angles in a right triangle if we know two lengths. In Figure 3–14 we need to find angle A. We know the length of the adjacent side and the length of the hypotenuse. We can use the cosine function to calculate the angle.

Note that the length of the adjacent side was divided by the hypotenuse. This is equal to the cosine of angle A. Then you look up .8261 in a cosine table or use a calculator with trigonometric calculations to calculate the angle. If we divide 1.9 by 2.3, we get .8261. This is the cosine of angle A. In this case, angle A equals 34.30 degrees.

In Figure 3–15 we need to find the coordinates for point A. We know an angle (10 degrees) and the length of the adjacent side (1.25 inches). We can use this to find the length of the opposite side.

The tangent formula was chosen. The length of the opposite side is .220. If we add .220 to the 1-inch dimension, we have the Y position for point A (1.22). The X position for point A is simply the length of the adjacent side (1.25) (Figure 3–16). Figure 3–17 contains a table that can be used to find unknown sides or angles. Note that we are assuming the lower left corner of the part to be X0, Y0.

$$\frac{\text{Hypotenuse}}{\text{Opposite}} = \text{Cosecant}$$

$$\frac{\text{Hypotenuse}}{\text{Adjacent}} = \text{Secant}$$

$$\frac{\text{Adjacent}}{\text{Opposite}} = \text{Cotangent}$$

FIGURE 3–13 *Three other formulas are useful in calculating unknown sides or angles. The choice of formula depends on which sides and angles you know.*

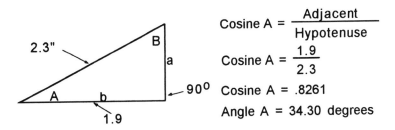

$$\text{Cosine A} = \frac{\text{Adjacent}}{\text{Hypotenuse}}$$

$$\text{Cosine A} = \frac{1.9}{2.3}$$

Cosine A = .8261

Angle A = 34.30 degrees

FIGURE 3–14 *How to calculate angle A when the lengths of the hypotenuse and side adjacents are known.*

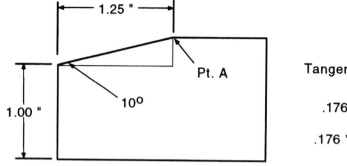

$$\text{Tangent} = \frac{\text{Opposite}}{\text{Adjacent}}$$

$$.176 = \frac{\text{Opposite}}{1.25}$$

$$.176 * 1.25 = \text{Opposite}$$

$$.220 = \text{Opposite}$$

FIGURE 3–15 *Calculating the coordinates for point A. The 10-degree angle is known, and the length of the adjacent side of the triangle is known. The tangent function was chosen to calculate the opposite side's length.*

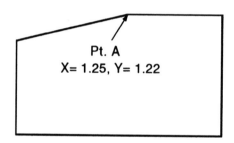

FIGURE 3–16 *Coordinates for point A. The X coordinate was found by adding the length of the opposite side (.220) to the 1.000 dimension. The Y dimension is simply equal to the length of the adjacent side (1.25). Note that we are considering the lower left corner to be X0, Y0.*

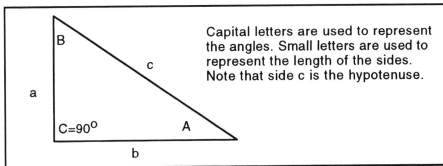

Sides and Angles Known	Formulas for Sides and Angles		
Side a & Hypotenuse c	$b=\sqrt{c^2 - a^2}$	$\text{sine } A = \dfrac{a}{c}$	$B = 90^0 - A$
Side a & Side b	$c=\sqrt{a^2 + b^2}$	$\tan A = \dfrac{a}{b}$	$B = 90^0 - A$
Side b & Hypotenuse c	$a=\sqrt{c^2 - b^2}$	$\text{sine } B = \dfrac{b}{c}$	$A = 90^0 - B$
Hypotenuse c & Angle B	$b = c \text{*sine } B$	$a = c \text{*cos } B$	$A = 90^0 - B$
Hypotenuse c & Angle A	$b = c \text{*cos } A$	$a = c \text{ * sine } A$	$B = 90^0 - A$
Side b & Angle B	$c = \dfrac{b}{\text{sine } B}$	$a = b \text{ * cot } B$	$A = 90^0 - B$
Side b & Angle A	$c = \dfrac{b}{\cos A}$	$a = b \text{ * tan } A$	$B = 90^0 - A$
Side a & Angle B	$c = \dfrac{a}{\cos B}$	$b = a \text{ * tan } B$	$A = 90^0 - B$
Side a & Angle A	$c = \dfrac{a}{\text{sine } A}$	$b = a \text{ * cot } A$	$B = 90^0 - A$

FIGURE 3–17 *This table can be used to choose the appropriate formula for trigonometric calculations. Look at the left column for the values you know and then choose the formula from that row that will calculate the value that you need.*

CHAPTER QUESTIONS

1. Calculate the length of side b.

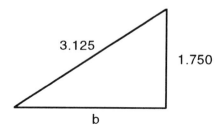

2. Calculate the length of side c (hypotenuse).

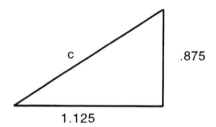

3. Calculate the length of side a.

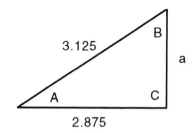

4. Calculate the missing lengths and angles.

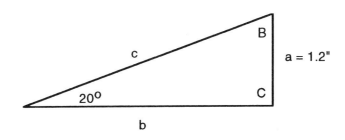

Sides		Angles	
b=	c=	B=	C=

5. Calculate the missing sides and angles.

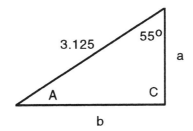

Sides		Angles	
a=	b=	A=	C=

6. Calculate the six hole locations for this part.

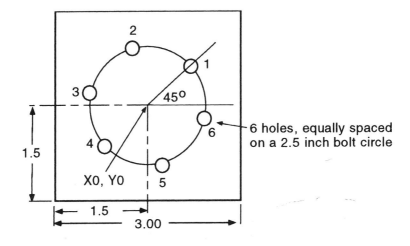

6 holes, equally spaced on a 2.5 inch bolt circle

7. The diagram in this question shows a portion of a part that needs to be machined. The programmer did not use offsets so he/she must now calculate the XY position of point A. A 1-inch cutter will be used.

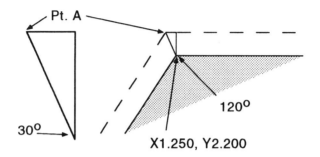

Chapter 4

··

CARBIDE FUNDAMENTALS

INTRODUCTION

The choice of which cutting tools to use on numerical control machines is an important one. Numerical control machines are highly productive and efficient machines, but if the cutting tools cannot hold up under extreme conditions, the machine's efficiency and accuracy will be greatly diminished.

OBJECTIVES

Upon completion of this chapter, the reader will be able to:

- *Describe how carbide inserts are made.*
- *State the two main characteristics of carbide.*
- *Describe the term "grade" as it applies to insert choice.*
- *Explain which factors to consider when selecting tool nose radius.*
- *Name three insert shapes in order of increasing strength.*
- *State two factors to consider when selecting insert shapes.*
- *Describe what the size of an insert is based on.*
- *State how insert size and depth of cut are interrelated.*
- *Describe the purpose and function of the different rake angles.*
- *Choose inserts for various applications.*
- *Describe how tool holders are identified.*
- *Explain the purpose of qualified tools.*
- *Troubleshoot typical machining problems.*

FUNDAMENTALS OF CARBIDE TOOLING

With today's CNC technology, machines can be operated with very little training; however, to be promoted to a CNC programmer, one must know the fundamentals of carbide cutting tools. Tooling is crucial to the success of a machine shop. A great machine with poor tooling will perform very poorly. Because good tooling techniques are vital to productivity, this chapter will concentrate on the fundamentals of carbide cutting tools.

CEMENTED CARBIDE

Cemented carbide, or tungsten carbide, is a form of powdered metallurgy. Fine powders consisting of tungsten carbide and other hard metals bonded with cobalt are pressed into required shapes and then sintered.

Sintering is the heating of the carbide materials to approximately 2500 degrees Fahrenheit. At this temperature the cobalt melts and flows around the carbide materials. Cobalt acts as the binder that holds the carbide particles together. After the carbide insert cools, the insert is almost as hard as a diamond.

The hardness and physical properties of cemented carbides allow them to operate at high cutting speeds and feeds with very little tool deformation.

The great hardness of carbide is also its Achilles' heel. Extremely hard materials are also very brittle, and this can cause problems under certain machining conditions. Through the use of different mixtures of hard materials, carbide manufacturers have come up with different grades of carbide materials. Selecting the proper grade for the machining application is important for economy and productivity.

SELECTION OF CARBIDE TOOL GRADE

Carbide tools come in a variety of grades. The grade is based on the carbide's wear resistance and toughness. As an insert becomes harder or more wear resistant, it becomes brittle (less tough).

If you used a very hard, wear-resistant insert on a material that has an uneven or interrupted surface (interrupted cut), the insert would most likely break. The ANSI/ISO standards organizations have devised systems of grading carbide based on the carbide insert's application and physical make up. These systems differ and can be quite confusing. Carbide manufacturers help clarify matters by putting cross-reference charts in their catalogues (see Figure 4–1).

COATINGS FOR CARBIDE INSERTS

Carbide is a very hard, durable cutting tool, but it still wears. The wear resistance of cemented carbide can be increased greatly by using coatings.

Grade selections

grades	composition and application	ISO
PVD TIN (titanium nitride) coated carbide grades		
KC710	**composition:** A PVD coated grade that has good toughness and thermal shock resistance with good crater wear resistance and resistance to buildup on the cutting edge. **application:** Greatly improved productivity when cutting a variety of steels and tool steels over a broad range of feeds at moderate to high speeds.	M15-M25 P15-P25
KC720	**composition:** A tough, durable PVD coated carbide grade. **application:** Developed for cutting high-temperature alloys, stainless steels, and low-carbon steels at low to moderate speeds. Its unique mechanical and thermal shock resistant properties, and resistance to edge buildup, enable KC720 to deliver superior performance and reliability on difficult operations, like interrupted cuts, and when milling high-temperature alloys with coolant.	K25-K35 M30-M40 P25-P45
KC730	**composition:** A PVD coated carbide grade. **application:** For milling cast and ductile irons, high-temperature alloys, aerospace materials, refractory metals, and 200 and 300 series stainless steels. The substrate offers superior thermal deformation resistance, depth of cut notch resistance, and edge strength. The uniformly dense PVD coating increases wear resistance, reduces problems with edge buildup, and provides an unusually good combination of properties for machining difficult-to-machine materials and aluminum.	K05-K15 M05-M15
New "M" multi-coating milling grades		
KC725M	**composition:** A patented, multi-layer, PVD coated, TiN/TiCN/TiN. **application:** A new milling grade engineered for high productivity wet milling of carbon, alloy and austenitic stainless steels. The high thermal shock resistance of the tough carbide substrate combined with the patented multi-layer coating provides long and reliable tool life in aggressive milling operations using coolant. It is the higher speed companion to KC720 in wet steel milling.	M15-M35 P20-P35
KC792M	**composition:** A PVD/CVD coated grade. **application:** Developed for 400-900 sfm milling of steels. Cobalt enriched, it is deformation resistant in interrupted cutting. The patent pending PVD over CVD coatings allow for thicker coatings than possible with other CVD coatings, as well as desirable compression stresses in the coatings to counteract thermal crack propagation. KC792M is the higher speed companion to KC710 for dry steel milling.	M25-M30 P30
KC992M	**composition:** A multi-layered ceramic CVD coated with TiCN/Al$_2$O$_3$ carbide. **application:** Designed to mill grey cast iron, with or without coolant, at medium speeds and feed rates with honed geometry inserts. Nodular irons with machinability index of 68-78 and BHN below 300 can also be machined up to 600 sfm and equivalent chiploads with T-land versions of insert geometries.	K10-K25
Uncoated carbide grades		
K313	**composition:** An unalloyed WC/Co fine-grained grade. **application:** Exceptional edge wear resistance, combined with very high edge strength and abrasion resistance, delivers high-speed metal removal rates with lighter chip loads when machining nonmetals and nonferrous metals including aluminum, stainless steels, and titanium materials.	K05-K15 M10-M20

FIGURE 4–1 *Kennametal's grade system chart gives grade selection choices for machining applications. If a depth of cut of 3/16 of an inch is required, an insert with a 3/8-inch inscribed circle should be used. Insert thickness also affects the feed rate. If an insert is going to be used for continuous, heavy-feed roughing cuts, an insert of greater thickness should be selected. (Courtesy of Kennametal, Inc.)*

Wear-resistant coatings can be applied to the carbide substrate (base material) through the use of plasma coating or vapor deposition. The coating that is deposited is very thin but very hard.

The most common types of coatings include titanium carbide (TiC), titanium nitride (TiN), and aluminum oxide (AlO). Aluminum oxide is a very wear-resistant coating used in high-speed finishing and light roughing operations performed on most steels and all cast irons. Titanium nitride coatings are very hard and have the strength characteristics to perform well under heavy rough-cutting conditions. All three coatings will perform well on most steels, as well as on cast iron.

DIAMOND COATED INSERTS

Coated cutting tools have been around for years. Titanium and boron nitride materials have driven the coated cutting tool industry. The newest material to make its presence known is the polycrystalline diamond, or PCD. PCD tools are becoming widely accepted as tooling solutions for difficult-to-machine materials. The PCD material has the hardness of a diamond and the friction coefficient of Teflon™. This combination has resulted in a remarkable increase in tool life. Diamond-coated tools are still being tested quite extensively, and manufacturers have not even published recommended speeds and feeds for their cutting tools. Diamond-coated tools are still in the infancy stage, but from all indications they could be the predominant cutting tool of the future.

TOOL-NOSE RADIUS

Although selecting the proper grade of insert is probably the most important, other factors such as nose radius are very important when selecting the proper tool for the application.

The nose radius of the tool directly affects tool strength and surface finish, as well as cutting speeds and feeds. The larger the nose radius, the stronger the tool. If the tool radius is too small, the sharp point will make the surface finish unacceptable, and the life of the tool will be shortened.

Larger nose radii will give a better finish and longer tool life, and will allow for higher feed rates to be used; however, if the tool nose is too large, it can cause chatter. It is always good machining practice to select an insert with a tool nose radius as large as the machining operation will allow.

INSERT SHAPE

Indexable inserts, also known as throw-away inserts, come in many shapes. Inserts are clamped in tool holders and provide cutting tools with multiple cutting edges. After the cutting edges are worn to a point where they can no longer be used, they are discarded.

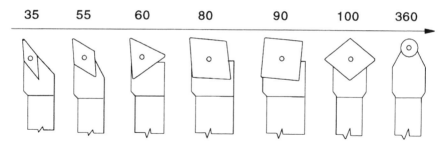

FIGURE 4–2 *The shape of the insert will have a great effect on the strength of the tool. Select the largest included angle that will cut the part.*

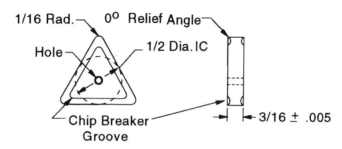

FIGURE 4–3 *This figure is a diagram of a typical triangular insert.*

When selecting an insert shape many factors must be taken into consideration:

What geometric features are required on the workpiece.

What lead angle can be used.

What operation needs to be performed.

How to get the maximum number of cutting edges.

What required strength is needed to do the job.

Figure 4–2 shows the characteristics of different shaped inserts in order of their strength. Round inserts have the greatest strength, as well as the greatest number of cutting edges, but the round configuration limits the operations that can be performed.

Square or 90-degree inserts have less strength and fewer cutting edges than round inserts, but are a little more versatile.

Triangular inserts (see Figure 4–3) are more versatile than square inserts, but as the included angle is reduced from 90 degrees to 60 degrees it becomes weaker and more likely to break under heavy machining conditions.

Diamond-shaped inserts are probably the most commonly used shape. Diamond-shaped inserts range from a 35 degree to an 80 degree included angle. Diamond-shaped inserts are much more versatile than square and round inserts. It is good machining practice to select the largest included angle insert that will properly cut the shape of the part because the insert will be stronger.

INSERT SIZE

The size of the insert is based upon the inscribed circle (IC, which is the largest circle that will fit inside the insert), the insert thickness, and the tool nose radius (see Figure 4–3).

The depth of cut possible with an insert depends greatly on the insert size. The depth of cut should always be as great as the conditions will allow. A good rule of thumb is to select an insert with an inscribed circle twice that of the depth of cut.

RAKE ANGLES

Rake angle, or back rake angle, is the angle at which the chips flow away from the cutting area (see Figure 4–4). There are three principal rake angles: negative, positive, and neutral.

When selecting the proper rake tool holder, it is essential to look at the machining conditions. Negative rake holders are economically a good choice because they hold neutral rake inserts. Neutral rake inserts have twice as many cutting edges as positive rake inserts because they can be turned over and used. Another advantage is that negative rake tool holders provide more support for the cutting edges of the insert. Under normal operating conditions negative rake inserts are also a little stronger because of the compressive strength of carbide. Negative rake holders should be used when the tool and the work are held very rigidly and when high machining speeds and feeds can be maintained. More horsepower is required to cut with negative rake tool holders, which is why there is an increasing trend toward the use of positive rake cutting.

Positive rake cutting is more of a shearing effect than the pushing effect generated by negative rake. Positive rake holders generate less cutting force and have less of a tendency to chatter.

Horsepower requirements are greatly reduced with positive cutting tools. Lower horsepower consumption is an important factor with today's smaller

Neutral Rake   Positive Rake   Negative Rake

FIGURE 4–4 *Side view of back rake angles.*

machines. Smaller machines are built with lower horsepower and less rigidity than the older manual machines.

The only drawback to positive rake cutting tools is their inability to stand up to harder materials. Eight to 10 years ago positive rake tool holders were considered unsuitable for machining steels because of cutting-edge weakness; however, recent advances in carbide technology have produced tougher substrate materials that provide greater edge strength. Some carbide companies have even gone as far as to recommend positive rake holders whenever possible.

Positive rakes should be used when machining softer materials because the chips are able to flow away from the cutting edge freely and the cutting action is more of a peeling effect. Positive rake cutting can be very successful on long slender parts or other operations that lack rigidity.

LEAD ANGLE

Lead angle, or side-cutting edge angle, is the angle at which the cutting tool enters the work (see Figure 4–5). The lead angle can be positive, negative, or neutral. The tool holder always dictates the amount of lead angle a tool will have.

Tool holders should always be selected to provide the greatest amount of lead angle that the job will allow. There are two advantages to using a large lead angle. First, when the tool initially enters the work, it is at the middle of the insert where it is strongest, instead of at the tool tip, which is the weakest point of the tool. Second, the cutting forces are spread over a wider area, reducing the chip thickness (see Figure 4–6).

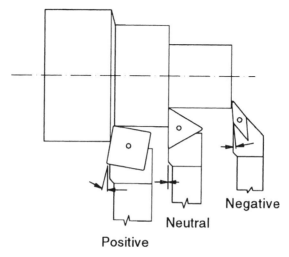

FIGURE 4–5 *Lead or side-cutting edge angle is determined by the tool holder type. The lead angle can be positive, neutral, or negative.*

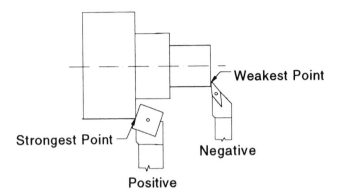

FIGURE 4–6 *The effect of the lead angle on the strength of the insert.*
Increasing the lead angle will greatly reduce tool breakage when
roughing or cutting interrupted surfaces.

INSERT SELECTION

Now that we have covered some aspects of carbide tool selection, let's look at
the questions that we need to answer when selecting the proper insert grade
and style. One of the first considerations is the material to be machined.

MACHINABILITY OF METALS

Machinability describes the ease or difficulty with which a metal can be cut.
Machining involves removing metal at the highest possible rate and at the
lowest cost per piece. Different materials' structures pose different problems
for the machinist. Materials that are easy to machine have high machinability
ratings and therefore cost less to machine. Materials that are difficult to machine have lower machinability ratings and cost more to machine.

The machinability of a material has a direct correlation to the material's hardness, or its ability to resist penetration or deformation. A number of tests measure the hardness of a material, but the most common test for machinability
is Brinell. Brinell hardness or BHN is stated as a number: the higher the BHN
number, the harder the material. Hardness, although a major factor affecting
machinability, is not the only factor that determines machinability.

STEELS

Steels are generally classified based on their carbon content and their alloying
elements. Plain carbon steels have only one alloy, carbon, mixed with iron.
Carbon has a direct effect on a steel's hardness. Plain carbon steel's machinability is directly reflected in its carbon content. Alloy steels, on the other
hand, have carbon and other alloying elements mixed with iron. These alloying

elements can give steel the characteristic of not only being hard, but also being tough. The major concern with machining alloy steels is their tendency to work harden, a phenomenon that occurs when too much heat from the cutting process is developed in the steel. The heat changes the properties of the steel, making it hard and difficult to machine. Great care must be taken when machining some alloy steels.

Plain carbon steel is divided into three categories: low carbon, medium carbon, and high carbon. Low-carbon steels have a carbon content of 0.10 percent to 0.30 percent and are relatively easy to machine. Medium-carbon steels have a carbon content of 0.30 percent to 0.50 percent. They are relatively easy to machine, but because of the higher carbon content, they have a lower cutting speed than that of low-carbon steel. High-carbon steels have a carbon content of 0.50 percent to 1.8 percent. When the carbon content exceeds 1.0 percent, high-carbon steel becomes quite difficult to machine.

STAINLESS STEEL

Stainless steels have carbon, chromium, and nickel as alloys. Stainless steels are a very tough, shock-resistant material and are difficult to machine. Work hardening can be a problem when machining stainless steels. To avoid work hardening, use lower speeds and increased feed rates. Chip control is sometimes a problem when machining stainless because of its toughness and its unwillingness to break.

CAST IRON

Cast iron is a broad classification for gray, malleable, nodular, and chilled-white cast iron. This grouping is in order of its machinability. Gray cast iron is relatively easy to machine, while chilled-white cast iron is sometimes unmachinable. Cast iron does not produce a continuous chip because of its brittleness.

The machinability of any material can be affected by factors such as heat treatment. Heat treating can either harden or soften a material. The condition of the material at the time of machining should be taken into consideration when deciding a material's machinability.

INSERT SELECTION PRACTICE

Figure 4–7 shows a typical lathe part. We can choose an appropriate insert for the part using the Kennametal grade system chart (Figure 4–1), the insert identification system shown in Figure 4–8, and the other information found previously in the chapter.

> **1.** *What type of material is being cut? Would you use a cast iron or a steel grade? Answer: The material used for the part is 1018 cold rolled steel, so a steel cutting grade would be appropriate.*

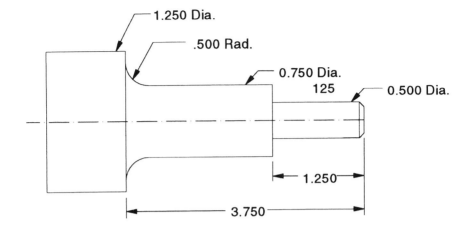

Material: 1018 Cold Rolled Steel, 200 BHN 125 Finish

FIGURE 4–7 *Typical lathe part.*

2. *How hard is the material? How does this affect the grade? Answer: The material is a low carbon alloy steel of only 200 Brinell hardness. A moderate hardness grade would be a good choice.*

3. *What is the condition of the material? Does the surface show evidence of scale or hard spots? How does this affect the selection of the grade, insert shape, rake angle, and nose radius? Answer: The material is cold rolled steel, which has little or no scale. Again, a general-purpose insert of moderate hardness and strength would be applicable.*

4. *What shape insert do we need to perform this job? Answer: For roughing this part, we would like to use a larger angled insert such as an 80-degree diamond. The finishing tool needs to have a little smaller angle to cut the radius. A 55-degree diamond or triangular insert would be a good choice for finishing the part.*

5. *How rigid is the machining setup? How does this affect the rake angles and nose radius? Answer: As you can see from the part print, the part has a small turned diameter on the end. This small diameter may tend to deflect and chatter. A positive rake insert with a 1/32 or 1/16 tool nose radius would probably be the best choice.*

6. *What are the surface finish requirements of the part? How does this affect the nose radius? Answer: The surface finish requirement of the part is 125. A 125 finish is a standard machine finish that can be held using a 1/32 or 1/16 tool nose radius. Slowing down the feed rate will also help to acquire the 125 finish requirement.*

Turning Insert Identification

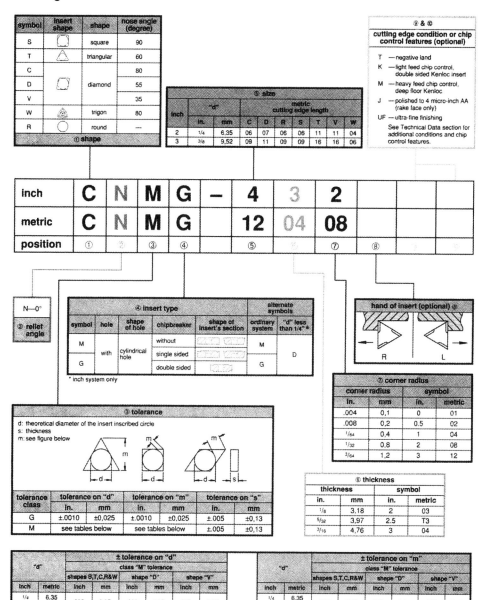

FIGURE 4–8 *Insert identification system. (Courtesy of Kennametal, Inc.)*

TOOL HOLDER STYLE & IDENTIFICATION

Carbide manufacturers and the American Standards Association have created a tool holder identification system for indexable carbide tool holders (see Figure 4–9). Because a huge variety of holders are available, we cannot include all of the holder types in this text.

QUALIFIED TOOLING

Tools that are used in numerical control machines are machined to a high level of accuracy. Qualified tools are typically guaranteed to be within .003 of an inch (see Figure 4–10).

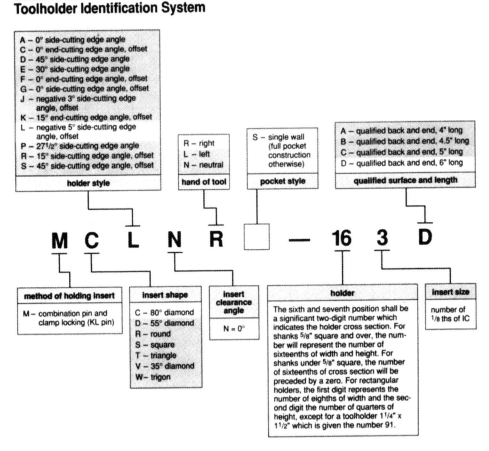

FIGURE 4–9 *Tool holder identification system.*
(Courtesy of Kennametal, Inc.)

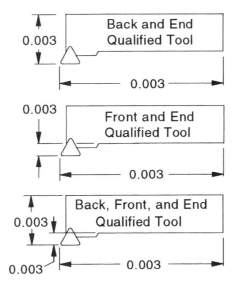

FIGURE 4–10 *Qualifications of tool holders.*

The accuracy of the cutting tip is referenced to specific points or datums located on the holders. This higher level of accuracy allows the operator to change inserts without having to remeasure the tools. The qualified measurements are located from two or three datums and are a measurement over the radius of the tool nose.

CHIP CONTROL

Chip control refers to the ability to control the outcome of chip formation. Chip control is important to operator safety, tool life, and chip handling. Numerical control machines typically have chip conveyers to automatically deposit chips into recycling containers. If chips are allowed to become long and stringy, they will clog up the chip conveyers. Straight, stringy chips will also wrap around the tool, the workpiece, and the work holding device, which can cause tool breakage and an especially dangerous situation for machine operators. Soft, gummy, and tough materials can wrap around spinning tools and workpieces. The chips begin whipping around, sending sharp, hot chips in every direction. For this reason, there has been considerable research in the area of chip control.

Tool manufacturers have provided us with two basic types of chip control devices for indexable carbide inserts: the mechanical chip breaker and the molded chip breaker (see Figure 4–11). The chip breaker is designed to redirect the flow of the chip, causing it to curl into a figure 6 or figure 9 (see Figure 4–12). A chip

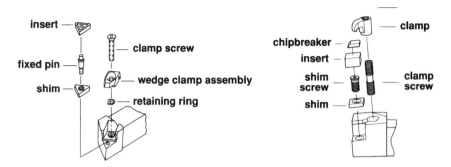

FIGURE 4–11 *Two styles of chip breakers: a molded chip breaker (left) and a mechanical chip breaker (right). (Courtesy of Kennametal, Inc.)*

of this configuration is said to be the perfect chip for steel cutting. When cutting cast iron, chip control is not a problem because iron is brittle and does not flow away from the cutting edge the way steel does.

MECHANICAL CHIP BREAKERS

The mechanical chip breaker is less common than the molded chip breaker. The mechanical chip breaker has more parts, making them more expensive to purchase. Changing an insert with a mechanical chip breaker takes more time than changing molded chip breaker inserts, and this adds to the cost of the workpiece.

MOLDED CHIP BREAKERS

Molded chip breakers use a molded groove to change the direction of the chip. Molded chip breakers are available in many different configurations. Some molded chip breakers are designed for certain materials, while others are designed for different feed rates and depths of cut.

FACTORS AFFECTING CHIP FORMATION

Chip breakers have greatly increased our ability to control chips; however, there are still factors that need to be addressed no matter which type of chip breaker is used. The three factors that affect chip control the greatest are feed rate, cutting speed, and tool shape. One of the quickest ways to eliminate long, stringy chips is to increase the feed rate: a thicker chip will curl and break easier than thin chips.

Decreasing the side-cutting edge angle or lead angle will also create a thicker chip. Sometimes increasing the speed will help the chip flow and curl easier. Long stringy chips are not the only chips that can cause problems in machining.

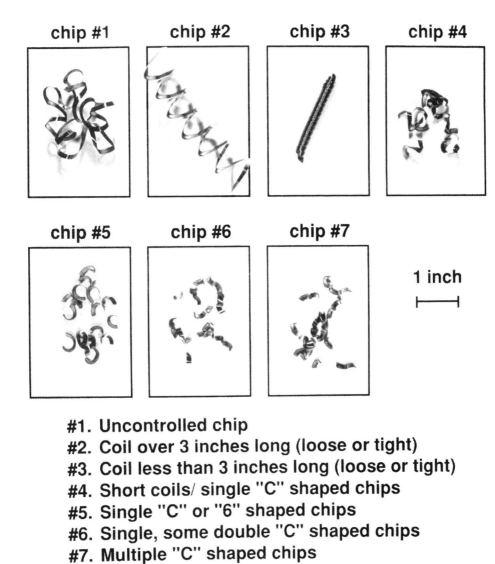

#1. Uncontrolled chip
#2. Coil over 3 inches long (loose or tight)
#3. Coil less than 3 inches long (loose or tight)
#4. Short coils/ single "C" shaped chips
#5. Single "C" or "6" shaped chips
#6. Single, some double "C" shaped chips
#7. Multiple "C" shaped chips

Chip forms 3,4,5,6 are acceptable   Chip forms 1,2, and 7 are unacceptable

FIGURE 4–12 *Different types of chips. (Courtesy of Kennametal, Inc.)*

Conduit chips are chips that are almost ready to break (see Figure 4–12). A conduit chip is a long curly chip, which is common when machining soft ductile materials. To remedy the problem, try increasing the feed rate. If using a mechanical chip breaker, move the chip breaker closer to the cutting edge or switch to a molded chip breaker.

Corrugated chips (see Figure 4–12) have a very tight curl and represent the opposite problem from stringy chips. These types of chips are being bent too much and are usually caused by excessive feed rates. Corrugated chips do not pose a chip control problem, but they are a sign of improper cutting action and should be dealt with immediately. If slowing the feed rate doesn't change the chip formation, move the mechanical chip breaker back. If using a molded chip breaker, use a narrower chip breaker to allow the chip to make a wider curl.

CHIP COLOR

As a machine operator, you should always be aware of the chips you are producing. Analyze the chip's shape and color. A deep blue steel chip indicates that the heat of the cutting action is being drawn away from the workpiece, as it should. A dark purple or black chip indicates excessive heat. In this case, reduce the cutting speed and any other machining conditions until the color of the chip is acceptable. Chips should always be clean and smooth on the underside, not torn and ragged. Proper chip formation is a balancing act of speeds, feeds, and chip breaker formation. Look to the chip for the information you need to balance these factors.

TROUBLESHOOTING

Carbide cutting tools, when operating as expected, are consistent, durable cutting tools; however, problems will sometimes result when using carbide cutting tools. It will be at this point the CNC operator will need to change certain factors or cutting conditions.

The first step is to diagnose the problem. Possible problems include premature failure of the insert, edge wear, crater wear, edge build up, depth of cut notching, chipping, thermal cracking, and thermal deformation.

CATASTROPHIC BREAKAGE

Premature failure or insert breakage is a problem that will be apparent even to the least experienced machine operator. If the insert breaks and continues to break even after being changed, there is a problem.

A possible cause of tool breakage is that the operating conditions are too high. Slow down the speed and especially the feed. If the grade that you have selected is too hard for the material or the condition of the material, select a tougher grade of insert. The lead angle may be too small. Select a tool holder that lends more support to the tool tip.

EDGE WEAR

Edge wear is more difficult to diagnose. Excessive edge wear is the unnatural wearing away of the insert along the side or flank of the cutting edge (see Figure 4–13).

Troubleshooting

Edge Wear[*]

Corrective Action
- Increase Feed Rate
- Reduce Speed (sfm)
- Use More Wear Resistant Grade
- Apply Coated Grade

Chipping

Corrective Action
- Utilize Stronger Grade
- Consider Edge Preparation
- Check Rigidity of System
- Increase Lead Angle

Heat Deformation (Up-Set)

Corrective Action
- Reduce Speed
- Reduce Feed
- Reduce Depth of Cut (doc)

Depth-of-Cut Notching

Corrective Action
- Change Lead Angle
- Consider Edge Preparation
- Apply Different Grade
- Adjust Feed

Thermal Cracking

Corrective Action
- Properly Apply Coolant
- Reduce Speed
- Reduce Feed
- Apply Coated Grade

Built-Up Edge

Corrective Action
- Increase Speed (sfm)
- Increase Feed Rate
- Apply Coated Grades or Cermets
- Utilize Coolant

Crater

Corrective Action
- Reduce Feed Rate
- Reduce Speed (sfm)
- Apply Coated Grades or Cermets
- Utilize Coolant

Catastrophic Breakage

Corrective Action
- Utilize Stronger Insert Geometry
- Reduce Feed Rate
- Reduce Depth of Cut (doc)
- Check Rigidity of System

*NOTE: Generally inserts should be indexed when .030 flank wear is reached. If it is a finishing operation, index at .015 flank wear or sooner.

FIGURE 4–13 *Common insert machining problems. (Courtesy of Kennametal, Inc.)*

The ability to recognize excessive edge wear comes with experience; however, if you believe that you are experiencing excessive edge wear, the probable cause is friction. Excessive friction causes heat to build along the cutting edge, which causes the binders to fail.

One possible cause is that the lead angle is too great. Choose a holder that reduces the lead angle. Check the tool height. A crash or bump of the tool turret may be causing the tool to be too high.

Another possible cause can be that the feed rate is too low. Increasing the feed rate will cause the chips to concentrate away from the cutting edge. Finally, it could be a grade selection problem. In that case, choose a harder grade of insert.

CRATER WEAR

Crater wear occurs when the binder is being replaced by the material you are cutting. When you are cutting steel, the constant passing of the chip over the insert causes the cobalt binder to be carried away by the chip, leaving the steel to act as the binder. The steel, not being a very good binder material, quickly wears away, leaving a crater (see Figure 4–13). Cratering is usually a grade selection problem or an extreme heat problem, caused by cutting conditions that are too high. To minimize cratering, reduce the speed and feed, use a harder grade of carbide, or use a coated carbide insert.

EDGE BUILD UP

Edge build up or adhesion occurs when metal deposits build up on the cutting edge (see Figure 4–13). Iron actually combines with the binder in the carbide substrate. Edge build up occurs when the cutting conditions are too slow. Carbide cuts best at high temperatures and will rapidly wear if these temperatures are not reached. Machine operators can reduce edge build up by increasing the speed and feed.

DEPTH-OF-CUT NOTCHING

Depth-of-cut notching is an unnatural chipping away of the insert right at the depth of cut line (see Figure 4–13). Depth-of-cut notching is usually a grade selection problem. If you are using an uncoated carbide, consider changing to a coated insert. If a coated insert is not available, try honing the edge of the insert. Honing should only be done on uncoated inserts. Honing is done at a 45-degree angle to the cutting edge. Proper honing just breaks the sharp edge of the insert. Depth-of-cut notching may also be solved by lowering the feed rate and/or by reducing the lead angle.

CHIPPING

Chipping is a common insert problem. Chipping occurs along the cutting edge and is sometimes mistaken for edge wear. The major causes of insert chipping are lack of rigidity, too hard of an insert grade, and low operating conditions. Carbide is very brittle and works best when it is well supported. By decreasing the overhang of the tool and supporting the work better, the operator can eliminate many carbide cutting tool problems. If rigidity is not the problem, use a softer or tougher grade of insert. When making roughing cuts through hard spots or sand inclusions, use a tougher, not harder, grade of carbide. If the operating conditions are too low, abnormal pressures may build up, causing chipping. Increasing the cutting speed will sometimes eliminate chipping.

THERMAL CRACKING AND THERMAL DEFORMATION

Two heat problems are commonly associated with carbide cutting tools: thermal cracking and thermal deformation. Thermal cracking will show up as small surface cracks along the cutting edge and tip of the insert. Cracking is caused by sudden changes in temperature. Thermal cracking can occur if coolant is being applied to the insert instead of in front of the insert. If coolant is applied in the middle of the cut, thermal cracking may occur.

The other heat-related problem associated with carbide cutting tools is thermal deformation. Thermal deformation is a melting away of the tool tip and is caused by operating conditions being too high. The excessive heat breaks down the binder materials in the carbide insert. There are two possible solutions to thermal deformation: reduce the cutting conditions or switch to a more heat-resistant grade of carbide.

When trying to diagnose problems concerning carbide cutting tools, remember that troubleshooting is not a shot in the dark and should be done systematically. Troubleshooting must be a methodical procedure. The first step is to determine the problem. The second step is to arrive at all of the possible solutions. The third step is to examine each of the possible causes, changing only one condition at a time.

Use Figure 4–14 to help diagnose your carbide cutting tool problems.

Problem	Remedy
Tool life too short, excessive wear	1. Change to a harder, more wear resistant grade. 2. Reduce the cutting speed. 3. Reduce the feed. 4. Increase the lead angle. 5. Increase the relief angle
Excessive cratering	1. Use a harder, more wear resistant grade. 2. Reduce the cutting speed. 3. Reduce the feed.
Cutting edge chipping	1. Increase the cutting speed. 2. Hone the cutting edge. 3. Change to a tougher grade. 4. Use a negative rake insert. 5. Increase the lead angle. 6. Reduce the feed.
Deformation of the cutting edge	1. Reduce the cutting speed. 2. Change to a grade with a higher red hardness. 3. Reduce the feed.
Poor surface finish	1. Increase the cutting speed. 2. Increase the nose radius. 3. Reduce the feed. 4. Use positive rake inserts.

FIGURE 4–14 *Troubleshooting chart. Find your machining problem on the left and the right column will list potential cures for the problem.*

CHAPTER QUESTIONS

1. State the two main characteristics of carbide.
2. What is meant by *insert grade?*
3. What is a coated carbide?
4. Name three types of coating that are applied to carbide inserts.
5. Name five different insert shapes in order of increasing strength.
6. What is one of the most common binding materials that holds the carbide particles together?
7. State the purpose of tool nose radius.
8. Name four factors to consider when selecting the proper shape carbide insert.
9. What is meant by *inscribed circle?*
10. Describe the three types of back rake angles.
11. What is lead angle?
12. What are qualified tool holders?
13. Study the part shown in Figure 4–15 and choose appropriate tool holders and inserts from the charts shown in Figures 4–1, 4–8, and 4–9.

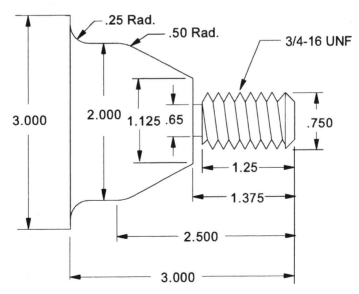

FIGURE 4–15 *Typical lathe part.*

Chapter 5

···

MACHINING CENTERS

INTRODUCTION

The machining center is a numerically controlled milling machine that is equipped with an automatic tool changer. The manual milling machine is a very versatile and productive machine tool, but when coupled with a computer control it becomes the production center of the machine shop. Repetitive operations, such as drilling, tapping, and boring, are perfect applications for the machining center.

OBJECTIVES

Upon completion of this chapter, the reader will be able to:

- *Describe the purpose and function of the machining center.*
- *Identify the major components of the machining center.*
- *Identify the axes and directions of motion on a typical machining center.*
- *Describe the function of the tool changer.*
- *Describe work-holding devices used on machining centers.*
- *Differentiate between climb and conventional milling.*
- *Calculate speeds and feeds for milling.*
- *Describe the different methods of manually moving the machine axes.*
- *Define terms such as "MDI" and "conversational."*

TYPES OF MACHINING CENTERS

Machining centers are identified by the direction in which the spindle lies. As with manual milling machines, machining centers are either horizontal or vertical spindle machines (see Figures 5–1 and 5–2).

FIGURE 5–1 *The horizontal spindle machining center is best suited for heavy machining operations on large workpieces. (Courtesy of Giddings & Lewis, Inc.)*

FIGURE 5–2 *Vertical spindle machining centers are very versatile machines. The quick setup of these machines makes them very popular in the machine shop. (Courtesy of Bridgeport Machines, Inc.)*

FIGURE 5–3 *Pallet systems allow operators the ability to load and unload parts while the machine is in operation. (Courtesy of Giddings & Lewis, Inc.)*

HORIZONTAL MACHINING CENTERS

Horizontal machining centers are typically the workhorses in the shop and are patterned after horizontal boring mills. The horizontal configuration of the spindle lends itself to heavy depths of cuts on large work pieces. Horizontal machining centers often have twin tables known as pallets (see Figure 5–3). While one pallet is within the machining envelope, the other pallet is swung free, allowing loading and/or unloading of parts. The two basic types of palletizing methods are the linear shuttle system and the rotary shuttle system.

VERTICAL MACHINING CENTER

The vertical machining center is probably the most versatile and common CNC machine found in the machine shop. The vertical configuration of the spindle lends itself to quick, easy workpiece setups. There are a variety of types and sizes of vertical machining centers. The type and size of the work done will determine which machine is best for your application.

PARTS OF THE MACHINING CENTER

The six main parts of the machining center are the column, saddle, bed, table, spindle, and tool changer (see Figure 5–4).

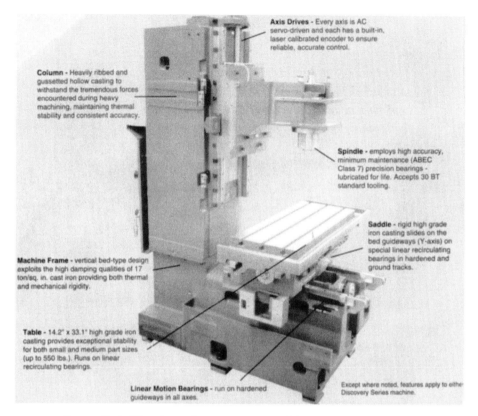

FIGURE 5–4 *The six main components of the vertical machining center. (Courtesy of Bridgeport Machines, Inc.)*

COLUMN

The column is the backbone of the machine. The column is typically mounted to the saddle and provides one of the axes or directions of travel. The rigid construction of the column will keep the machine from twisting during the machining operation.

SADDLE

The saddle provides the major axis of travel on the machining center, typically the X axis of travel on the horizontal machining center. The saddle is mounted to the bed.

BED

The bed is one of the more integral parts of the machining center. The bed is typically produced from high-quality cast iron, which absorbs the vibration of the machining operation. Hardened and ground slideways are mounted to the bed to provide alignment and support for the machining center.

TABLE

The table is mounted on the bed and the work or a work-holding device is mounted to the table. The table has T-shaped slots milled in it for mounting the work or work-holding device. The machining center equipped with a pallet-changing system may have more than one table.

SPINDLE

The spindle holds the cutting tool and is programmable in revolutions per minute.

TOOL CHANGERS

Tool changers are an automatic storage and retrieval system for the cutting tools. An automatic tool changer makes the CNC milling machine a machining center.

Tool changers come in two types: carousel and chain. Carousel-type tool changers are spindle direct tool changers, meaning they do not use auxiliary arms to change tools (see Figure 5–5). The carousel can be mounted on the back or side of the machine. Carousel tool changers are typically found on vertical machining centers. When a tool change is commanded, the machine moves to the tool change position and puts the current tool away. The carousel then rotates to the position of the new tool and picks it up.

FIGURE 5–5 *Carousel-type tool changing systems are usually found on vertical machining centers. The spindle positions itself over the tool and then goes down and clamps the tool in the spindle. (Courtesy of Bridgeport Machines, Inc.)*

FIGURE 5–6 *Chain-type tool changers typically hold more tools than carousel-type tool changers. (Courtesy of Giddings & Lewis, Inc.)*

Chain-type tool changers are found on horizontal machining centers (see Figure 5–6). Chain-type tool changers typically hold the tool in a horizontal position and usually, but not always, incorporate a pivot arm (see Figure 5–7).

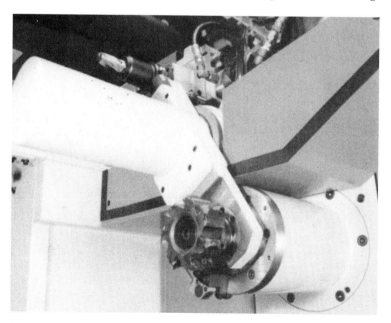

FIGURE 5–7 *The tool changing pivot arm can remove a tool from the spindle and place a new tool in the spindle, all in one motion. (Courtesy of Yamazaki Mazak Corp.)*

The arm rotates and picks up the new tool and removes the old tool from the spindle, all in one motion. The chain may be located on the side or top of the machine tool.

Both tool-changing mechanisms typically have bi-directional capabilities, which allows for quicker tool changes. Tool-changing cycle time is very important to a machining center's productivity. Tool changing time is nonproductive time. There is no machining taking place while a tool is being changed.

AXES OF MOTION

The linear axes or directions of travel of the machining center are defined by the letters X, Y, and Z (see Figure 5–8). Axis designation letters appear with positive or negative signs for direction of travel. The Z axis always lies in the same direction as the spindle. A negative Z (–Z) movement always moves the cutting tool closer to the work. The X axis is usually the axis with the greatest amount of travel. On a vertical machining center the X axis would be the left/right travel of the table. The Y axis on the vertical machining center would be the travel toward and away from the operator. On the horizontal machining center the Y axis would be the up and down travel of the spindle head.

ROTATIONAL AXES

More complex machining centers are capable of rotary cutting motions. A horizontal machining center that is equipped with a rotary table is capable of four axes of motion. The fourth axis is known as the C axis (see Figure 5–9). The C axis is a rotational axis about the Z axis. Machining centers that are

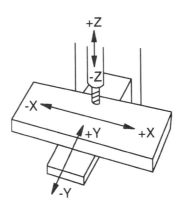

FIGURE **5–8** *The three linear axes of the vertical machining center are identified as X, Y, and Z. The Z axis always lies in the same plane as the spindle.*

FIGURE 5–9 *The rotary table adds a fourth axis of travel to the machining center. The rotary table can be used to do helical-type machining or it can be used to reposition the workpiece. (Courtesy of Bridgeport Machines, Inc.)*

FIGURE 5–10 *The additional capabilities of tilting and rotating the spindle give the machining center five axes of travel. (Courtesy of Giddings & Lewis, Inc.)*

equipped with a rotary table and tilting, contouring spindle are said to have five axes: three linear and two rotary (see Figure 5–10). Four-and five-axis machines are used to machine parts with complex surfaces such as mold cavities or rotary turbines.

WORK-HOLDING DEVICES

Work-holding techniques play a very important role in the setup and operation of the machining center. Before any machining can be done, the operator or setup person must make sure that the part or work-holding device is properly positioned and fastened to the table. Some setups may be as simple as placing the part in a vise, but some setups may take considerable ingenuity and time. Whatever the case, it is important to remember to make your setup as safe as possible.

VISES

The vise may be the most common work holding device in the machine shop (see Figure 5–11). The plain vise is used for holding work with parallel sides and is bolted directly to the table using the T-slots in the machine table. Air or hydraulically operated vises can be used in high-production operations to increase productivity.

ANGLE PLATE

Work that needs to be held at a 90-degree angle to the axis of travel is best held on an angle plate (see Figure 5–12). An angle plate is an L-shaped piece of cast iron or steel that has tapped holes or slots to provide a way to clamp parts to it.

FIGURE 5–11 *The standard milling machine vise is used to hold relatively small parts that have a square or rectangular shape.*

FIGURE **5–12** *Angle plates come in a variety of sizes and are typically bolted directly to the machine table. (Courtesy of Giddings & Lewis, Inc.)*

DIRECT WORKPIECE MOUNTING

Work that is too big or has an odd configuration is customarily bolted directly to the table (see Figure 5–13). This method of work holding takes the most ingenuity and expertise. There are a number of accessories that can be used to aid the setup person. A variety of clamp styles are commercially available for directly mounting workpieces to the machine table (see Figure 5–14). Figure 5–15 shows some clamping practices. Following are some tips for clamping work directly to the table:

1. *Tables should be protected from abrasive materials, such as cast iron, by placing plastic or aluminum shims between the work and the table.*
2. *Clamps should be located on both sides of the workpiece if possible.*
3. *Clamps should always be located over supports to prevent distortion or breakage of parts.*
4. *Clamps and supports should be placed at the same height.*
5. *Screw jacks should be placed under parts for support to prevent vibration and distortion.*

FIXTURES

Fixtures are tools built exclusively to hold and accurately position a part. They are typically found in a production atmosphere and can be built to hold one

FIGURE 5–13 *Parts that are clamped directly to the table are typically of an odd configuration, such as a casting. Setting up these types of workpieces takes some ingenuity.*

FIGURE 5–14 *Strap clamps are used to fasten work to the machine table, fixture, or angle plate. Strap clamps are usually supported by step blocks. T-bolts should be placed as close to the workpiece as possible.*

part or thousands of parts, depending on the application. The fixtured part is usually of an odd configuration, one that cannot be held in something as simple as a vise (see Figure 5–16). Fixtures are used quite extensively in the machining industry. They should be kept simple to allow for quick loading and unloading of parts because loading time is unproductive time. Fixtures also need to be designed and built in a fool-proof manner so that the part can be loaded in only one way. A well-designed fixture will lower the cost of producing parts and thus create a higher profit for your company.

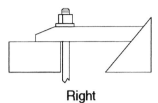

Right

Place clamp stud close to the workpiece

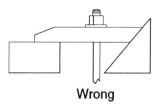

Wrong

Do not place the clamp stud
closer to the support

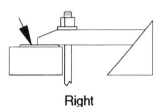

Right

Use shims between finished surfaces
of workpieces and clamps

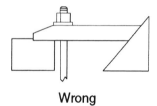

Wrong

Work-holding clamps in contact
with finished surfaces will mar th
workpiece

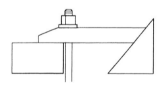

Right

Clamps that are level or with a
slight decline toward the
workpiece equalize
pressure on the workpiece

Wrong

Angling clamps incorrectly puts
pressure on the support, not the
workpiece, and tend to push
the piece out of position

FIGURE 5–15 *Study these acceptable clamping practices carefully.*

TOOLS AND TOOL HOLDERS

Milling and drilling tools will make up the majority of the types of tools used
on the machining center. This section will discuss standard types of tools and
tool holders used on milling machines.

Carbide tools, the most common tools used on CNC equipment, will be covered in this section. Carbide tooling is crucial to the productive use of CNC
machines. You may wish to review the selection and application of carbide
tools in Chapter 4.

FIGURE 5–16 *Fixtures are used to hold and accurately position workpieces. Repeatability in locating the part is of great importance when using fixtures. (Courtesy of Giddings & Lewis, Inc.)*

HIGH-SPEED STEEL DRILLS

The two basic types of high-speed drills are the twist drill and the spade drill. The high-speed steel twist drill is the most commonly used tool for producing holes.

Twist drills are great for rapidly producing holes that do not have to be very accurate in size or position. If the holes must be very accurate in size they are drilled to a smaller size and either reamed, milled, or bored to size. If the position of the hole must be very accurate, the drilled hole must be milled or bored on location.

Twist drills are made with two or more flutes and come in a variety of styles (see Figure 5–17). Twist drills have either a straight or tapered shank. Straight-shank drills are common up to 1/2 inch in diameter and are held in drill chucks.

FIGURE 5–17 *Twist drills are the most common hole-producing tools in the machine shop. (Courtesy of Kennametal, Inc.)*

FIGURE 5–18 *The tang on the end of the tapered shank drives the drill. (Courtesy of Kennametal, Inc.)*

Larger drills typically have a tapered shank with a tang on the end (see Figure 5–18). The tang keeps the tapered shank drill from slipping under the higher torque conditions associated with drilling large holes.

CENTER OR SPOTTING DRILLS

When drilled holes need to be accurately located, it is advisable to center or spot drill the holes prior to drilling. This spotting or centering is achieved with center or spotting drills (see Figure 5–19). These drills are short, stubby, and rigid and do not flex or deflect, as do longer drills. The spot drill produces a small startpoint that is accurately located. When the hole is drilled, the drill point will follow the starting hole that the spot drill made. This method can produce holes that are reasonably accurate in location.

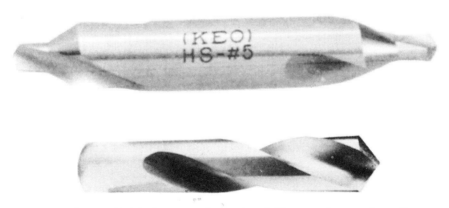

FIGURE 5–19 *Center and spotting drills come in a variety of sizes. The short, stubby design of the drills allows them to accurately locate holes. (Courtesy of Kennametal, Inc.)*

SPADE DRILLS

The spade drill has a flat blade with sharpened cutting edges (see Figure 5–20). The spade cutting tool is clamped in a holder and can be resharpened many times. Spade drills typically are used for drilling very large diameter holes. They can lower tooling costs because standard blade holders can hold a variety of sizes of blades. Designed to drill holes in one pass, spade drills require approximately 50 percent more horsepower than twist drills. Spade drilling also requires a rigid machine and setup.

CARBIDE DRILLS

Carbide-tipped twist drills have been around for many years. They are basically carbon steel drills with a piece of tungsten carbide brazed into them. They look similar to a spade drill, but are usually made in smaller diameters. Solid carbide drills are just that, a solid carbide cutting tool in a twist configuration. Solid carbide drills are typically found in small diameters because of the cost of the carbide materials.

One of the newer innovations in carbide drilling technology is carbide insert drills (see Figure 5–21). These drills incorporate indexable or replaceable inserts and can remove metal four to ten times faster than a high-speed steel drill. Carbide insert drills need a very rigid setup and a machine with substantial horsepower.

FIGURE 5–20 *Spade drills are a two-piece tool consisting of the blade and the holder. Spade blades and holders increase productivity because one holder can be used to drill many different diameter holes simply by changing blades. (Courtesy of Kennametal, Inc.)*

FIGURE **5–21** *Carbide insert drills allow you to drill hard materials at feeds and speeds much higher than those of conventional drills. When the carbide insert drill becomes dull, the carbide inserts can be indexed or replaced. (Courtesy of Kennametal, Inc.)*

AUXILIARY HOLE-PRODUCING OPERATIONS

Drilling may be the most common method of producing holes, but it is not the most accurate. In some cases, holes may need a very accurate size, location, and/or finish. If an accurate-size hole is needed, reaming may be the quickest method.

A reamer is a cylindrical tool similar in appearance to the drill (see Figure 5–22). Reamers produce holes of exact dimension with a smooth finish. Reaming can be done only after a slightly smaller hole has been drilled in the part. The reamer follows the drilled hole, so inaccuracies in location cannot be corrected by reaming. If accurate location is needed as well as accurate size, boring may be necessary. Reamers are a quick way of producing accurately sized small holes; however, boring can produce holes of any size and in the exact location.

BORING

Boring is done with an offset boring head and cutting tool. Figure 5–23 illustrates a variety of boring heads that can use high-speed steel tools, as well as

FIGURE **5–22** *Reamers consistently produce accurately sized holes. Reamers should be run at half the speed and twice the feed of the same size drill. (Courtesy of Kennametal, Inc.)*

FIGURE 5–23 *The offset boring head can be adjusted to cut any size hole within its size range. (Courtesy of Kennametal, Inc.)*

FIGURE 5–24 *Boring bars are tools used to do the cutting in a boring operation. Carbide boring tools need a rigid setup to operate properly. (Courtesy of Kennametal, Inc.)*

carbide tools. The offset boring head holds the tool and can be adjusted to cut any size hole within its range. The boring head is rotated and fed down into the piece, removing material (see Figure 5–24). Boring accurately produces holes of any size and in the exact location.

Boring, similar to reaming, can be done only on previously prepared holes. As a general rule, for best results, the boring tool should be as short and as large in diameter as possible. When using a high-speed steel tool, the diameter-to-length ratio should be no greater than 5-to-1. Example: If a 1-inch diameter boring bar is used, no more than 5 inches should be sticking out. Carbide has a 3-to-1 recommended ratio. Reducing the ratio insures against chatter because as the tool overhang becomes greater, the amount of force it takes to flex the boring bar decreases greatly.

TAPPING

Tapping is the process of producing internal threads by using a pre-formed threading tool known as a tap. Tapping can be done only on previously drilled holes. There are many different types of taps (see Figure 5–25). The most common type of tap used on the machining center is the spiral pointed gun tap which is especially useful for tapping holes that go through the workpiece or holes with sufficient space for chips. Chip clearance is especially important when tapping. If chips clog the hole, a broken tap often results.

There are two ways to tap on numerical control or CNC machines. One way uses a special tapping head. The tapping head is spring loaded, and the lead of the tap provides the primary feed (see Figure 5–26). The secondary or programmed feed need only be approximate because the spring-loaded head allows the tap to float up and down at the lead rate of the thread. The lead of the tap is the distance that the thread travels in one revolution (see Figure 5–27).

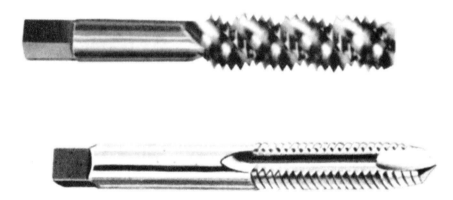

FIGURE 5–25 *The two most common types of machine taps used on CNC machines are the spiral fluted tap and the gun tap. The spiral flutes on each allow lubricant or coolant to reach the end of the tap. Gun taps push the chips ahead of the tap, so you must consider the amount of chip clearance available. (Courtesy of Kennametal, Inc.)*

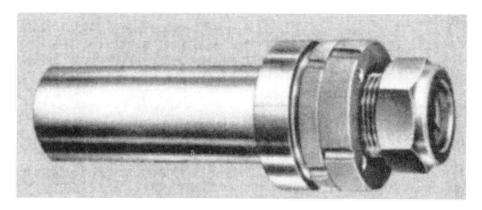

FIGURE 5–26 *The spring-loaded tapping head allows the tap to feed down at its own rate. (Courtesy of Kennametal, Inc.)*

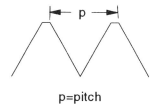

p=pitch

FIGURE 5–27 *The lead for a single lead thread is the same as the pitch of the thread. The pitch is the distance from a point on one thread to the same point on the next thread.*

The second type of tapping is called rigid tapping. This doesn't need special holders, but precise feed and RPM synchronization are needed to insure undamaged threads. Some machines are not capable of such synchronization. In either case the feed of the tap needs to be calculated. The feed for tapping is calculated by dividing 1 inch by the number of threads per inch. The quotient is then multiplied by the revolutions per minute of the spindle. For example: 1/4-20 UNC tap running at 250 RPM. 1/2 * 250 = 0.05 * 250 RPM = 12.5 inches per minute feed rate. Most modern machining centers are equipped with tapping cycles, which feed the tap down to the programmed depth and then automatically reverse the spindle and feed up, unscrewing the tap from the hole.

TOOLS FOR MILLING

Milling cutters are classified by the type of relief they have ground on the cutting edges and by the method in which they are mounted. When selecting the best milling cutter for a particular operation, four things must be taken into consideration; the kind of cut to be made, the material to be cut, the number of parts to be machined, and the type of machine available.

ENDMILLS

One of the most frequently used tools on a machining center is the endmill (see Figure 5–28). Endmills are made from two types of materials: solid carbide and high-speed steel. Endmills are ground with a relief on the sides and ends just behind the cutting edges. They come in two or more flutes. Two fluted or end-cutting endmills can be used for plunging. The teeth on the end come together much like those of a drill. Two considerations determine the number of flutes a milling cutter should have: does the endmill need to be end cutting for a plunging operation, and what is the depth of cut going to be? Increasing the number of teeth on the endmill greatly reduces the chip clearance area to prevent clogging. Endmills can be used for profile cutting, slotting, cavity cutting, or facemilling, although facemilling is usually done with a shell milling cutter. Endmills are typically held in solid-type holders with set screws for positive holding (see Figure 5–29).

Two Flute – Double End – Regular

Two Flute – Double End – Regular

Multi-Flute – Center Cutting

FIGURE 5–28 *Endmills are manufactured with two or more flutes. The two-flute double endmill is used for plunging and profiling. Ball endmills have a radius ground on the end of the tool and are used for milling radii in slots or contouring the bottom surfaces of mold cavities. The multi-fluted roughing endmill, or hog mill, has scallops or grooves around the body of the tool and can remove three times as much material as standard endmills. (Courtesy of Kennametal, Inc.)*

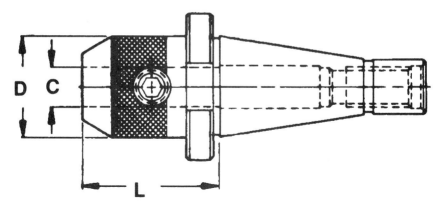

FIGURE 5–29 *Endmill tool holders use set screws to positively locate on the flats that are located on endmills. This positive locking-style holder keeps the endmill from slipping during machining operations. (Courtesy of Kennametal, Inc.)*

SHELL-ENDMILLS

Shell-endmilling cutters are designed to remove large amounts of material. They are sometimes referred to as facemills because of their ability to take large facing cuts. Shell-endmills range in size from 1-1/4 to 6 inches and up. They have a hole for mounting on an arbor and a key way to receive a driving key (see Figure 5–30). Shell-endmills are available in high-speed steel, as well as carbide. The carbide shell mill has indexable carbide inserts that can be indexed or thrown away when they become dull.

Shell-endmills represent a potential savings to machine shops because several different cutter sizes and materials can be used with one mounting arbor. This reduces the number of arbors required.

CLIMB AND CONVENTIONAL MILLING

In milling there are two basic directions you can feed: into the rotation of the cutter or with the rotation of the cutter (see Figure 5–31). Feeding with the rotation of the cutter is known as *climb milling*. In climb milling, the cutter is attempting to climb onto the workpiece.

On most manual machines, not equipped with backlash eliminators, climb milling is unheard of. If there is any slack or backlash between the screw and nut driving the table, the workpiece would be pulled into the cutter, possibly causing tool breakage, a spoiled workpiece, and possible serious injury to the operator; however, CNC machines are equipped with ball screws, and backlash is all but eliminated, thus allowing climb milling.

FIGURE 5–30 *Shell-endmills are designed to remove large amounts of material. Facing operations are usually performed with these types of tools, which are available in high-speed steel or indexable carbide. (Courtesy of Kennametal, Inc.)*

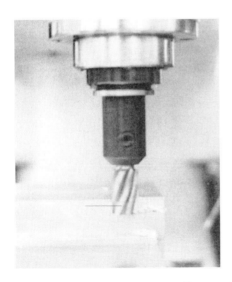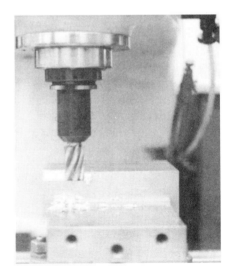

FIGURE 5–31 *Climb milling and conventional milling represent the two directions of feed associated with milling. When climb milling, the outer scale of the material is cut first. In conventional milling, the inside of the material is cut first.*

Climb milling is desirable in most cases because it takes less horsepower to cut in this fashion. Other benefits of climb milling are better surface finishes, less tool deflection, and extended tool life, and the chips are discarded away from the cutter. Feeding against the rotation of the cutter is known as *conventional milling*. The conventional milling chip has no thickness at the beginning, but builds in size toward the exit of the cutter. Conventional milling is recommended on materials with a hard outer scale, such as cast steels or forgings. If the tool needs to extend out of the holder a greater-than-normal length, it may be best to conventional cut. This will cause the tool to flex and stay in the flexed position, avoiding chatter.

CUTTING SPEED, FEED, AND DEPTH OF CUT

Cutting speeds, feed rates, and depths of cut are the three factors that have to be determined prior to machine operation. These cutting conditions will greatly affect production rates, as well as milling cutter life.

DEPTH OF CUT

The depth of cut is the first cutting condition that needs to be determined. It has the least effect on tool life, so to maximize production the depth of cut should be as great as possible. Rigidity of setup, rigidity of machine, and horsepower availability are the deciding factors when selecting the maximum depth of cut. The rule of thumb for endmills says that the depth of cut should not exceed two-thirds of the diameter of the tool. For example, if you have a 1.00-inch endmill, you should not exceed a 0.666-inch depth of cut. Depth of cut for carbide cutting tools is covered in the carbide cutting tool chapter.

FEED RATE

Feed rate, the next cutting condition determined, is the speed in which the material is advanced into the cutter. Feed rate on a milling machine is expressed in inches per minute, although feed per tooth or chipload per tooth is needed to determine the feed rate. As the material advances into the cutter, the rate of that advancement puts a strain on each tooth of the milling cutter, known as chip load. If the chip load becomes too great, the edges of the cutter can chip or break. On the other hand, if the chip load or feed rate is too small, the cutter may suffer premature dulling. Chip load factors are found on charts distributed by milling cutter manufacturers or in resource texts. Selecting a feed rate should not be left to guesswork. The formula for selecting the proper feed rate should be memorized. This formula is

$$IPM = CL * N * RPM$$

Where:

IPM = Feed rate in inches per minute.

CL = Chip load or feed per tooth.

N = Number of teeth on the cutter.

RPM = Revolutions per minute of the cutter.

Figure 5–32 gives you some standard chip load factors for high-speed steel end-mills.

CUTTING SPEEDS

Cutting speeds may have the greatest effect on cutting tool life. The cutting speed in milling is the speed at which the tool rotates past a given point on the workpiece in a given period of time. Cutting speed is controlled primarily, but not exclusively, by the hardness of the material. Harder materials generate more heat because of the difficulty in penetrating the harder surfaces. Cutting temperatures need to be kept within acceptable ranges, or cutting tools will break down and become dull prematurely.

Recommended Feed in Inches per Tooth (chip load)	End Mills		
Material	Sizes in Inches		
	1/2	3/4	1&up
Plain Low Carbon Steel	.001	.003	.003
Plain Medium Carbon Steel	.001	.002	.003
Plain High Carbon Steel	.001	.002	.002
Cast Iron, Soft	.001	.003	.003
Cast Iron, Medium	.001	.002	.003
Cast Iron, Hard	.001	.002	.002
Alloy Steel, Normalized	.001	.002	.002
Aluminum	.003	.004	.005
Stainless Steel, Free Machining	.001	.002	.003

FIGURE 5–32 *Feed rate table.*

Material	Cutting Speed, SFPM	
	HSS	Carbide
Plain Low Carbon Steel	90-120	270-450
Plain Medium Carbon Steel	70-100	225-375
Plain High Carbon Steel	30-70	145-300
Cast Iron, Soft	100-150	325-500
Cast Iron, Medium	70-120	225-400
Cast Iron, Hard	30-100	145-375
Alloy Steel, Normalized	30-90	145-350
Aluminum	200-300	500-950
Stainless Steel, Free Machining	65-90	200-300

FIGURE 5-33 *Cutting speeds for common materials.*

Cutting speed is expressed in surface feet per minute (SFPM). The RPM (revolutions per minute) of the cutter is a direct function of cutting speed. Each material has its own cutting speed. To correctly calculate the RPM setting for the machining center, follow the formula

RPM = (CS * 4)/D

Where:

RPM = Revolutions per minute of the cutter.

CS = Cutting speed of the material in feet per minute.

D = Diameter of the cutter in inches.

Figure 5-33 shows cutting speeds of some commonly used materials.

For more information concerning carbide selection and carbide cutting speeds, see Chapter 4.

MACHINING CENTER OPERATION

The machining center operator works on the machining center day in and day out. The operator gets to know the machine tool better than anyone. Duties usually include loading and unloading parts, inspecting machined parts, and changing offsets and tools when he/she deems it necessary. Machine operation is the first step in the process of becoming a programmer or setup person. To become a good programmer, you must have a good understanding of the operations and functions of the machine controls. An operator who learns how to operate the machining center and pays close attention to the setups done by the setup person will typically find the transition from operator to programmer to be an easy one.

SAFETY

Before operating any machine, remember that no one has ever thought he/she was going to be injured. But it happens! It can happen in a split second when you are least expecting it. An injury can affect you for the rest of your life. You must be safety minded at all times. Please get to know your machine before operating any part of the machine control and please keep these safety precautions in mind.

1. *Wear safety glasses and side shields at all times.*

2. *Do not wear rings or jewelry that could get caught in a machine.*

3. *Do not wear long sleeves, ties, loose fitting clothes, or gloves when operating a machine. These can easily get caught in a moving spindle or chuck and cause severe injuries.*

4. *Keep long hair covered or tied back while operating a machine. Many severe accidents have occurred when long hair became entangled in moving tooling and machinery.*

5. *Keep hands away from moving machine parts.*

6. *Use caution when changing tools. Many cuts occur when a wrench slips.*

7. *Stop the spindle completely before doing any setup or piece loading and unloading.*

8. *Do not operate a machine unless all safety guards are in place.*

9. *Metal cutting produces very hot, rapidly moving chips that are very dangerous. Long chips are especially dangerous. You should be protected from chips by guards or shields. You must also always wear safety glasses with side shields to prevent chips from flying into your eyes. Shorts should not be worn because hot chips can easily burn your legs. Hot chips that land on the floor can easily burn though thin-soled shoes.*

10. *Many injuries occur during chip handling. Never remove chips from a moving tool. Never handle chips with your hands. Do not use air to remove chips. They are dangerous when blown around and can also be blown into areas of the machine where they can damage the machine.*

11. *Securely clamp all parts. Make sure your setup is adequate for the job.*

12. *Use proper methods to lift heavy materials. A back injury can ruin your career. It does not take an extremely heavy load to ruin your back; bad lifting methods are enough.*

13. *Safety shoes with steel toes and oil-resistant soles should be worn to protect your feet from dropped objects.*
14. *Watch out for burrs on machined parts. They are very sharp.*
15. *Keep tools off the machine and its moving parts.*
16. *Keep your area clean. Sweep up chips and clean up any oil or coolant that people could slip on.*
17. *Use proper speeds and feeds. Reduce feed and speed if you notice unusual vibration or noise.*
18. *Dull or damaged tools break easily and unexpectedly. Use sharp tools and keep tool overhang short.*

MACHINE CONTROL FEATURES

Machining center controls come in all styles and levels of complexity, but they all have many of the same features. If you can get a good understanding of one machine, the next control will be that much easier. The newer, more capable controls are typically easier to operate because the control features are more straightforward. Most functions are at your fingertips. Control functions are divided into two distinct areas: manual controls and program controls.

MANUAL CONTROL

Manual control features are those buttons or switches that control machine movement (see Figure 5–34).

EMERGENCY STOP BUTTON

The emergency stop button is the most important component on the machine control. This button has saved more than one operator from disaster by shutting down all machine movement. This big red button with the word "Reset" or "E-stop" on the front should be used when it is evident that a collision or tool breakage is going to occur. Emergency stop buttons are located in more than one area on the machine tool and should be located prior to doing any machine operations.

MOVING THE AXES OF THE MACHINE

Manual movement of the machine axes is done in a number of different ways. Most controls are equipped with a pulse-generating hand wheel (see Figure 5–35).

The hand wheel has an axis selection switch that allows the operator to choose which axis he/she wants to move. The handle sends a signal or electronic pulse

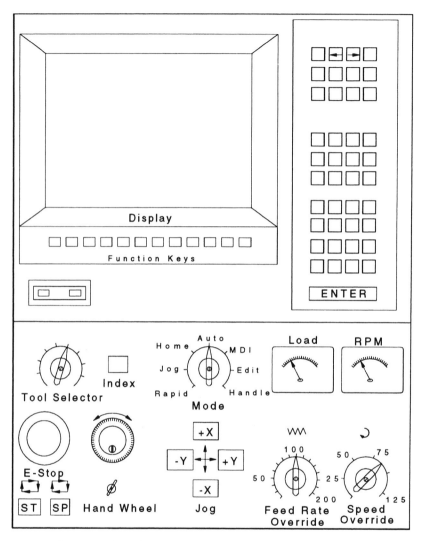

FIGURE 5–34 *The manual machine controls are usually located toward the bottom of the control.*

to the motors, which move the table or the spindle head. If the handle of the Y axis is turned in the negative direction, the tool moves toward the operator. If the Z axis hand wheel is moved in a negative direction, the tool moves toward the table.

Some machines are equipped with jog buttons (see Figure 5–36). When the jog button for a certain axis is pressed the axis moves. The distance or speed at which the machine moves is selected by the operator prior to the move and is controlled by a jog mode selector switch (see Figure 5–37). Selection options include rapid traverse, selected feed rate, or incremental distance.

Hand wheel

FIGURE 5–35 *Some controls have a hand wheel for each axis, while other controls have one hand wheel and a switch to select which axis the operator will move.*

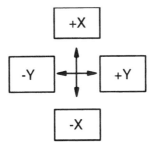

FIGURE 5–36 *Jog buttons are used for machine axis movement. The amount of movement per push of the jog button is controlled by the mode selector switch.*

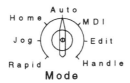

FIGURE 5–37 *The jog mode selector switch can control the mode of travel, such as rapid or controlled feed, and the amount of incremental travel.*

CYCLE START/FEED HOLD BUTTONS

The two most commonly used buttons on the control are the cycle start and feed hold buttons (see Figure 5–38). The cycle start button is used to start execution of the program. The feed hold will stop execution of the program without stopping the spindle or any other miscellaneous functions. By pushing the cycle start, the operator can restart the execution of the program.

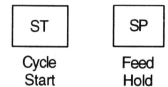

FIGURE **5–38** *The cycle start and feed hold buttons are located adjacent to one another. These buttons will typically light up when activated. The feed hold button will only stop the axis from traveling; it will not stop the spindle or any other miscellaneous function.*

HOME OR ZERO RETURN

The home or zero return button, when selected, will return all of the axes of the machine to the home position. Every CNC machine has an assigned stationary position known as home or reference zero. Home position is usually defined as a position at the extreme travel limits of the three main axes. The zero or home position is set by using switches and encoders. When the machine is home the sensors are triggered and the axes indicator lights illuminate. The home return button is used when the operator wants to load or unload parts and when he/she wants to start the program from the home position. Before the operator shuts the machine down, the axes of the machine should be brought back to the home position. This will insure quick and easy home positioning of the machine upon startup. The home position is very important because the workpiece coordinate or program zero is set from this position.

WORKPIECE COORDINATE SETTING

The workpiece coordinate or program zero is the point or position from which all of the programmed coordinates are established. For example, when the programmer looks at the part print and notices that all of the dimensions come from the lower left-hand corner and the top of the part, these datums are then used to establish the program zero or workpiece coordinate (see Figure 5–39).

The part origin is the X0, Y0, Z0 location of the part in the rectangular or Cartesian coordinate system. In absolute programming, all of the tool movements would be programmed with respect to this point. If all of the dimensions were located from the center of the bored hole, then that point would become the program zero. During part setup the X and Y zero position of the part has to be located. Using the hand wheels or other manual positioning devices and an edge finder or probe, the setup person locates the point at which

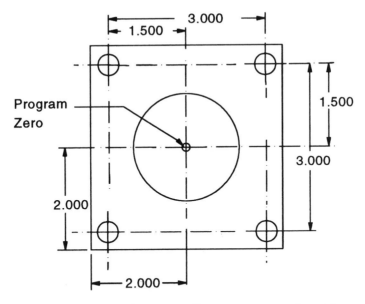

FIGURE 5–39 *The workpiece coordinate system can be set from any datum feature. Pick the feature that would allow the programmer to do the least amount of calculations.*

the center of the spindle and the part origin are the same (see Figure 5–40). The "home zero" is then entered as a G-code in the appropriate area of the program or in an offset table.

The setup person must then measure and enter the values of the tool lengths in the offset table for each tool being used in the program. Each tool used in the program has a different length.

The control must then be told to compensate for the difference in the lengths. The tool length offsets are typically the distance from Z at home zero to the Z position of the part zero (see Figure 5–41). The tool length offsets are stored as an offset in the control (see Figure 5–42). This process may seem a little confusing, but once you have done it, it really is quite straightforward. We will also talk more about workpiece coordinates and tool length offsets in Chapter 6.

SPINDLE SPEED AND FEED RATE OVERRIDE SWITCHES

Spindle speed and feed rate overrides are used to speed up or slow down the feeds and speeds of the machine during cutting operations (see Figure 5–43). The override controls are typically used by the operator to adjust to changes in cutting conditions, such as hard spots in the material. Feed rates can typically be adjusted from 0 to 150 percent of the programmed feed rate. Spindle speeds can be adjusted from 0 to 200 percent of the programmed spindle speed.

FIGURE 5–40 *It is extremely important to accurately locate the workpiece zero point; otherwise, all of the machined features will be shifted out of location. An edge finder or probe is usually used to position the center of the spindle on the home zero location. (Courtesy of Kennametal, Inc.)*

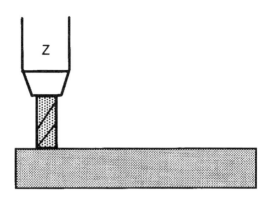

FIGURE 5–41 *The operator touches the tool off on the Z0 point of the part. He/she then takes the Z axis distance from the Z0 point of the part and the machine home position. The distance is recorded in the tool length offset table. This must be done for all of the tools used in the program.*

FIGURE 5–42 *When the program calls for the tool length offset, the control accesses the register located in the tool offset area. The control uses this to compensate for the tool length.*

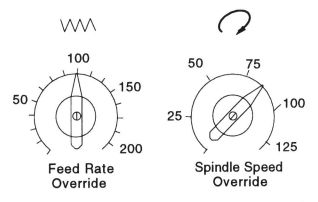

FIGURE 5–43 *The spindle speed and feed rate override switches give the operator a greater amount of control over the machine.*

SINGLE BLOCK OPERATION

The single block option on the control is used by the operator to advance through the program one block or line at a time. When the single block switch is on, the operator presses the button each time he/she wants to execute a program block. When the operator wants the program to run automatically, he/she can turn the single block off and press cycle start, and the program will run through without stopping. The purpose of the single block switch is to allow the operator to watch each operation of the program carefully. It is typically used on unproven programs.

OTHER CONTROL FEATURES

Some control features do not control the machine directly the way the manual control features do. Programming controls are used as information input devices.

MANUAL DATA INPUT

Manual data input or MDI is a means of inputting data. MDI can be used to enter a simple command, such as starting the spindle, or entering an entire program, and is done through the alphanumeric keyboard located on the control (see Figure 5–44).

PROGRAM EDITING

After the part program is loaded into the control, it always needs some modification. The need for the changes usually shows up on the shop floor. The operator or programmer can make the needed changes using the program edit mode. The programmer uses the display screen to find the program errors and the keyboard to correct the errors.

CATHODE RAY TUBE

The cathode ray tube (CRT) is commonly referred to as the screen. The screen displays information such as the written program or part graphics. In most

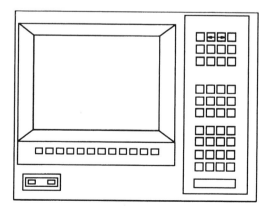

FIGURE 5–44 *The alphanumeric keyboard lets the operator edit or enter a program at the machine control.*

cases the program is too long to fit on the screen, so it is separated into pages. The page or cursor button allows the operator to move through consecutive parts of the program. Graphics can also be displayed on the screen if you have purchased graphics capabilities from the machine manufacturer. Graphics are a representation of the part and the tool path, which would be generated by the active program. Graphics are a safe way to prove out part programs. The simulation of the tool path will allow you to see program errors quickly, such as misplaced decimal points or missing minus signs. Graphics are used extensively in conversational programming.

CONVERSATIONAL PROGRAMMING

Conversational programming is a built-in feature that allows the programmer to respond to a set of questions that are displayed on the graphics screen. The questions guide the programmer through each phase of the machining operation. First, the operator might input the material to be machined. This information will be used by the control to calculate speeds and feeds. The operator can override these. Next, the operator would choose the operations to be performed and input the geometry. Operations such as pocket milling, grooving, drilling, and tapping can be performed.

With each response to the questions, more questions are presented until the operation is complete. By answering the questions, the programmer is filling in variables in a canned or preprogrammed cycle. This type of programming is quicker than methods using machine code language. There is no standard conversational part programming language, and each system can be quite different; however, once the programmer completes the conversational program, many CNC controls convert the conversational language into standard EIA/ISO machine language. That is why we will concentrate heavily on EIA/ISO programming language in this textbook.

DIAGNOSTICS

The diagnostics mode consists of several routines that detect errors in the machine system. An error number and message will be displayed on the screen. If any error is found in the CNC operation or servo system, the error message will prompt the operator or service technician to the cause of the problem.

This chapter has given you some basic insights into the machining center. Use this information as a place to start; take some time with an operator or instructor and get to know the operation of the control on your machining center.

CHAPTER QUESTIONS

1. Describe the function of the tool changer.
2. What are the three major axes associated with the machining center?
3. Which axis always lies in the same plane as the spindle?
4. What purpose does a pallet-changing system perform?
5. Describe the six main components of the machining center.
6. What is the purpose of a rotary table?
7. What type of work is held in a vise?
8. What type of work-holding device would you use to hold a workpiece at a 90-degree angle to the axis of travel?
9. How are large workpieces held?
10. What types of work-holding devices are typically used in production atmospheres?
11. What is the most common hole-producing tool?
12. What keeps tapered shank drills from slipping during drilling?
13. What is the purpose of a center drill?
14. How can using spade drills lower the cost of producing holes?
15. What is a reamer?
16. What components are used in boring?
17. What is the feed for tapping a 1/2-13 UNC hole?
18. Describe the two types of tapping done on machining centers.
19. What type of end mill would be used for plunge cutting?
20. What is a carbide face milling cutter?
21. Describe climb milling.
22. Calculate the RPM for a 1/2-inch, high-speed steel end mill cutting low carbon steel.
23. Calculate the feed rate in inches per minute for a 1 inch, four-flute, high-speed steel endmill cutting stainless steel.
24. Describe two methods of manually moving the axes of the machining center.
25. Describe the workpiece coordinate or zero point.

Chapter 6

PROGRAMMING MACHINING CENTERS

INTRODUCTION

This chapter will take you through the steps necessary to properly plan, set up, and program a machining center. The programmer will need to rely on the information provided in previous chapters. Chapter 2 presented the information necessary to write simple programs; now we are going to take you a step further and discuss fixed cycles. Fixed or canned cycles will reduce the amount of programming necessary for repetitive machining operations such as drilling, boring, reaming, and tapping.

OBJECTIVES

Upon completion of this chapter, the reader will be able to:

- *Demonstrate an understanding of acceptable machining center programming practices.*
- *Describe the sequence of operations for machining centers.*
- *Describe the steps necessary to properly plan a program.*
- *State the purpose of a setup sheet.*
- *Program a machining center using linear moves, circular moves, and canned cycles.*

PLANNING THE PROGRAM

The first step in preparing any program is to plan carefully all of the steps necessary to make the part. This preparation is called a *process plan*. The process plan outlines all of the steps and tooling considerations necessary to manufacture the part. The first step in planning the program is to study the part drawing carefully.

THE PART DRAWING

The part drawing gives the programmer detailed information on how the part looks. The shape of the part, the tolerances, material requirements, surface finishes, and the quantity required all have an impact on the program. From the part drawing the programmer will determine what type of machine tool is required, work-holding considerations, and part datum (workpiece zero) location. Once these questions have been answered, the programmer can develop a process plan.

DECIDING ON THE MACHINE

The type of machine to use is based on its compatibility with the type of part or parts to be manufactured. The machine will be selected based upon its size, horsepower, accuracy, tooling capacity, and the number of axes of travel required. The size of the machine is based on the amount of travel of the axes.

Will we be able to machine the whole part without moving and resetting the part? Setup time is wasted time. When the machine is not making chips, we are not making money.

Does the part safely fit on the machine? Machine tools have a weight capacity. It can be unsafe to overload the capacity of the machine. What are the machining power requirements? The size of the tool, depth of cut, and part material have a direct effect on the horsepower requirements of the machine.

Is the machine rigid enough to withstand heavy rough machining operations? The rigidity of the machine affects the depth of cut, feed and speed rates, and surface finishes. To maximize production rates, choose a machine that is rigid enough to handle the task.

What are the part tolerances? Is our machine capable of the accuracy required? How many different tools will be needed to manufacture this part? Will the machine's tool carousel accept that many tools? Will the machine be available when you are ready to make these parts? By answering these questions you have made an informed decision on the machine that best fits the needs of your part.

WORK HOLDING

Will fixtures to hold the part be required? Fixtures need to be planned early in the process. This will ensure that the proper lead time is allotted for purchasing or manufacturing this type of holding-device. Small square or rectangular parts can be held in a vise.

For large parts, a setup using strap clamps may by used. If the part is small but has a complicated shape, soft vise jaws may be machined to hold the part. Some typical methods of work holding were shown in Chapter 5. Whenever possible use standard work-holding devices to cut down on the cost to manufacture the part.

PART DATUM LOCATION

Deciding on the part datum location is based upon how the part designer drew the part. When dimensioning a part, the designer typically uses a part feature that has a direct influence on how the part is used. If the designer dimensions the majority of the part features from the corner of the part, then we need to use that corner as the part datum. If the designer uses the center of a hole as the datum feature, then we need to use this as the programming datum.

SELECTING THE PROPER TOOLING

Decide the tooling requirements of the part ahead of time. If specially ground tools are needed, send them out for grinding. When standard tools are going to be used, make sure you have enough of them on hand to complete the job. Waiting for tooling can be very costly.

Deciding on the types of tools that are going to be used greatly influences the way you write your program. Carbide or specially coated tools can greatly speed up the machining process and should be chosen whenever possible.

Cost is the major factor in deciding on the tooling. If using a higher priced carbide tool will allow you to make the part faster, it may be the one to use. When considering the cost of the tool, remember to take into account the cycle time to produce the part, the part tolerance, surface finish, and quantity of parts needed.

THE PROCESS PLAN

Process planning involves deciding when certain machining operations will take place. Primary machining operations will take place on the CNC machine. Although the part configuration will have a decided effect on the sequence of operations, there are some general rules to follow when deciding on machining operations.

The recommended procedures for machining are as follows:

1. *face mill top surface*
2. *rough machine the profile of the part*
3. *rough bore*
4. *drill and tap*
5. *finish profile surfaces*
6. *finish bore*
7. *finish reaming*

You should notice that all roughing operations took place first, then finishing operations were done. This minimizes the effect of high-pressure operations moving or stressing the part.

Sometimes secondary operations may be done more economically on other types of machines. For example, if a large plate needs to have a series of holes drilled and tapped, it may be more economical to drill the holes on a machining center and tap the holes on a less costly machine using a tapping head. These secondary operations may be done by the same operator while another part is running. This keeps the expensive machine making parts while the operator does the less demanding secondary operations on a simpler, less expensive machine.

This approach to process planning is usually done by the operator/programmer in smaller job shop settings. In large shops the process plan would come down from the engineering area and would include information for each step in the manufacturing of the part. In a small job-shop setting, the operator, in a way, becomes the manufacturing engineer.

The operator must determine the best, most economical way to produce the part. The operator must determine the operating sequence, types of cutting tools, cutter path, work-holding devices, and the machining conditions (cutting speed, feed rate, and depth of cut). The operator in this type of setting must have a good background in the machine tool field.

Whether process planning is done by the manufacturing engineer or the operator, a plan for each operation needs to be developed. The process plan is done with paper and pencil on a process planning sheet (see Figure 6–1). A well organized planning sheet will be the programmer's ready reference.

THE SETUP SHEET

The CNC setup sheet is a detailed explanation of how the part will be set up, which fixture is to be used, where the part datum is located, and which tools are to be used (see Figure 6–2).

Setup sheets tell setup personnel exactly what the programmer had in mind when he/she programmed the part. In small job shops the programmer and the operator are usually the same person, but the setup sheet can be a useful review if the same or similar parts are programmed in the future. The setup sheet should contain all of the necessary information to prepare for the job and may include sketches of parts and fixtures. Setup sheets can drastically reduce setup times.

PROGRAMMING THE PART

After all of the preliminary steps are done, it is time to write the program. If you have done a good job of planning the job, programming the part should be fairly simple and straightforward. When writing the program by hand, a form can be

Fox Valley Techical College - Process Plan				Part No.
Operation	Tool #	Tool Description	RPM	Feed Rate

FIGURE 6–1 *The process plan is an outline of the machining steps to be done on the part.*

Fox Valley Technical College - Setup Sheet	Part No.
	Sheet \

Machine:	Prepared by:
Machining Operations	**Setup Sketch**

NO.	Description	
Comments:		

FIGURE 6–2 *The CNC setup sheet usually includes a sketch for clarity. Setup sheets help assure consistent quality, rapid setups, and consistent job setups and machining.*

Fox Valley Technical College - Numerical Control Programming Sheet										
Part Name				Part No.			Page /			
Machine				Programmer			Date			
N Seq No.	G Prep. Funct	Axis Coordinate			Arc Center Definition			F Feed Rate	M Misc Funct	S RPM	H D L
		X	Y	Z	I	J	K				

FIGURE 6–3 *Programming sheets can reduce errors and keep the program orderly.*

used to keep the program organized (see Figure 6–3). Programming sheets keep track of the sequence of operation, coordinates, tool numbers, and miscellaneous functions. Programming sheets, usually available from the machine manufacturer, are available for both machining centers and turning centers.

CANNED CYCLES FOR MACHINING CENTERS

Canned cycles (fixed cycles) simplify the programming of repetitive machining operations such as drilling, tapping, and boring. Canned cycles are a set of preprogrammed instructions that eliminate the need for many lines of programming. Programming a simple drilled hole, without the use of a canned cycle, can take four or five lines of programming. Think of the lines that are needed to produce a hole:

1. *position the X and Y axis to the proper coordinates with a rapid traverse move (G00),*
2. *position the Z axis to a clearance plane,*
3. *feed the tool down to depth,*
4. *rapid position the tool back to the clearance point.*

That is one drilled hole! By using a drilling canned cycle, a hole can be done with one line of programming. Standard canned cycles, or fixed cycles, are common to most CNC machines. Figure 6–4 lists the most commonly used canned cycles for machining centers.

The most commonly used canned cycle is the G81 canned drilling cycle. This cycle will automatically do all of the things necessary to drill a hole in one line of programming (see Figure 6–5). The Z position is very important when you call the canned cycle. The present position will become the Z initial plane. The machine will normally rapid back to the Z initial position before a rapid to the next hole. If the tool is 12 inches above the work when the canned cycle is called, that will be the Z initial plane, and the machine will

G Code	Function	Z Axis	At Depth	Z Axis Return
G81	Drill	Feed	–	Rapid Traverse
G83	Peck Drill	Feed With Peck	–	Rapid Traverse
G84	Tap	Feed	Reverse Spindle	Feed
G85	Bore/Ream	Feed	Stop Spindle	Cutting Feed
G86	Bore	Feed	Stop Spindle	Rapid Traverse

FIGURE **6–4** *Canned cycles for machining centers.*

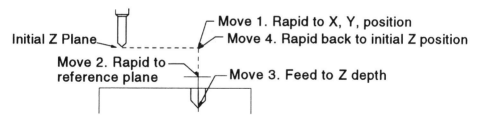

FIGURE **6–5** *The G81 canned drilling cycle.*

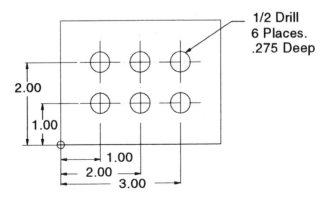

FIGURE 6–6 *Canned drilling cycle example.*

rapid up to 12 inches above the work between each hole. This would be a very slow program.

As you can see from Figure 6–5, the canned drilling cycle consists of four moves. Those four moves are controlled by one line of programming. A typical G81 canned drilling cycle would look like this:

N0010 G81 X4.500 Y2.250 Z-.375 R.100 F2.5;

N is the line number, G81 specifies which canned cycle, X and Y are the coordinates of the hole, Z is the depth of the hole, and F is the feed rate in inches per minute.

The G81 canned cycle is modal, which means that it will stay active until it is canceled by a G80. If we were drilling a series of holes, we would only need to specify the coordinates for the next hole. Figure 6–6 and the program that follows are a drilling example using a G81 canned drilling cycle.

N0010 G90 G70; (Absolute programming)
N0020 M06 T01; (Tool change, tool #1, 1/2 inch drill)
N0030 G54 X-10.500 Y-6.750 Z-16.564; (Workpiece zero setting)
N0040 M03 S1000; (Spindle start clockwise, 1000 RPM)
N0050 G00 X1.00 Y1.00; (Rapid to hole position #1)
N0060 Z1.0; (Rapid to initial level)
N0070 G81 Z-0.275 R0.100 F3.00; (Drill hole #1 .275 inches deep)
N0080 X2.00; (Drill hole #2)
N0090 X3.00; (Drill hole #3)
N0100 Y2.00; (Drill hole #4)
N0110 X2.00; (Drill hole #5)
N0120 X1.00; (Drill hole #6)

N0130 G80; (Cancel drilling cycle)

N0140 G28; (Return all axes to home position)

N0150 M05; (Spindle stop)

N0160 M30; (Rewind program, reset the control, and end the program)

PECK DRILLING CYCLE (G83)

When the part calls for drilling deep holes (holes that are three to four times deeper than the diameter of the drill), the programmer needs to use a special drilling cycle. The peck drilling cycle (G83) is very similar to the G81 drilling cycle, but it uses an extra word address (Q) to specify the depth of each drill peck. After the drill reaches the depth of the peck, it rapids out of the hole, clearing the hole of chips (see Figure 6–7). The canned peck drilling cycle will rapid position the tool to the hole coordinates and to the Z position specified by the R plane value. After the R positioning move, the tool will feed down the distance indicated by the Q address word, retract to the R plane, rapid back into the hole, and feed an additional Q distance until the hole is drilled to the programmed Z depth. When the hole depth is reached, the drill rapid positions up to the R plane value and repeats the routine at the next commanded hole position.

The G83 canned peck drilling cycle is modal. This means that it will stay active until it is canceled by a G80. If we were drilling a series of deep holes, we would only need to specify the coordinates for the next hole, just like we did with the standard G81 cycle. Figure 6-8 is a drilling example similar to Figure 6–6 using a G83 canned peck drilling cycle.

Retract Position

Two G-codes control the Z axis retract position. The G98 default code tells the control to retract the Z axis to the initial plane after each drilled hole. This is useful when there are clamps or other obstacles above the part surface that we

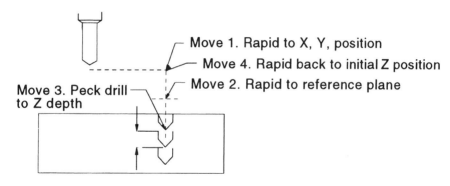

FIGURE 6–7 *Canned peck drilling cycle.*

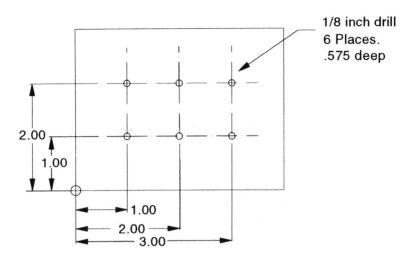

FIGURE 6–8 *Canned peck-drilling cycle example.*

need to avoid. Because the Z position when the canned cycle is called establishes the initial plane, the programmer would move to a Z position safely above any obstacles and then call the canned cycle.

As mentioned before, this is important. You should rapid the Z axis to an appropriate position above the workpiece before any canned cycle is called to establish the Z initial plane. An inappropriate Z initial plane could waste time or cause the tool to run into clamps or other obstacles.

A G99 can be used when there are no obstacles involved. A G99 tells the control to retract the Z axis to the R plane after drilling. Then it will rapid to the next hole. This shorter move up is faster. To use this code, simply put it in the line before or in the line where the canned cycle is called, as follows:

    N0060 G99 G00 Z1.0; (Return to R plane code and rapid to Z1.0)
    N0070 G81 Z-0.275 R0.100 F3.00; (Drill canned cycle)

The G98 and G99 apply to other canned cycles as well.

    N0010 G90 G70; (Absolute programming)
    N0020 M06 T01; (Tool change, tool #1, 1/8 inch drill)
    N0030 G54 X-10.500 Y-6.750 Z-16.564; (Workpiece zero setting)
    N0040 M03 S1000; (Spindle start clockwise, 1800 RPM)
    N0050 G00 X1.00 Y1.00; (Rapid to hole position #1)
    N0060 Z1.0; (Rapid to initial level)
    N0070 G83 Z-0.275 R0.100 Q.2 F3.00; (Drill hole #1 .575 inches deep)
    N0080 X2.00; (Drill hole #2)
    N0090 X3.00; (Drill hole #3)
    N0100 Y2.00; (Drill hole #4)

N0110 X2.00; (Drill hole #5)

N0120 X1.00; (Drill hole #6)

N0130 G80; (Cancel drilling cycle)

N0140 G28; (Return all axes to home position)

N0150 M05; (Spindle stop)

N0160 M30; (Rewind program, reset the control, and end the program)

CANNED TAPPING CYCLE (G84)

When threaded or tapped holes are needed, a G84 tapping cycle can be used. The tapping cycle is commanded much like the previous canned cycles. The difference occurs when the tap reaches the programmed depth. The spindle stops and reverses itself and automatically feeds the tap out of the hole. The feed rate of the tapping canned cycle must be coordinated with the spindle. To do this we must multiply the lead of the tap (lead equals 1 inch divided by the number of threads per inch) by the spindle RPM. When the G84 tapping cycle is commanded, the tap rapid positions to the specified coordinates and from the Z initial position to the R plane. The tap then feeds down to the specified depth, cutting the threads. At the programmed depth, the spindle automatically reverses, and the tap is fed back to the R plane (see Figures 6–9 and 6–10). Because the spindle reverses and feeds up simultaneously, use a floating tap holder to reduce the possibility of tap breakage.

The G84 canned tapping cycle is modal and will stay active until it is canceled by a G80. If we were tapping a series of holes, we would need only to specify the co-ordinates for the next tapped hole, as with the standard drilling canned cycles.

N0010 G90 G70; (Absolute programming)

N0020 M06 T03; (Tool change, tool 3, 1/4-20 UNC tap)

N0030 G54 X-10.500 Y-6.750 Z-16.564; (Workpiece zero setting)

N0040 M03 S800; (Spindle start clockwise, 800 RPM)

N0050 G00 X1.00 Y1.00; (Rapid to hole position #1)

N0060 Z1.0; (Rapid to initial level)

N0070 G84 Z-0.250 R0.100 F4.00; (Tap hole #1 .250 inches deep)

N0080 X2.00; (Tap hole #2)

N0090 X3.00; (Tap hole #3)

N0100 Y2.00; (Tap hole #4)

N0110 X2.00; (Tap hole #5)

N0120 X1.00; (Tap hole #6)

N0130 G80; (Cancel tapping cycle)

N0140 G28; (Return all axes to home position)

N0150 M05; (Spindle stop)

N0160 M30; (Rewind program, reset the control, and end the program)

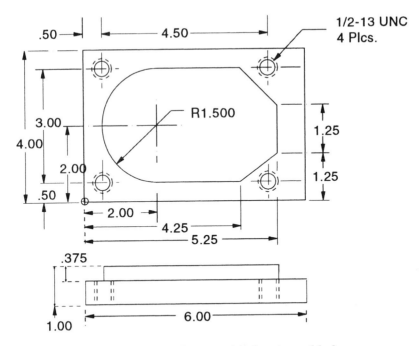

FIGURE 6–9 *Sample part with four tapped holes.*

Tool #	Operation	Tool	Speed (RPM)	Feed (IPM)
1	Mill profile	.750 End mill	150	6.00
2	Tap drill	.422 Drill	850	3.00
3	Tap	1/2-13	450	3.4

FIGURE 6–10 *Tools for the tapping example.*

HELICAL INTERPOLATION

Helical milling can be used to produce large internal or external threads (see Figure 6–11) or helical pockets. Helical milling involves circular interpolation in two axes (usually X and Y) plus a linear feed in the third axis (usually Z). This is especially useful for large threads.

A special type of milling cutter, called a *thread hob*, is used to mill threads. It looks like an end mill with teeth shaped like the desired thread. The hob is

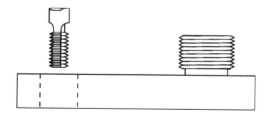

FIGURE 6–11 *Examples of helical milling. The left part of the workpiece is an internal thread that could be produced by helical milling. The right is an external thread that could be produced by helical milling.*

sent to a start position. It is then fed into the workpiece and does a helical interpolation three times around the part. With each interpolation it feeds the Z axis down an amount equal to the lead of the thread. Lead, which is how far a thread advances in one revolution, can be calculated by dividing 1 inch by the number of threads per inch. For example, if we had a 2-inch diameter thread with 10 threads per inch, the lead would be 1 inch/10 or .1 inch.

Helical interpolation can be done in the X/Y plane (G17), X/Z plane (G18), or Y/Z plane (G19). The format is as follows:

    X/Y plane—G17 G02/G03 X Y I J Z F

    X/Z plane—G18 G02/G03 X Y I K Z F

    Y/Z plane—G19 G02/G03 X Y J K Z F

G17, G18, and G19 select the plane to be used. G02 or G03 selects the direction of the helical interpolation (clockwise or counterclockwise). X, Y, and Z are the endpoint coordinates (depending on which plane is chosen). I, J, and K are the arc centerpoint coordinates, and F sets the feed rate.

Imagine that we need to cut an external 1-inch thread with 20 threads per inch (similar to the thread shown on the right side of the workpiece in Figure 6–11). The following program would perform that operation. It rapids the cutter close to the part, feeds the cutter into the part, and makes three circular moves around the part while moving the Z axis down to machine the threads. Diameter and length offsets have been ignored to simplify the program.

    N0100 G00 X1.8 Y.0 S750 M03;

Line N0100 rapids the cutting tool to X1.8 Y0.0 and turns on the spindle at 750 RPM.

    N0110 G00 Z-.818 M08;

Line N0110 rapids the Z axis to Z-.818 and turns on the coolant.

N0115 G01 X.969 F5.;

Line N0115 feeds the X axis to X.969 at a rate of 5 inches per minute. This moves the thread hob into the work to the proper thread depth.

N0120 G17 G02 X.969 Y0. Z-.868 I-1. J0.;

Line N0120 calls for XY helical interpolation and clockwise circular interpolation. The endpoint is X .969, Y0., and the vector from the start to the center of the arc is described by the I and J values. The Z is the endpoint (in Z) of this move. In one revolution, the cutter will move from Z-.818 to Z-.868. This is .050 and is equal to the lead of the thread.

N0125 G02 X.969 Y0. Z-.918 I-1. J0.;

Line N0125 remains in XY helical interpolation (the G17 is modal), clockwise circular interpolation. The endpoint is X .969 Y0., and the vector from the start to the center of the arc is described by the I and J values. The Z is the endpoint (in Z) of this move. In one revolution, the cutter will move from Z-.868 to Z-.918. This is .050 and is equal to the lead of the thread.

N0130 G02 X.969 Y0. Z-.968 I1. J0.;

Line N0130 remains in XY helical interpolation (the G17 is modal), clockwise circular interpolation. The endpoint is X .969 Y0., and the vector from the start to the center of the arc is described by the I and J values. The Z is the endpoint (in Z) of this move. In one revolution, the cutter will move from Z-.918 to Z-.968. This is .050 and is equal to the lead of the thread.

N0135 G01 X1.8;

Line N0135 feeds the X axis away from the workpiece to X 1.8.

Now that you have an understanding of canned cycles and how they are used, you need to put this knowledge to work. Program the parts after the chapter questions. Keep in mind the process plan and the setup sheets.

CHAPTER QUESTIONS
••

1. What type of information does the programmer get from the part drawing?

2. What must be taken into consideration when deciding on the machine to be used?

3. What is a process plan?

4. Describe the setup sheet.

5. What are the advantages of using a canned cycle?

6. State three of the more commonly used canned cycles.

7. Calculate the feed rate for a 1/2-13 tap running at 600 RPM.

8. Using the tool data tables shown in Figure 6–12, develop a process plan (use Figure 6–13) and then write a program to execute the profile milling and the hole operations for the part drawing shown in Figure 6–14. The part has been previously squared and milled to the proper thickness. Use absolute programming and cutter compensation.

9. Write a program to machine the part shown in Figure 6–15. Use canned cycles where appropriate.

10. Write a program to machine the part shown in Figure 6–16. Use canned cycles where appropriate.

11. Write a program to machine the part shown in Figure 6–17. Use canned cycles where appropriate.

12. Write a program to machine the part shown in Figure 6–18. Use canned cycles where appropriate.

Tool #	Operation	Tool	Speed (RPM)	Feed (IPM)
1	Mill profile	.750 End mill	150	6.00
2	Tap drill	5/16 Drill	1000	3.00
3	1/4 drill	1/4 Drill	1200	2.5
4	Tap	3/8-16 Tap	500	3.12

FIGURE 6–12 *Tools available to program the part shown in Figure 6–14.*

Fox Valley Techical College - Process Plan				Part No.
Operation	Tool #	Tool Description	RPM	Feed Rate

FIGURE 6–13 *Process planning sheet for question 8.*

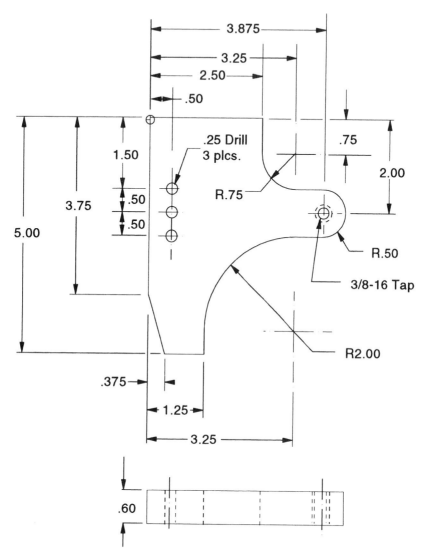

FIGURE 6–14 *Part for question 8.*

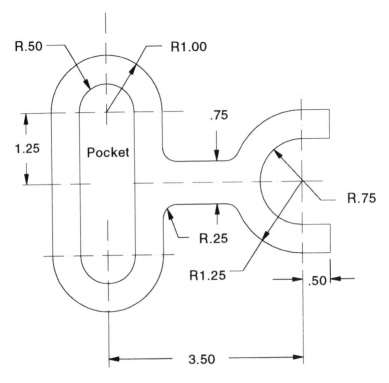

Operations: Profile & Pocket Milling   .50 Deep

FIGURE **6–15** *Part for question 9.*

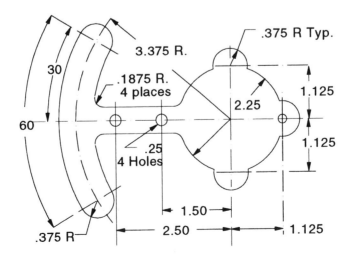

Profile .500 deep drill .75 deep

FIGURE **6–16** *Part for question 10.*

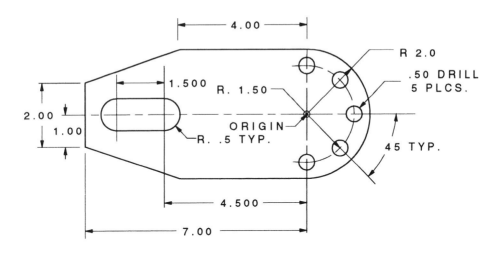

Operations:

1. Mill Profile .50 Deep
2. Mill Slot .500 Deep
3. Drill .50 Holes .500 Deep

FIGURE **6–17** *Part for question 11.*

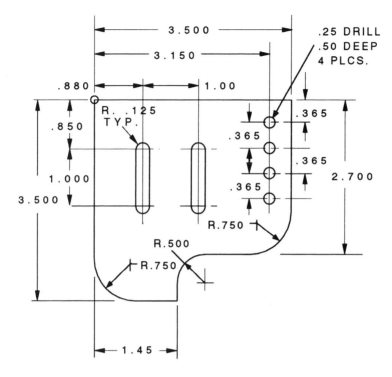

Operations:

1. Mill Profile .500 Deep
2. Mill Slots .500 Deep
3. Drill .25 holes .500 Deep

FIGURE **6–18** *Part for question 12.*

Chapter 7

INTRODUCTION TO MAZATROL PROGRAMMING

COMPETENCY

Perform a variety of basic Mazatrol programming operations using either the machine control or the Griffo Brothers programming software. Demonstrate a working knowledge of the Mazak machining center by properly setting up and machining a series of specified parts.

INTRODUCTION

Mazatrol is the conversational programming format of the Mazak machine control. Mazak is a very popular machine tool manufacturer. One of the reasons why Mazak is so popular is because of the Mazatrol programming language, which resides on Mazak machine tools. Mazatrol is a powerful yet straightforward programming language. In this unit we will be using a tutorial type format to familiarize you with the Mazatrol programming language. To aide us in producing this tutorial we are using a software called Griffo Brothers. Griffo Brothers is a programming package that closely emulates a variety of Mazak controls. Although Griffo Brothers resides on a personal computer this software very closely emulates the Mazatrol controller on the machine tool (see Figure 7–1). The differences between the actual control and the Griffo Brothers emulation package are very subtle.

OBJECTIVES

Upon completion of this chapter, the reader will be able to:

- *Initiate a new program*
- *Program a face machining operation*
- *Perform line machining operations*
- *Program point machining operations*
- *Describe arbitrary shapes in Mazatrol programming language*

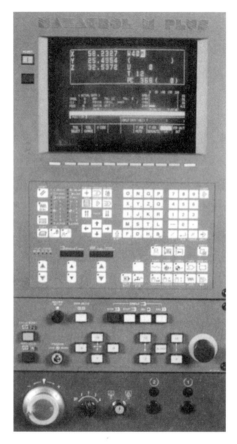

FIGURE 7–1 *The Mazatrol M2 control is a very popular interactive CNC control.*

- *Confirm programming accuracy using Shape Check*
- *Program pocket milling routines*
- *Program pocket mountain machining operations*
- *Program circle milling routines*

BASIC PROGRAMMING

Figure 7–2 is the part print of our first programming exercise. In this exercise we will facemill the top of the workpiece, cut a step around the outside of the part, and drill four holes.

Mazatrol programming is a conversational-type programming language that uses a dialogue box. The dialogue box (Figure 7–3) is located in the lower right-hand corner of the screen.

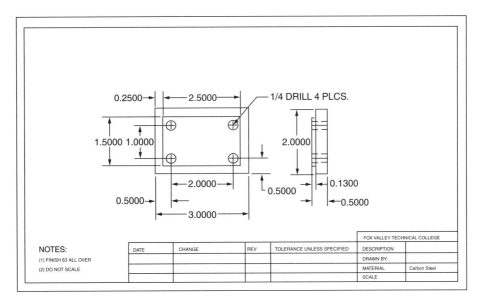

FIGURE 7–2 *Part print for the first programming exercise.*

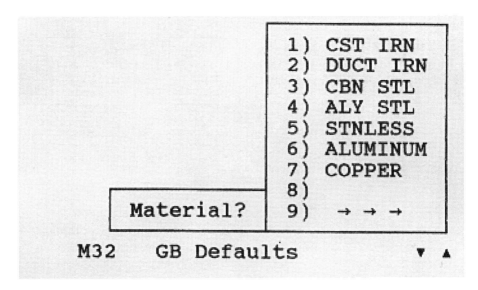

FIGURE 7–3 *Dialogue box.*

The dialogue box acts as an interface between the programmer and the control. The dialogue box contains the current question that needs to be answered by the programmer. Once the current question is answered, the control will continue to prompt the programmer for additional information pertaining to the part to be manufactured.

STARTING A NEW PROGRAM

Double click on the CAMLINK icon to get started using the Griffo Brothers software. Once the machine tool or software has been initialized we need to start a new program. In the Griffo Brothers emulation software we need to select **Data Entry**. On the Mazak machine control starting a new program begins by selecting **Program** from the Main Menu. You are now ready to start a new program. Select **Start a New Data File** now. At the prompt for a work number enter the last four digits of your Social Security number. Entering this number will help you remember what your first program number is.

Upon entering the actual programming mode the initial question that will need to be answered deals with the raw material.

In the dialogue box **Material?** appears. This refers to the type of material to be used to make the workpiece. According to our print the material we will be using is carbon steel. Use the mouse, or type in the number of the material we will be using. Select **3) CBN STL**.

Initial Point Z—Clearance? This is the point above the top of the part or workpiece coordinate location (WPC) setting that is the destination for the tool's rapid traverse. See Figure 7–4 for more information. Typically, we will rapid traverse to within .100 of the top of the part, unless there are clamps or some type of work holding device sticking up. Type in **.1** and press Enter.

Zero Return<Z.X+Y:0, X+Y+Z: 1>? Zero return (Figure 7–5) refers to how you would like the tools to return to the home position after machining has been completed or when tool changing is taking place between machining units.

The <Z.X+Y:0 return method is an input of 0. This return method brings the Z back to home position first then return the X and Y axes back simultaneously.

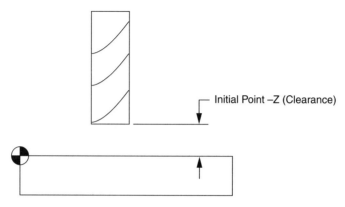

FIGURE 7–4 *This figure shows the initial point–Z clearance.*

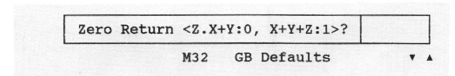

FIGURE 7–5 *Zero return dialogue box.*

This is the safest method. The X+Y+Z: 1> ? return method is an input of 1. This return method brings the X and Y and Z back to home position simultaneously. This is the fastest, but the most dangerous, method of returning the axes back to home. Type in a **0** and press Enter.

Multi Mode (Figure 7–6) would be used for a multiple fixture or vise setup. We will have only one setup. To turn multi mode off select **1) Multi Off**.

You will notice that everything in Unit 0 (Figure 7–7) deals with the setup of the job; the material, the tool changing mode, the rapid clearance planes and the type of setup.

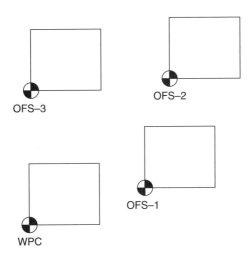

FIGURE 7–6 *Multi Mode would be used for a multiple fixture or vice setup.*

Griffo	File	WkNo	View	Edit	Copy	Search		464M		
UNO MAT		INITIAL-Z	ATC MODE		MULTI MODE		MULTI FLAG	PITCH-X		PITCH-Y
0					OFF		♦	♦		♦

FIGURE 7–7 *Everything in Unit 0 deals with the setup of the job.*

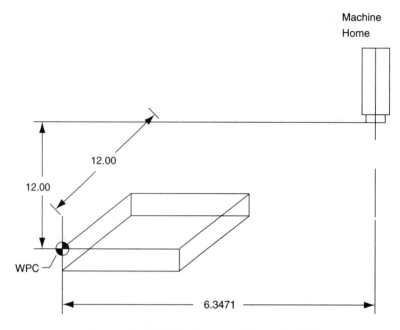

FIGURE 7–8 *Workpiece coordinate (WPC).*

Unit 1 is always the **WPC** or workpiece coordinate (Figure 7–8). This is the equivalent of the G54 or G92 in the EIA/ISO word address programming format. This tells the computer where the part zero is located on the machine table. Select **5) WPC**.

This is the first or number 1 WPC. Type in a number **1** and press Enter.

We don't know where the workpiece is going to be located on the machine so we will have to input a zero for each of the axes. In the dialogue box you will see the prompt:

Workpiece Coordinate WPC - X?

Type in a **0** and press Enter.

Workpiece Coordinate WPC - Y?

Type in a **0** and press Enter.

Workpiece Coordinate WPC - Θ?

This is for an angle of rotation of the workpiece or holding device. Type in a **0** and press Enter.

Workpiece Coordinate WPC - Z?

Type in a **0** and press Enter.

FACE MACHINING

Unit 2 is the unit in which we begin to describe actual machining operations. Looking back at the part print in Figure 7–2 you will note that first machining operation we want to perform is to facemill the top of the part. We will do this to cleanup and finish the top surface of the piece. From the initial Machining Unit menu (Figure 7–9) select **Face Machining** from the dialogue box.

Face Machining is used to machine the top of the part surface (Figure 7–10). Face machining can be used for a variety of different machining methods. Cutting steps, cutting pockets, and cutting slots are only a sample of the machining methods you will find in the Face Machining unit. These will be used later on in this manual. For the part we are doing now we need to face the top of the part. Select **1) Face Mil** now.

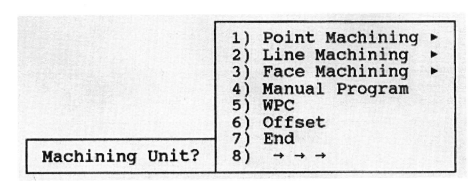

FIGURE 7–9 *Machining Unit menu.*

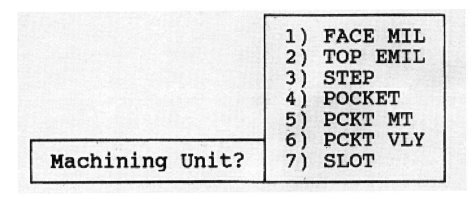

FIGURE 7–10 *Machining options.*

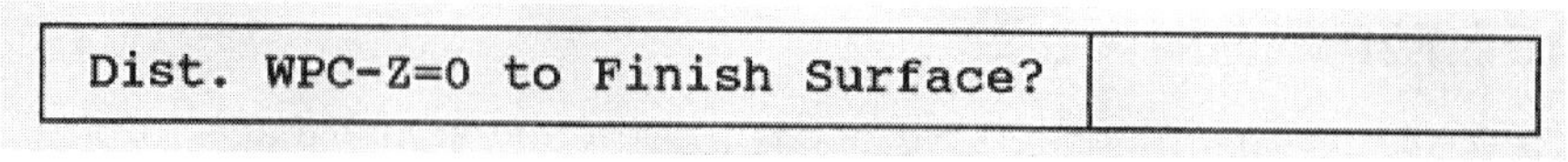

```
┌─────────────────────────────────────────────────────┐
│ Dist. WPC-Z=0 to Finish Surface?  │                 │
└─────────────────────────────────────────────────────┘
```

FIGURE 7–11 *Dialog box asking for the distance from the workpiece Z zero point to the finish surface of the part.*

```
┌──────────────────────────────┬──────────────────────┐
│ Z-Axis Stock Removal?        │ .03                  │
└──────────────────────────────┴──────────────────────┘
```

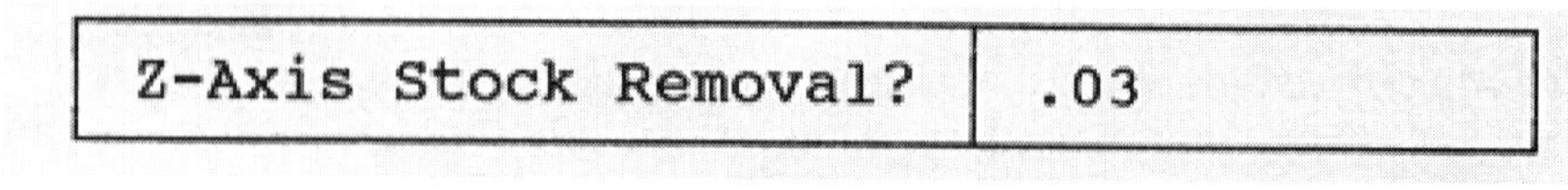

FIGURE 7–12 *Dialogue box asking for the amount of material to be removed in the Z axis.*

Dist. WPC-Z=0 to Finish Surface? What this question in the dialogue box (Figure 7–11) refers to is the distance from the workpiece Z zero point to the finish surface of the part. We want the top finished surface of our workpiece to be equal to Z0.0. Type in a **0** and press Enter.

Z-Axis Stock Removal? This is the amount of material the tool will be removing in the Z axis (Figure 7–12). This would equate to the depth of cut on the Z axis. When we set our WPC-Z (Figure 7–11) we are going to touch off with the tool on the top of the workpiece and tell it to remove .030 inches when facemilling the top of the workpiece. If you needed to take a deeper cut along the surface of the part to completely clean up this surface, you would adjust your depth accordingly. Type in **.03** and press Enter.

Bottom Roughness? This question in the dialogue box is asking for the surface finish requirements for this particular machining unit. The higher the number, the better the surface finish (Figure 7–13). The machine will adjust the speed, feed, depth of cut, and the amount of material that should be left for finishing according to what is input for surface requirements. Select **1)** for bottom roughness.

Finish Allowance –Z? How much stock would you like to leave for finishing on the Z axis or the top or face of this part? The machine will recommend or "default" an amount based on the selection(s) you have made previously. Either type in a **0** or accept the machine default of zero by pressing Enter (Figure 7–14).

The next area of this unit is called the SNO area or sequence area. In this area we will be describing the cutting tool, how it gets to a part, and the cutting conditions (speeds and feeds).

Which Type Of Tool? The dialogue box gives you three choices (Figure 7–15). The control is defaulting to Facemill because of the prior choices you

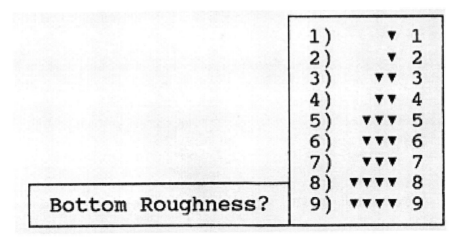

FIGURE 7–13 *Dialogure box asking for the surface finish requirements.*

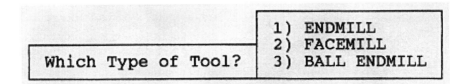

FIGURE 7–14 *Dialogue box asking how much stock you would like to leave for finishing on the Z axis.*

Which Type of Tool?	1) ENDMILL
	2) FACEMILL
	3) BALL ENDMILL

FIGURE 7–15 *Tool choices.*

have made. You can accept the default of Facemill by pressing Enter or you can select **Facemill** by using the arrow keys to highlight your choice, then press Enter to accept this choice.

Nominal Diameter? What size cutting tool will we be using? Type **3.0** and press Enter for a 3-inch facemill.

Tool I.D. Code? The machine will pick the first 3.0-inch facemill it sees in the tool carousel. If you want to use a certain type of cutter, e.g., a roughing endmill, you would attach an I.D. code to this tool. We will accept any 3-inch facemill. Select **1)** for no tool I.D. code.

Machining Priority Number? The Mazatrol control will allow you to prioritize how you want the part to machined. Typically, we write the program in the sequence we would like it to be machined; however, there are times when we would like to change the sequence of the machining units. This priority number allows you to prioritize the sequence of machining units. We want to machine the part in the sequence it was written. Select **1) No Priority**.

Approach Point X? Approach point X is the precise point on the X axis where the tool will rapid traverse to just prior to the start of the machining process. In the dialogue box you will see a question mark for unknown (Figure 7–16). If you would like the control to calculate for you the approach point type in a ?. Otherwise, you can input the exact coordinate in the box and press Enter. The ? feature is one of the major advantages of this control. Type a ? and press Enter.

Approach Point Y? Approach point Y is the precise point on the Y axis where the tool will rapid traverse to just prior to the start the machining process. In the dialogue box you will see a question mark for unknown. If you would like the control to calculate for you the approach point type in a ?. Otherwise, you can input the exact coordinate in the box and press Enter. The ? feature is one of the major advantages of this control. Type a ? and press Enter.

Cutting Direction? The cutting direction refers to how you want the piece machined. Look at Figure 7–17, you are given six choices on how you want the part machined. Figure 7–18 will help to explain what these choices mean.

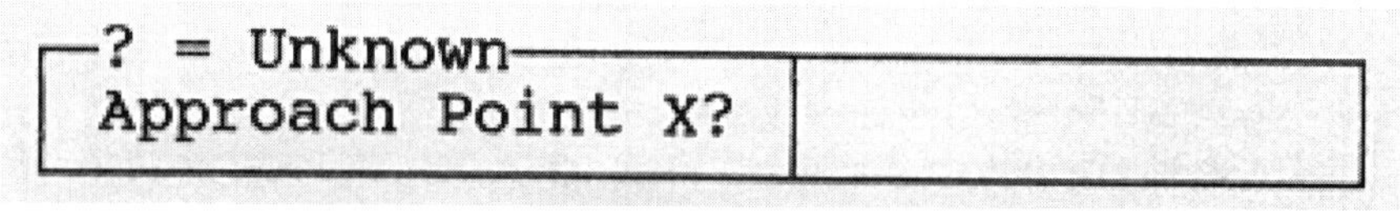

FIGURE **7–16** *Dialogue box asking for the approach point X.*

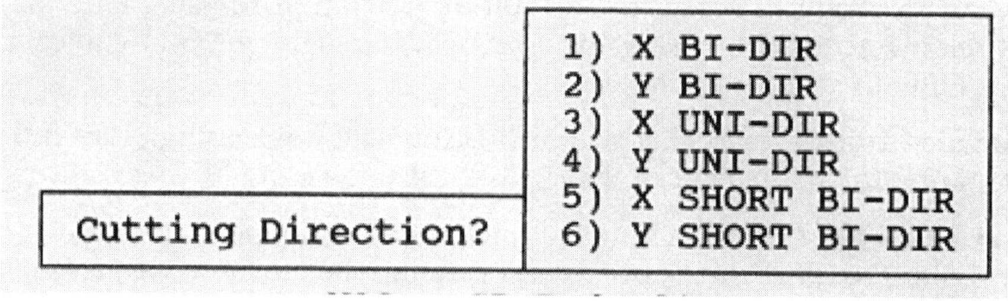

FIGURE **7–17** *Cutting direction choices.*

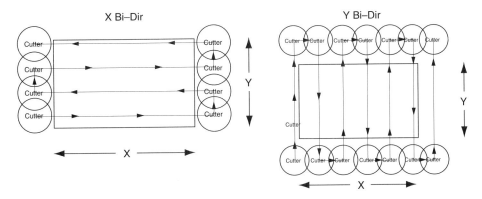

FIGURE 7–18 *Details of the cutting choices.*

X BI - DIR - In Figure 7–18 notice how the cutter path relates to the name of the cut direction. X BI - DIR defined means that the X axis is the major axis of the cutter path and BI - DIR means that the cutter path travels in both directions along the X axis.

Y BI - DIR - In Figure 7–18 notice how the cutter path relates to the name of the cut direction. Y BI - DIR defined means that the Y axis is the major axis of the cutter path and BI - DIR means that the cutter path travels in both directions along the Y axis.

X UNI - DIR - XBI - DIR defined means that the X axis is the major axis of the cutter path and UNI -DIR means that the cutter path travels in only one direction along the X axis. Once the cutter has reached the end of the cut in that axis it picks up and moves back to the start of the next cut.

Y UNI - DIR - YBI - DIR defined means that the Y axis is the major axis of the cutter path and UNI - DIR means that the cutter path travels in only one direction along the Y axis. Once the cutter has reached the end of the cut in that axis it picks up and moves back to the start of the next cut.

X SHORT BI - DIR - X SHORT BI - DIR defined means that the X axis is the major axis of the cutter path and BI - DIR means that the cutter path travels in both directions along the X axis. Note in Figure 7–18 that the tool, at the end of each pass, is completely past the end of the part. "Short" tells us that the tool will never completely travel past the end of the part.

Y SHORT BI - DIR - Y SHORT BI - DIR defined means that the Y axis is the major axis of the cutter path and BI - DIR means that the cutter path travels in both directions along the Y axis. Note in Figure 7–18 that the tool, at the end of each pass, is completely past the end of the part. "Short" tells us that the tool will never completely travel past the end of the part.

For the purposes of our face machining we will use the X axis bi-directional cutting direction. Select **1)** and press Enter.

Depth Of Cut? The depth of cut should be defaulting to .030. Remember we put this in under Z axis stock removal. Accept .03 by pressing Enter. If your program is not defaulting to **.03** type it in and press Enter.

Width Of Cut? The width of cut is the amount the tool is going to step over on each pass when facing off the part. The control will automatically calculate the amount to step-over based on the tool diameter. The stepover distance is approximately two-thirds the diameter of the cutting tool. Accept the default by pressing Enter.

Cutting Speed? The operator needs to look up the cutting speed in surface feet per minute for the workpiece material and the cutting tool material. Type in **400** and press Enter.

Feed Rate? The operator needs to look up the feed rate in feed per tooth (chip load) for the workpiece material and the cutting tool. Type in **.015** and press Enter.

M-Code? This entry will allow us to add Miscellaneous functions to our machining unit. With Mazatrol programming you don't have to turn on the spindle or put in tool changes using M-codes; this is done already. What we will use the M-code entry for is to turn on the coolant. To turn on the coolant type in an **8** for flood coolant on and press Enter.

The next area of this machining unit is called the Figure area or FIG. In this area we will actually be describing the figure we want machined. Keep in mind that we are going to be using the lower left-hand corner of the part as our workpiece coordinate zero position (Figure 7–19).

Point Cutting Pattern? The point cutting pattern gives a general shape description of the portion of the part we want to be face machined in this unit. Your selection choices (Figure 7–20) are in the dialogue box.

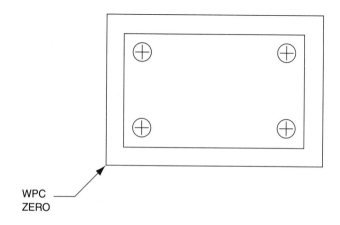

WPC
ZERO

FIGURE 7–19 *The left-hand corner will be used for the WPC zero.*

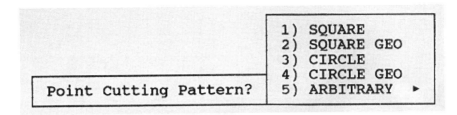

FIGURE 7–20 *Dialogue box showing point cutting pattern choices.*

Square will include any four-sided figure with parallel opposing sides. In other words, a rectangle is actually a square under these circumstances. Square Geo is not available with this software version.

Circle is a circle.

Circle Geo is not available with this software version.

Arbitrary shapes are any shapes other than circles or squares.

The shape of our part is a square. Select the choice of **Square**.

Corner 1 Coordinate X? This is the X coordinate of the first corner of the square (Figure 7–21). A square can be described by the coordinates of the two corners: corner one and corner three.

Corner one is our workpiece X zero point. Type in **0.0** and press Enter.

Corner 1 Coordinate Y? This is the Y coordinate of corner one of the square figure. This is our workpiece Y zero point. Type in **0.0** and press Enter.

Corner 3 Coordinate X? This is the X coordinate of corner three of the square figure. Looking back at our part print in Figure 7–2, the X axis of corner three will be 3.00. Type in **3.0** and press Enter.

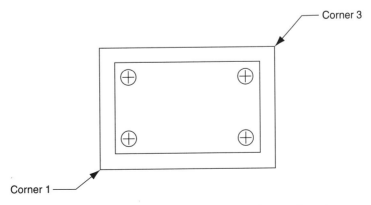

FIGURE 7–21 *Corner 1 and corner 3 coordinates locations.*

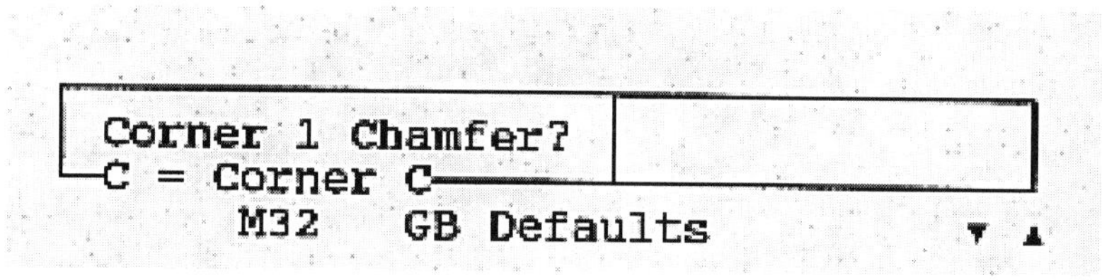

FIGURE 7–22 *Dialogue box asking for chamfer values.*

Corner 3 Coordinate Y? This is the Y coordinate of corner three of the square figure. Looking back at our part print in Figure 7–2, the Y axis of corner 3 will be 2.00. Type in **2.0** and press Enter.

Corner 1 Chamfer? If the corner or corners of the part have chamfers or radii, they can be input now. We have no corner chamfer or radii, so what we need to do is to skip the rest of the questions in this unit. As you look in lower right-hand corner of the screen, notice the up and down arrows (Figure 7–22). Move the cursor to the down arrow using the mouse. Click on the down arrow using the left mouse button. You have just moved down to the start of the next unit. We are done with the face machining unit so select **Shape End**.

The arrows we just used to move down to the next unit can also be used to move throughout the program, unit by unit.

LINE MACHINING

Unit 3 is where we describe the step around the outside of the part. From the dialogue box choose **Line Machining** from the initial Machining Unit menu (Figure 7–23). Select **Line Machining** from the dialogue box now.

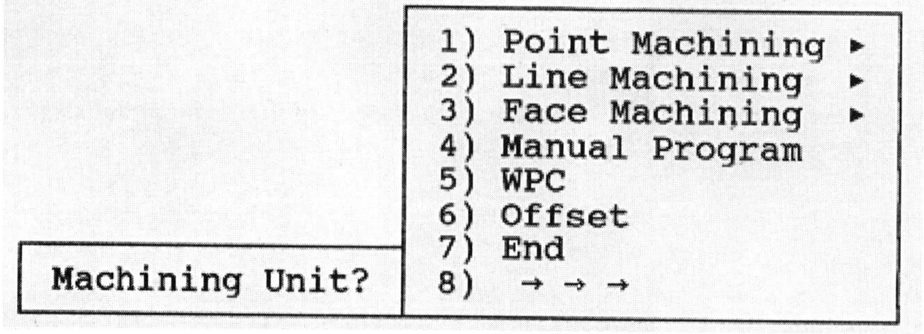

FIGURE 7–23 *Dialogue box showing the machining unit menu.*

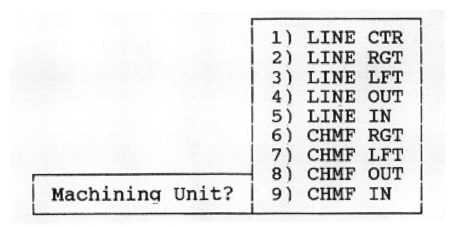

```
                          ┌──────────────────────┐
                          │ 1) LINE  CTR         │
                          │ 2) LINE  RGT         │
                          │ 3) LINE  LFT         │
                          │ 4) LINE  OUT         │
                          │ 5) LINE  IN          │
                          │ 6) CHMF  RGT         │
                          │ 7) CHMF  LFT         │
┌─────────────────────────┤ 8) CHMF  OUT         │
│ Machining Unit?         │ 9) CHMF  IN          │
└─────────────────────────┴──────────────────────┘
```

FIGURE **7–24** *Dialogue box listing the line machining unit options.*

Line Machining is a machining unit that can be used to perform single-path cutting operations (Figure 7–24).

Line Left - To determine which line machining unit to use look at Figure 7–25. If we want to climb cut the step around the part we will use the Line Left machining unit. If we wanted to go around the part in a counterclockwise direction we would use the Line Right machining unit. The choices of line left, line right, and line center work in the same manner as cutter compensations did in Word Address/EIA/ISO programming. The other choices in the Line Machining units will be discussed in later units of this manual. Select **Line Left** at this time to place the cutter on the left side of the programmed path.

Dist. WPC-Z=0 to Finish Surface? What this question in the dialogue box is asking is the distance from workpiece Z zero to the finish surface of the part. We want the step around the outside our workpiece to be equal to Z.130. Type in a **.130** and press enter.

Z-Axis Stock Removal is the amount of material the tool will be removing in the Z axis. This would equate to the depth of cut on the Z axis. When we set our WPC-Z we said that the top of our work is Z – zero. The step is will be removing .130 inches of material on the Z axis. Type in **.130** and press Enter.

X/Y Axis Stock Removal is the actual amount of material the tool will be removing on the radius of the cutter. This would equate to the width of cut or the width of the endmill, whichever is the actual amount. The step is .250 inches wide. We will be using a .500 inch endmill. The actual amount of material being removed is .250 inches. Type in **.250** and press Enter.

Surface Roughness in the dialogue box is asking for the surface finish requirements for this particular machining unit. The higher the number, the

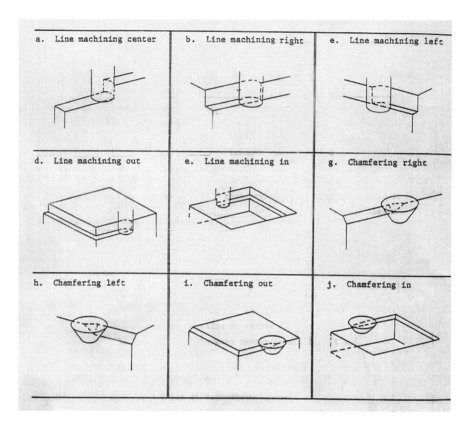

FIGURE 7–25 *Line machining unit cuts.*

better the surface finish (Figure 7–26). The machine will adjust the speed, feed, depth of cut, and the amount of material that should be left for finishing according to what is input for surface requirements. Select **1)** for bottom roughness.

Finish Allowance –Z? How much stock would you like to leave for finishing on the Z axis or the top of the step? The machine will recommend or "default" an amount based on the selection(s) you have made previously. Either type in a **0** or accept the machine default of zero by pressing Enter (Figure 7–27).

Finish Allowance –R? How much stock would you like to leave for finishing on X and Y axes. This is the side of the step. The machine will recommend or "default" an amount based on the selection(s) you have made previously. Either type in a **0** or accept the machine default of zero by pressing Enter.

The next area of this unit is called the SNO area or sequence area. In this area we will be describing the cutting tool, how it gets to the part, and the cutting conditions, such as speeds and feeds, as they relate to the material type.

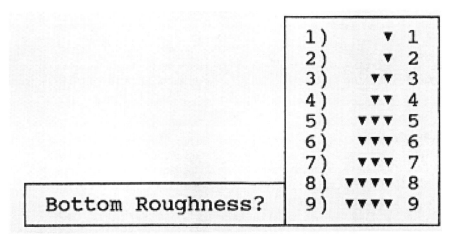

FIGURE **7–26** *Dialoue box asking for surface finish requirements.*

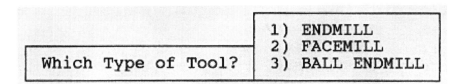

FIGURE **7–27** *Dialogue box asking for the amount of finish allowance on the Z axis.*

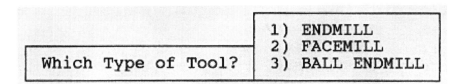

FIGURE **7–28** *Dialogue box showing tool choices.*

Which Type Of Tool? The dialogue box gives you three choices (Figure 7–28). The control is defaulting to Endmill because of the prior choices you have made. You can accept the default of Endmill by pressing Enter or you can select **Endmill** by using the arrow keys to highlight Endmill and then pressing the Enter key.

Nominal Diameter? What size cutting tool will we be using? Type **.500** and press Enter for a .500-diameter endmill.

Tool I.D. Code? The machine control will pick the first .500 diameter end-mill it sees in the tool carousel. If you want to use a certain type of cutter, e.g., a roughing endmill, you would attach an I.D. code to this tool. We will accept any .500 diameter endmill. Select **1)** for no tool I.D. code.

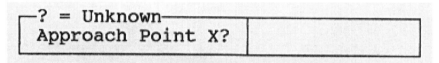

FIGURE **7–29** *Dialogue box asking for the approach point X.*

Machining Priority Number? The Mazatrol control will allow you to prioritize how you want the part to machined. Typically, we write the program in the sequence we would like it to be machined; however, there are times when we would like to change the sequence of the machining units. This priority number allows you to prioritize the sequence of machining units without changing the programmed sequence. We want to machine the part in the sequence it was written. Select **1) No Priority**.

Approach Point X? Approach point X is the precise point on the X axis where the tool will rapid traverse to just prior to the start the machining process. In the dialogue box you will see a question mark (Figure 7–29). If you would like the control to calculate for you the approach point type in a ?. Otherwise, you can input the exact coordinate in the box and press Enter. The ? feature is one of the major advantages of this control. Type a **?** and press Enter.

Approach Point Y? Approach point Y is the precise point on the Y axis where the tool will rapid traverse to just prior to the start the machining process. In the dialogue box you will see a question mark. If you would like the control to calculate for you the approach point type in a ?. Otherwise, you can input the exact coordinate in the box and press Enter. The ? feature is one of the major advantages of this control. Type a **?** and press Enter.

Feed Rate-Z? In the very beginning we set up an approach point. This approach point was the height above the part to which we wanted to rapid traverse. Feed rate Z is asking how we want to get from the approach point down the cut depth. 1) G01 says we want to feed down. 2) G00 says we want to rapid down to depth. If the approach point on the X and Y axes is completely off of the part we can rapid traverse, otherwise we need to feed down. Select **1) G01** to accept a controlled Z feed down to depth.

Depth Of Cut? The depth of cut should be defaulting to .130. Remember we put this in under Z axis stock removal and we are not leaving anything for finishing on the Z axis. Accept **.13** by pressing Enter. If your program is not defaulting to .13 type it in and press Enter.

Cutting Speed? The operator needs to look up the cutting speed in surface feet per minute for the workpiece material and the cutting tool material. Type in **100** and press Enter.

Feed Rate? The operator needs to look up the feed rate in feed per tooth (chip load) for the workpiece material and the cutting tool. Type in **.005** and press Enter.

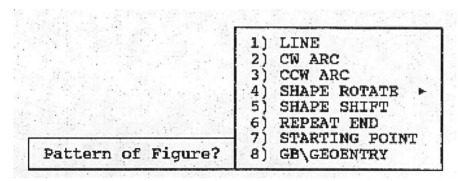

FIGURE 7–30 *Dialogue box showing part geometry choices.*

M-Code? This entry will allow us to add Miscellaneous functions to our machining unit. With Mazatrol programming you don't have to turn on the spindle or put in tool changes using M-codes because it is done already. We turned on the coolant in the face machining unit so we can bypass the M-code selection.

The next area of this machining unit is called the Figure area or FIG. In this are we will actually be describing the step we want machined. Keep in mind that we are going to be using the lower left-hand corner of the part as our workpiece coordinate zero.

Pattern of Figure? Pattern of figure describes the part geometry by using arbitrary shape description choices such as lines, arcs, etc. (Figure 7–30).

Whenever we describe a part or figure shape we must start at some point. The first selection we make when beginning to describe a part shape is to select **Line.** The first selection of Line is actually the Start Point of the figure. You will notice that this first Line selection will be a different color than choices that come after the first Line. Select **Line** at this point.

Coordinate X of Figure? What is the X coordinate of the start point of the step around the outside of the part. Look back at the part print on Figure 7–2. We will be starting at the lower left-hand corner of the step (see Figure 7–31).

The X coordinate of the start point of the figure is .25. Type in **.25** and press Enter.

Coordinate Y of Figure? What is the Y coordinate of the start point of the step around the outside of the part. Look back at the part print on Figure 7–2. We will be starting at the lower left hand corner of the step. The Y coordinate of the start point of the figure is also .25. Type in **.25** and press Enter. To describe a line, or in this case a start point, it takes only an X and Y coordinate. We don't need to input anything more on this line. To jump down to the next unit, use the mouse to select the down arrow in the lower right-hand corner of the screen (Figure 7–32). Pick the **down arrow** key now.

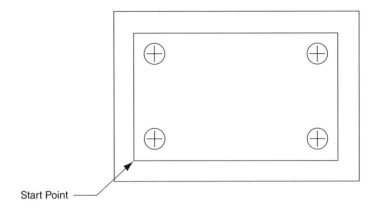

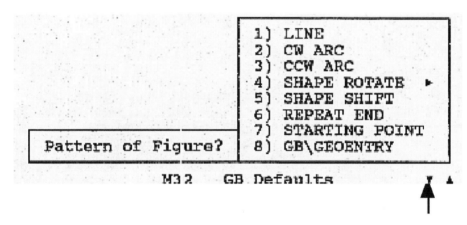

Start Point

FIGURE 7–31 *Start point on part.*

FIGURE 7–32 *Pattern of figure menu.*

Pattern of Figure? The next pattern of the figure is a line. Select **Line**.

Coordinate X of Figure? What is the end point of the line that starts at X.25 and Y.25 (Figure 7–33).

The **X** end point of the line is still .25. Type **.25** and press Enter.

Coordinate Y of Figure? The Y coordinate of the end point of the first line is 1.75. Type **1.75** and press Enter. Pick the **down arrow** key now.

Pattern of Figure? The next pattern of the figure is again a line. Select **Line**.

Coordinate X of Figure? What is the end point of the line that starts at X.25 and Y1.75. The X end point of the line is 2.75. Type **2.75** and press Enter.

Coordinate Y of Figure? The Y coordinate of the end point of the line is still 1.75. Type **1.75** and press Enter. Pick the **down arrow** key now.

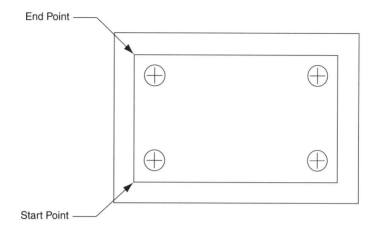

FIGURE 7–33 *Start point and end point of the part.*

Pattern of Figure? The next pattern of the figure is again a line. Select **Line**.

Coordinate X of Figure? What is the end point of the line that starts at X2.75 and Y1.75. The X end point of the line is still 2.75. Type **2.75** and press Enter.

Coordinate Y of Figure? The Y coordinate of the end point of the line is back to .25. Type **.25** and press Enter. Pick the **down arrow** key now.

Pattern of Figure? The next pattern of the figure is again a line. Select **Line**.

Coordinate X of Figure? What is the end point of the line that starts at X2.75 and Y.25. The X end point of this line is going to be back to our original start point. Type **.25** and press Enter.

Coordinate Y of Figure? The Y coordinate of the end point of the line is back to .25, which is our original start point. Type **.25** and press Enter. Pick the **down arrow** key now.

This concludes the step or frame around the outside of the part. Select **Shape End** from the dialogue box.

Let's check to make sure that we have input everything correctly by using the **Shape Check** option. Use the mouse to move the cursor to the Menu Bar along the top of the screen. Select **Griffo**. Under the Griffo menu select **Shape Check**. Under the Shape Check menu select **Check Cont**. On the graphics screen you should now see two boxes, one inside the other. The outside box depicts the face machining and the inside box depicts the step frame we just finished. If your screen doesn't show this, go back and check the numbers you input. Your inputs should look exactly like these in Figure 7–34. To go back to the main programming screen select **End Check**.

```
 Griffo  File  WkNo  View  Edit  Copy  Search                    123M
UNO UNIT        DEPTH    SRV-Z    SRV-R    BTM   WAL  FIN-Z     FIN-R
  2 FACE MIL   0.      0.03       ✦        1    ✦   0.          ✦
SNO   TOOL  NOM-φ  NO  APRCH-X  APRCH-Y TYPE ZFD DEP-Z  WID-R C-SP    FR     M M
R 1 E-MILL  3.           ?        ?      XBI  ✦  0.03   2.1    400 0.015    8
FIG PTN  P1X/CX   P1Y/CY   P3X/R     P3Y      CN1      CN2       CN3      CN4
  1 SQR   0.       0.       3.       2.

UNO UNIT        DEPTH    SRV-Z    SRV-R    RGH   CHMF  FIN-Z     FIN-R
  3 LINE LFT   0.13     0.13     0.25       1    ✦   0.        0.
SNO   TOOL  NOM-φ  NO  APRCH-X  APRCH-Y TYPE ZFD DEP-Z  WID-R C-SP    FR     M M
R 1 E-MILL  0.5          ?        ?      ✦  G01 0.13    ✦      100 0.005
FIG PTN     X         Y          R/θ      I        J         P        CNR
  1 LINE   0.25      0.25
  2 LINE   0.25      1.75
  3 LINE   2.75      1.75                      ┌─────────────────────────┐
  4 LINE   2.75      0.25                      │ 1) Point Machining  ▶   │
  5 LINE   0.25      0.25                      │ 2  Line Machining   ▶   │
                                               │ 3  Face Machining   ▶   │
UNO UNIT                                       │ 4  Manual Program       │
  4 ███████                                    │ 5  WPC                   │
                                               │ 6  Offset                │
                              ┌──────────────┐ │ 7  End                   │
                              │Machining Unit?│ │ 8   → → →                │
                              └──────────────┘ └─────────────────────────┘
```

FIGURE 7–34 *Input values for this exercises.*

POINT MACHINING

Machining Unit? We will be establishing a new machining unit. The new machining unit will be used to put the four 1/4-inch drilled holes. Drilling is found under the Point Machining unit. Select **Point Machining**.

Point Machining, as you can see from your screen or Figure 7–35, includes a number of different machining methods.

We will be discussing a number of these units as we continue in this module. Select the **Drilling** unit from the dialogue box.

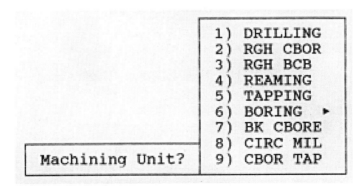

FIGURE 7–35 *Dialogue box listing machining unit options.*

Hole Diameter? The hole is 1/4-inch diameter. Type **.25** and press Enter.

Hole Depth? The holes go through a .500-inch thick part. Type **.550** and press Enter. We allow .050 to assure the drill goes all the way through the part.

Chamfer Width? There is no chamfer. Type **0.0** and press Enter.

Type of Tool? You will see a multitude of tool types now appearing on the screen. We really only need to drill the holes. They holes don't need to be center drilled. Select **Drill** from the tool menu in the dialogue box.

Tool Diameter? Accept the **.25** default by pressing Enter.

Tool I.D. Code? The machine will pick the first .25 drill it sees in the tool carousel. If you want to use a certain type of cutter, e.g., a short drill, you would attach an I.D. code to this tool. We will accept any 1/4-inch diameter drill. Select **1)** for no tool I.D. code.

Machining Priority Number? The Mazatrol control will allow you to prioritize how you want the part to machined. Typically, we write the program in the sequence we would like it to be machined; however, there are times when we would like to change the sequence of the machining units. This priority number allows you to prioritize the sequence of machining units. We want to machine the part in the sequence it was written. Select **1) No Priority**.

Hole Diameter? Accept the **.25** default by pressing Enter.

Hole Depth? Accept the **.55** default by pressing Enter.

Feed Rate Reduction Distance? Mazatrol allows you to adjust the cutting conditions if a previously drilled hole is present. When drilling large holes you sometimes want to start with a smaller hole. Mazatrol will recognize the prepared hole and make adjustments in the cutting speeds, feed rates, and approach points. The .25-inch holes we are drilling have not been previously prepared. Type a **0** and press Enter.

Bottom Cutting Feed Rate (%)? Mazatrol will recognize the prepared hole and make adjustments in the feed rate as it approaches the part of the hole that has not been previously prepared, such as in spot drilling. We have not center drilled or spot drilled. Type a **0** and press Enter.

Drilling Type? The control is now trying to find out if you want to peck drill the holes or drill directly to depth without stopping (see Figure 7–36).

The Drilling Cycle drills the hole straight to depth without stopping or pecking. The Drilling Cycle is typically used for center drilling or for shallow drilled holes.

The Pecking Cycle 1 drills to an incremental programmed peck depth and stops the feed to break the chip, but doesn't retract until it reaches the programmed depth.

DRILL CYCLE   PECKING CYCLE 2   PECKING CYCLE 1

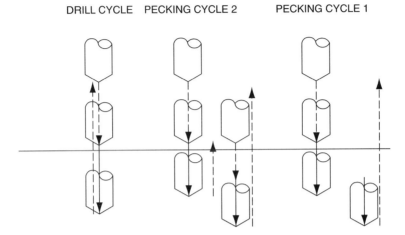

FIGURE 7–36 *Different drilling cycles.*

The Pecking Cycle 2 drills to an incremental programmed peck depth and stops, then it retracts to a clearance plane above the part to clear the chips. The drill then rapids back down the hole to a predetermined clearance position and begins to feed again.

The drilled holes in our part are fairly shallow. Select **1) Drilling Cycle** and press Enter.

Depth of Cut? This is an automatic default for drilling. This is based upon a percentage of the drill diameter. This is the amount of penetration in one operation. Accept the **.125** default by pressing Enter.

Cutting Speed? The operator needs to look up the cutting speed in surface feet per minute for the workpiece material and the cutting tool material. Type in **100** and press Enter.

Feed Rate? The operator needs to look up the feed rate in feed per tooth (chip load) for the workpiece material and the cutting tool. Type in **.0015** and press Enter.

M-Code? Arrow down by using the arrow key on the keyboard to jump down to the next unit. This is another method of moving around the programming screen.

End Sequence? We have input all of the tool information that is needed for this particular operation. Select **End Sequence** at this time to move on to the next step in programming a drill-hole pattern.

Point Cutting Pattern? What drilling pattern are we going to be programming. From Figure 7–37, you can get a little better idea what point cutting patterns are available. This is the real strength of Mazatrol programming.

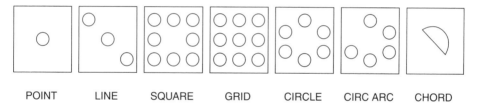

POINT   LINE   SQUARE   GRID   CIRCLE   CIRC ARC   CHORD

FIGURE 7–37 *Available point cutting patterns.*

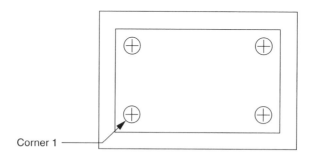

Corner 1 ⎯⎯⎯

FIGURE 7–38 *Corner 1 of the part.*

Looking back at the part print (Figure 7–2), note that the pattern of drilled holes we need is a square. Select **Square** from the dialogue box.

Z Value of the Work Surface? What is the Z coordinate value of the top of the surface where the drilled hole resides. It is the top of the part or Z0.0. Type **0.0** and press Enter.

Corner 1 Coordinate X? This is the X coordinate of the first corner of the square drill pattern figure (Figure 7–38).

Corner one is our workpiece X zero point. Type in **.500** and press Enter.

Corner 1 Coordinate Y? This is the Y coordinate of corner one of the square drilling figure. Type in **.500** and press Enter.

Angle of Start Line From X Axis? When you program a point cutting pattern such as square, the square can be described with two lines. This is possible because the control can calculate the opposite two lines (Figure 7–39).

In Figure 7–39, if we describe line 1 and line 2 the control will automatically calculate the lines 3 and 4. The start line is line 1. The angle of the start line on the X axis is therefore going to be 0. This will typically be true unless the square drilled holes are askew as in a parallelogram. Type in a **0** and press Enter.

Angle Between Two Lines? The angle between line 1 and line 2 in this case is 90 degrees. Type in **90** and press Enter.

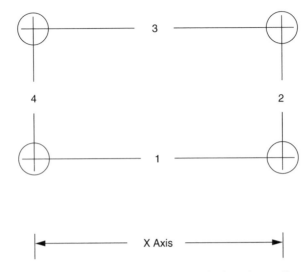

FIGURE 7–39 *A square can be described with two lines.*

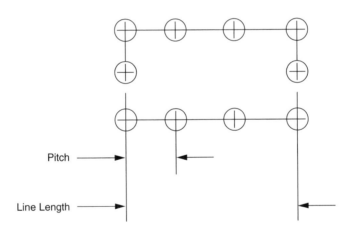

FIGURE 7–40 *Comparison of pitch and line length.*
Pitch is the distance between holes.

Pitch/Length of Pattern AN1? Pitch is the distance from one hole to the next. Length indicates the length of line 1. In our program the distance from the first hole to the second hole is the same as the length of line 1. It will not always be this way. Take a look at Figure 7–40. Type in **2.00**. This is the distance from the first hole to the second hole on our part.

Pitch/Length of Pattern AN2? Pitch is the distance from one hole to the next. Length indicates the length of the line 2. In our program the distance from the second hole to the third hole is the same as the length of line 2. Type in **1.00** and press Enter.

T1&T2<Pitch:0, Line Length:1>? The control needs to know what those numbers were that we just input for Pitch/Length. Did the input values represent the pitch of the holes or the total line length? For the example we are using, a four-hole square pattern, the pitch is the same as the line length. Type in either a **0** or a **1** and press Enter.

Number of Holes in Line AN 1? How many holes in line 1? Type in a **2** and press Enter.

Number of Holes in Line AN 2? How many holes in line 2? Type in a **2** and press Enter.

Omit 4 CNR Ex Spt<Y:1, N:0>? It is possible to program a square pattern of drilled holes and have the control skip the holes in the four corners. We need the holes in the corners. All of our holes are corner holes. Type in a **0** for no omission. Press Enter.

Omit Spt Machining <Y:1, N:0>? It is possible to program a square pattern of drilled holes and have the control skip the first hole or Start Point in the pattern. We need all the holes in the pattern. Type in a **0** for no omission. Press Enter.

Return Position <INIT:0, R:1>? After the control has drilled the last hole in this process, where do you want the drill to return to on the Z axis? If we tell it 0 for INIT or initial point, this value was set on the first line of our program. It was called Initial Point Z clearance. We had input .100. The R plane is set by a parameter within the control. The R plane is typically around .100. We would use the R plane if we planned on drilling more holes with the same drill. Type in a **0** for initial point clearance and press Enter.

We have reached the end of the drilling cycle. At the next question in the dialogue box select **9) END SHAPE**.

We have also reached the end of the program. At the next question in the dialogue box select **7) END**.

Continue < Y:1, N:0 >? No, we don't wish to continue. Type in a **0** and press Enter.

Parts Counter < Y:1, N:0 >? No, we don't wish for parts counting. Type in a **0** and press Enter.

If you are using Griffo Brothers off-line programming software we now need to save our program. If you are programming at the machine control, the program is automatically saved. In Griffo Brothers, save the program by using the mouse to move the cursor to the Menu Bar along the top of the screen. Select File. Under the File menu select **Save**.

Save Data File? Type a **Y** for yes or accept the default of yes by pressing Enter.

Let's check to make sure that we have input everything correctly by using the **Shape Check** option. Use the mouse to move the cursor back to the Menu Bar along the top of the screen. Select **Griffo**. Under the Griffo menu select

Shape Check. Under the Shape Check menu select **Check Cont**. On the graphics screen you should now see two boxes, one inside the other and the drilled holes. The outside box depicts the face machining and the inside box depicts the step frame we just finished and the circles are the drilled holes. If your screen doesn't show this, go back and check the numbers you input. To go back to the main programming screen select **End Check**.

ARBITRARY SHAPE MILLING

Figure 7–41 is the part print of our second programming exercise. In this exercise we will machine a path around the outside of the part.

Once the machine tool or software has been initialized we need to start this new program. In the Griffo Brothers emulation software we need to select **Data Entry**. On the Mazak machine control starting a new program begins by selecting **Program** from the Main Menu. Griffo Brothers now prompts you to **Start a New Data File**. Make this selection now. At the prompt enter a work number to represent the name of this program.

Upon entering the programming mode the initial question that will need to be answered deals with the raw material. In the dialogue box **Material?** appears. According to our print the material we will be using is cast iron. Use the mouse or type in the number of the material we will be using. Select **1) CST IRN**.

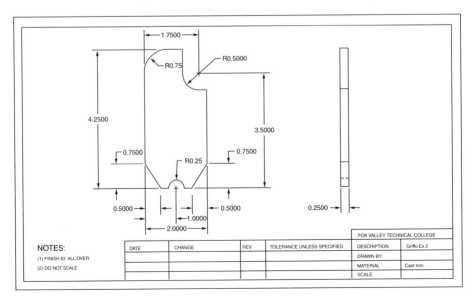

FIGURE 7–41 *Part print for the second programming exercise.*

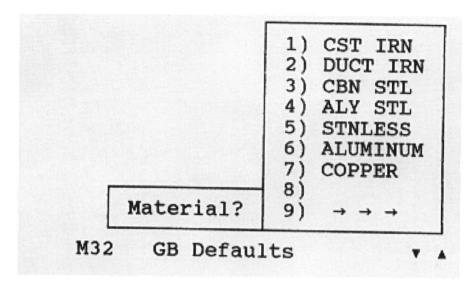

FIGURE 7–42 *Dialogue box showing material options.*

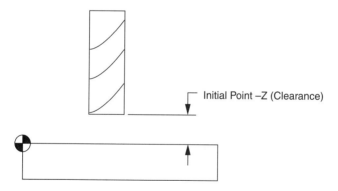

FIGURE 7–43 *This figure shows that the initial point is the place where the machine will rapid to. Make sure you consider clamps and obstructions when you decide your initial point.*

Initial Point Z - Clearance? This is the point above the top of the part or workpiece coordinate location (WPC) setting. This is the point that is the destination for the tool's rapid traverse. See Figure 7–43 for more information. Typically we will rapid traverse to within .100 of the top of the part, unless there are clamps or some type of work-holding device sticking up. Type in **.1** and press Enter.

```
┌─────────────────────────────────────────────┐  ┌──────────────┐
│  Zero Return <Z.X+Y:0, X+Y+Z:1>?            │  │              │
└─────────────────────────────────────────────┘  └──────────────┘
              M32    GB Defaults                    ▼  ▲
```

FIGURE 7–44 *Zero return dialogue box.*

Zero Return<Z.X+Y:0, X+Y+Z: 1>? Zero return (Figure 7–44) refers to how you would like the tools to return to the home position after machining has been completed or when tool changing is taking place between machining units.

The <Z.X+Y:0 return method is an input of 0. This return method brings the Z back to home position first then return the X and Y axes back simultaneously. This is the safest method. The X+Y+Z: 1> ? return method is an input of 1. This return method brings the X and Y and Z back to home position simultaneously. This is the fastest, but the most dangerous, method of returning the axes back to home. Type in a **0** and press Enter.

Multi Mode (Figure 7–45) would be used for a multiple fixture or vise setup. We will have only one setup. To turn multi mode off select **1) Multi Off**.

You will notice that everything in Unit 0 (Figure 7–46) deals with the setup of the job; the material, the tool changing mode, the rapid clearance planes and the type of setup.

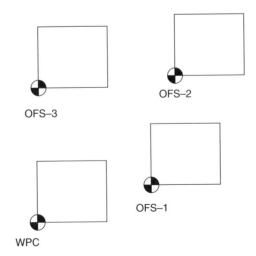

FIGURE 7–45 *Multi Mode would be used to set up locations for multiple parts on the table.*

```
    Griffo  File  WkNo  View  Edit  Copy  Search            464M
 UNO MAT       INITIAL-Z  ATC MODE  MULTI MODE    MULTI FLAG   PITCH-X   PITCH-Y
     0                              OFF               ♦          ♦         ♦
```

FIGURE 7–46 *Everything in Unit 0 deals with the setup of the job.*

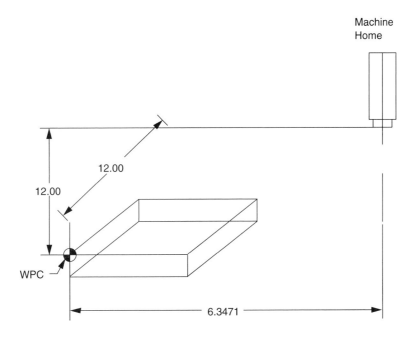

FIGURE 7–47 *Workpiece coordinate (WPC).*

Unit 1 is always the **WPC** or workpiece coordinate (Figure 7–47). This is the equivalent of the G54 or G92 in the EIA/ISO word address programming format. This tells the computer where the part zero is located on the machine table. Select **5) WPC**.

This is the first or number 1 WPC. Type in a number **1** and press Enter.

We don't know where the workpiece is going to be located on the machine so we will have to input a zero for each of the axes. In the dialogue box you will see the prompt:

Workpiece Coordinate WPC - X?

Type in a **0** and press Enter.

Workpiece Coordinate WPC - Y?

Type in a **0** and press Enter.

Workpiece Coordinate WPC - ⊖?

This is for an angle of rotation of the workpiece or holding device. Type in a **0** and press Enter.

Workpiece Coordinate WPC - Z?

Type in a **0** and press Enter.

LINE MACHINING AN ARBITRARY SHAPE

Unit 2 is the unit in which we begin to describe actual machining operations. Looking back at the part print in Figure 7–41, note that only machining operation we want to perform is to contour mill the profile of the part. We will do this to cleanup and finish the wall of the cast piece. From the initial Machining Unit menu select **Line Machining** from the dialogue box.

Line Machining is a machining unit that can be used to perform single-path cutting operations (Figure 7–48).

Line Left - To determine which line machining unit to use look at Figure 7–49. If we want to climb cut the step around the part we use the Line Left machining unit. If we would want to go around the part in a counterclockwise direction we would use the Line Right machining unit. The choices of line left, line right, and line center work in the same manner as cutter compensations did in Word Address/EIA/ISO programming. Select **Line Left** at this time to place the cutter on the left side of the programmed path.

Dist. WPC-Z=0 to Finish Surface? This question in the dialogue box refers to the distance from workpiece Z zero to the finish surface of the part. We

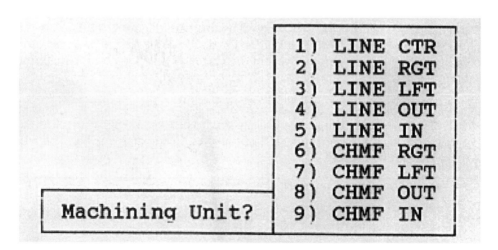

FIGURE **7–48** *Dialogue box showing machining unit options.*

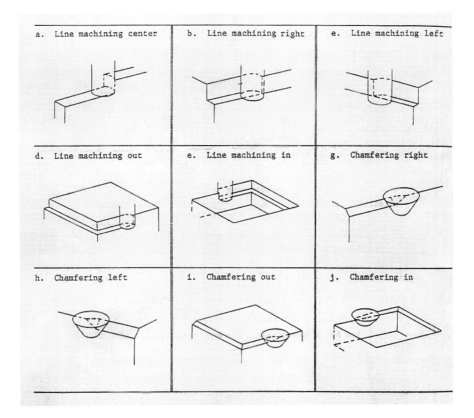

FIGURE 7–49 *Line machining unit cuts.*

want the step around the outside of our workpiece to be equal the thickness of the casting. Type in a **.250** and press Enter.

Z-Axis Stock Removal? is the amount of material the tool will be removing in the Z axis. This would equate to the depth of cut on the Z axis. When we set our WPC-Z we said that the top of our work is Z - zero. The step is will be removing .250 inches of material on the Z axis. Type in **.250** and press Enter

X/Y Axis Stock Removal? is the actual amount of material the tool will be removing on the radius of the cutter. This would equate to the width of cut or the width of the endmill, whichever is the actual amount. We have to see approximately how much material there is left on the casting. We will be using a .375-inch endmill. The actual amount of material being removed is about .050 inches. Type in **.050** and press Enter.

Surface Roughness? in the dialogue box is asking for the surface finish requirements for this particular machining unit. The higher the number, the better the surface finish (Figure 7–50). The machine will adjust the speed, feed, depth of cut, and the amount of material that should be left for finishing according to what is input for surface requirements. Select **1)** for bottom roughness.

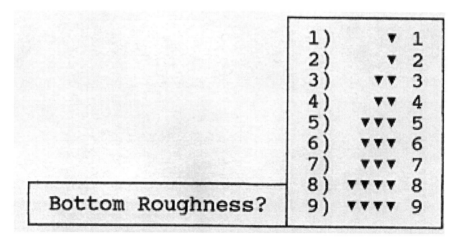

FIGURE **7–50** *Dialogue box asking for surface finish requirements.*

Finish Allowance-Z?	0.

FIGURE **7–51** *Dialogue box asking for the amount of finish allowance on the Z axis.*

Finish Allowance - Z? How much stock would you like to leave for finishing on the Z axis or the top of the step? The machine will recommend or "default" an amount based on the selection(s) you have made previously. Either type in a **0** or accept the machine default of zero by pressing Enter (Figure 7–51).

Finish Allowance - R? How much stock would you like to leave for finishing on X and Y axes. This is the side of the step. The machine will recommend or "default" an amount based on the selection(s) you have made previously. Either type in a **0** or accept the machine default of zero by pressing Enter.

The next area of this unit is called the SNO area or sequence area. In this area we will be describing the cutting tool, how it gets to the part, and the cutting conditions, such as speeds and feeds, as they relate to the material type.

Which Type Of Tool? The dialogue box gives you three choices (Figure 7–52). The control is defaulting to Endmill because of the prior choices you have made. You can accept the default of Endmill by pressing Enter or you can select **Endmill** by using the arrow keys to highlight Endmill and then pressing the Enter key.

Nominal Diameter? What size cutting tool will we be using? Type **.38** and press Enter for a .375-inch diameter endmill. Only 2 decimal places are allowed for the endmill size.

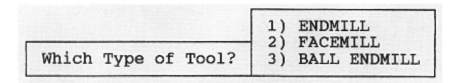

FIGURE **7–52** *Dialogue box showing tool choices.*

Tool I.D. Code? The machine control will pick the first .375 diameter end-mill it sees in the tool carousel. If you want to use a certain type of cutter, e.g., a roughing endmill, you would attach an I.D. code to this tool. We will accept any .375 diameter endmill. Select **1)** for no tool I.D. code.

Machining Priority Number? The Mazatrol control will allow you to pri-oritize how you want the part to machined. Typically, we write the program in the sequence we would like it to be machined; however, there are times when we would like to change the sequence of the machining units. This priority number allows you to prioritize the sequence of machining units without changing the programmed sequence. We want to machine the part in the sequence it was written. Select **1)** No Priority.

Approach Point X? Approach point X is the precise point on the X axis where the tool will rapid traverse to just prior to the start the machining process. In the dialogue box you will see a question mark (Figure 7–53). If you would like the control to calculate for you the approach point type in a ?. Otherwise, you can input the exact coordinate in the box and press Enter. The ? feature is one of the major advantages of this control. Type a **?** and press Enter.

Approach Point Y? Approach point Y is the precise point on the Y axis where the tool will rapid traverse to just prior to the start the machining process. In the dialogue box you will see a question mark. If you would like the control to calculate for you the approach point type in a ?. Otherwise, you can input the exact coordinate in the box and press Enter. The ? feature is one of the major advantages of this control. Type a **?** and press Enter.

Feed Rate - Z? In the very beginning we set up an approach point. This ap-proach point was the height above the part to which we wanted to rapid traverse. Feed Rate Z is asking how do we want to get from the approach point down the cut depth. 1) G01 says we want to feed down. 2) G00 says we want

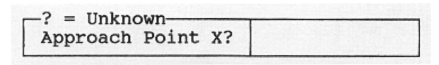

FIGURE **7–53** *Dialogue box asking for the approach point X.*

to rapid down to depth. If the approach point on the X and Y axes is completely off of the part we can rapid traverse, otherwise we need to feed down. Select **1) G01** to accept a controlled Z feed down to depth.

Depth Of Cut? The depth of cut should be defaulting to .250. Remember, we put this in under Z axis stock removal and we are not leaving anything for finishing on the Z axis. Accept **.25** by pressing Enter. If your program is not defaulting to .25 type it in and press Enter.

Cutting Speed? The operator needs to look up the cutting speed in surface feet per minute for the workpiece material and the cutting tool material. Type in **125** and press Enter.

Feed Rate? The operator needs to look up the feed rate in feed per tooth (chip load) for the workpiece material and the cutting tool. Type in **.005** and press Enter.

M-Code? This entry will allow us to add Miscellaneous functions to our machining unit. With Mazatrol programming you don't have to turn on the spindle or put in tool changes using M-codes, this is done already. We don't typically use coolant when cutting cast iron so we can by pass the M-code selection.

The next area of this machining unit is called the Figure area or FIG. In this are we will actually be describing the step we want machined. Keep in mind that we are going to be using the lower left-hand corner of the part as our workpiece coordinate zero.

Pattern of Figure? Pattern of figure describes the part geometry by using arbitrary shape description choices such as lines, arcs, etc. (Figure 7–54).

When we describe a part or figure shape we must start at some point. The first selection we make when beginning to describe a part shape is to select **Line**.

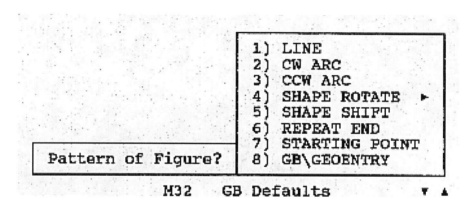

FIGURE 7–54 *Dialogue box showing pattern choices.*

The first selection of line is actually the start point of the figure. You will notice that this first line selection will be a different color than choices that come after the first Line. Select **Line** at this point.

Coordinate X of Figure? What is the X coordinate of the start point of the cut around the outside of the part. Look back at the part print on Figure 7–41. We will start defining our part profile at the point where the angled line meets the vertical line. (see Figure 7-55). Note where the workpiece coordinate setting is going to be. This is the point from which all of the programming coordinates will emanate.

The X coordinate of the start point of the figure is 0.0. Type in **0.0** and press Enter.

Coordinate Y of Figure? What is the Y coordinate of the start point of the profile of the part. Look back at the part print on Figure 7–41. The Y coordinate of the start point of the figure is .75. Type in **.75** and press Enter. To describe a line or, in this case, a start point, it takes only an X and Y coordinate. We don't need to input anything more on this line. To jump down to the next unit, use the mouse to select the down arrow in the lower right-hand corner of the screen. Pick the **down arrow** key now.

Pattern of Figure? The next pattern of figure is a line. Select **Line**.

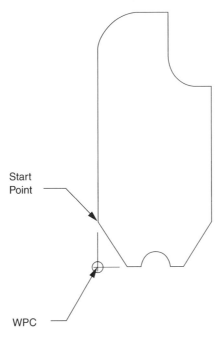

FIGURE 7–55 *Figure showing that dimensions come from the WPC.*

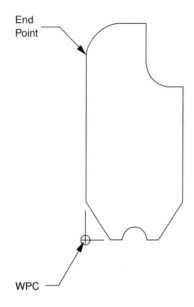

FIGURE 7–56 *Figure showing end point vs. WPC.*

Coordinate X of Figure? What is the end point of the line that starts at X0.0 and Y.75 (Figure 7–56).

The X end point of the line is still 0.0. Type **0** and press Enter.

Coordinate Y of Figure? The Y coordinate of the end point of the first line is 4.25 - R.75. Type **3.5** and press Enter. Pick the **down arrow** key now.

Pattern of Figure? The next pattern of the figure is a clockwise arc. Select **CW arc**.

Coordinate X of Figure? What is the end point of the arc that starts at X0.0 and Y3.50. The **X** end point of the clockwise arc is .75. Type **.75** and press Enter.

Coordinate Y of Figure? The Y coordinate of the end point of the arc is 4.25. Type **4.25** and press Enter.

Radius R? The radius value of the arc is .75. Type **.75** and press Enter.

Circle Interpolation Center X? Mazatrol needs only three pieces of information to create arcs. We have three pieces of information on this line. Input a question mark and Mazatrol will fill in the answer later. Type in **?**.

Circle Interpolation Center Y? Mazatrol needs only three pieces of information to create arcs. We have three pieces of information on this line. Input a question mark and Mazatrol will fill in the answer later. Type in **?**.

Pick the **down arrow** key now.

Pattern of Figure? The next pattern of the figure is again a line. Select **Line**.

Coordinate X of Figure? What is the end point of the line that starts at X4.25 and Y.75. The **X** end point of the line is 1.25. Type **1.25** and press Enter.

Coordinate Y of Figure? The Y coordinate of the end point of the line is still 4.25. Type **4.25** and press Enter.

Pick the **down arrow** key now.

Pattern of Figure? The next pattern of the figure is again a line. Select **Line**.

Coordinate X of Figure? What is the end point of the next line. The **X** end point of this line is going to be the same as the last X axis end point. Type **1.25** and press Enter.

Coordinate Y of Figure? The Y coordinate of the end point of the line is 3.50. This is the tangent point of the arc. Type **3.50** and press Enter.

Pick the **down arrow** key now.

Pattern of Figure? The next pattern of the figure is a counterclockwise arc. Select **CCW arc**.

Coordinate X of Figure? What is the end point of the counterclockwise arc. The **X** end point of the arc is 1.75. Type **1.75** and press Enter.

Coordinate Y of Figure? The Y coordinate of the end point of the arc is 3.0. Type **3.0** and press Enter.

Radius R? The radius value of the arc is .50. Type **.50** and press Enter.

Circle Interpolation Center X? Mazatrol needs only three pieces of information to create arcs. We have three pieces of information on this line. Input a question mark and Mazatrol will fill in the answer later. Type in **?**.

Circle Interpolation Center Y? Mazatrol needs only three pieces of information to create arcs. We have three pieces of information on this line. Input a question mark and Mazatrol will fill in the answer later. Type in **?**.

Pick the **down arrow** key now.

Pattern of Figure? The next pattern of figure is again a line. Select **Line**.

Coordinate X of Figure? What is the end point of the line that starts at X1.75 and Y3.0. The X endpoint of the line is 2.0. Type **2.0** and press Enter.

Coordinate Y of Figure? The Y coordinate of the end point of the line is still 3.0. Type **3.0** and press Enter.

Pattern of Figure? The next pattern of the figure is again a line. Select **Line**.

Coordinate X of Figure? What is the end point of the vertical line that starts at X2.0 and Y3.0. The **X** end point of the line is 2.0. Type **2.0** and press Enter.

Coordinate Y of Figure? The Y coordinate of the end point of the line is .75. Type **.75** and press Enter.

Pattern of Figure? The next pattern of the figure is again a line. Select **Line**.

Coordinate X of Figure? What is the end point of the angled line. The **X** end point of the line is 1.50. Type **1.50** and press Enter.

Coordinate Y of Figure? The Y coordinate of the end point of the line is still 0.0. Type **0** and press Enter.

Pattern of Figure? The next pattern of the figure is again a line. Select **Line**.

Coordinate X of Figure? What is the end point of the short horizontal line that starts at X1.5 and Y0. The **X** end point of the line is 1.25. Type **1.25** and press Enter.

Coordinate Y of Figure? The Y coordinate of the end point of the line is still 0. Type **0** and press Enter.

Pattern of Figure? The next pattern of the figure is a counterclockwise arc. Select **CCW arc**.

Coordinate X of Figure? What is the end point of the counterclockwise arc. The **X** end point of the arc is .75. Type **.75** and press Enter.

Coordinate Y of Figure? The Y coordinate of the end point of the arc is 0. Type **0** and press Enter.

Radius R? The radius value of the arc is .25. Type **.25** and press Enter.

Circle Interpolation Center X? Mazatrol needs only three pieces of information to create arcs. We have three pieces of information on this line. Input a question mark and Mazatrol will fill in the answer later. Type in **?**.

Circle Interpolation Center Y? Mazatrol needs only three pieces of information to create arcs. We have three pieces of information on this line. Input a question mark and Mazatrol will fill in the answer later. Type in **?**.

Pick the **down arrow** key now.

Pattern of Figure? The next pattern of the figure is again a line. Select **Line**.

Coordinate X of Figure? What is the end point of the next line. The **X** end point of the line is .50. Type **.50** and press Enter.

Coordinate Y of Figure? The Y coordinate of the end point of the line is still 0.0. Type **0** and press Enter.

Pattern of Figure? The next pattern of the figure is again a line. Select **Line**.

Coordinate X of Figure? What is the end point of the angle line. This end point gets us back to where we began. The **X** end point of the line is 0. Type **0** and press Enter.

Coordinate Y of Figure? The Y coordinate of the end point of the line is .75. Type **.75** and press Enter.

This concludes the profile mill around the outside of the part. Select **Shape End** from the dialogue box.

Let's check to make sure that we have input everything correctly by using the **Shape Check** option. Use the mouse to move the cursor to the Menu Bar along the top of the screen. Select **Griffo**. Under the Griffo menu select **Shape Check**. Under the Shape Check menu select **Check Cont**. On the graphics screen you should now see two boxes, one inside the other. The outside box depicts the face machining and the inside box depicts the step frame we just finished. If your screen doesn't show this, go back and check the numbers you input. Your inputs should look exactly like those in Figure 7-57. To go back to the main programming screen select **End Check**.

UNO	MAT	INITIAL-Z	ATC MODE	MULTI MODE		MULTI FLAG	PITCH-X	PITCH-Y		
0	CST IRN	0.1	0	OFF		✦	✦	✦		

UNO	UNIT	ADD. WPC	X	Y	⊕	Z	4		
1	WPC- 1		0.	0.	0.	0.	0.		

UNO	UNIT	DEPTH	SRV-Z	SRV-R	RGH	CHMF	FIN-Z	FIN-R		
2	LINE LFT	0.25	0.25	0.05	1	✦	0.	0.		
SNO	TOOL	NOM-φ NO	APRCH-X	APRCH-Y	TYPE	ZFD DEP-Z	WID-R C-SP	FR	M	M
R 1	E-MILL	0.38	?	?	✦	G01 0.25	✦ 125	0.005		
FIG	PTN	X	Y	R/⊕	I	J	P	CNR		
1	LINE	0.	0.75							
2	LINE	0.	3.5							
3	CW	0.75	4.25	0.75	0.75	3.5				
4	LINE	1.25	4.25							
5	LINE	1.25	3.5							
6	CCW	1.75	3.	0.5	1.75	3.5				
7	LINE	2.	3.							
8	LINE	2.	0.75							
9	LINE	1.5	0.							
10	LINE	1.25	0.							
11	CCW	0.75	0.	0.25	1.	0.				
12	LINE	0.5	0.							
13	LINE	0.	0.75							

UNO	UNIT	CONTI.	NUMBER	ATC	X	Y	Z	4	ANGLE
3	END	0	0	0	0.	0.	0.	0.	0

FIGURE 7-57 *Input values for this exercise.*

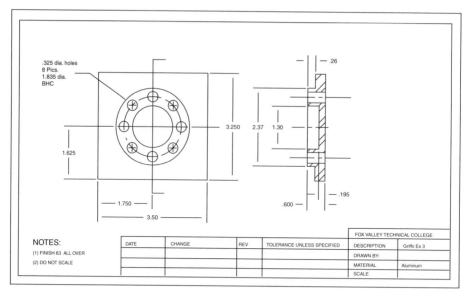

FIGURE 7–58 *Part print for the third programming exercise.*

Figure 7–58 is the part print of third programming exercise. In this exercise we will facemill the top of the workpiece, cut a circular pocket in the middle of the part, cut a circular mountain from a square pocket, and circle mill a series of holes in a bolt-hole circle.

Once the machine tool or software has been initialized we need to start a new program. In the Griffo Brothers emulation software we need to select **Data Entry**. On the Mazak machine control starting a new program begins by selecting **Program** from the Main Menu. Griffo Brothers now prompts you to **Start a New Data File**. Make this selection now. At the prompt for a work number enter the last four digits of your Social Security number. Entering this number will help you remember what your program number is.

Upon entering the actual programming mode the initial question that will need to be answered deals with the raw material. In the dialogue box **Material?** appears. This refers to the type of material to be used to make the workpiece. According to our print the material we will be using is aluminum. Use the mouse, or type in the number of the material we will be using. Select **6) ALUMINUM**.

Initial Point Z - Clearance? Typically we will rapid traverse to within .100 of the top of the part, unless there are clamps or some type of work holding device sticking up. Type in **.1** and press Enter.

Zero Return<Z.X+Y:0, X+Y+Z: 1>? Zero return refers to how you would like the tools to return to the home position after machining has been completed or when tool changing is taking place between machining units.

The <Z.X+Y:0 return method is an input of 0. This return method brings the Z back to home position first then return the X and Y axes back simultaneously. This is the safest method. The X+Y+Z: 1> ? return method is an input of 1. This return method brings the X and Y and Z back to home position simultaneously. This is the fastest, but the most dangerous, method of returning the axes back to home. Type in a **0** and press Enter.

Multi Mode would be used in a multiple fixture or vise setup. We will have only one setup. To turn multi mode off select **1) Multi Off**.

Unit 1 is always the **WPC** or workpiece coordinate. The WPC tells the computer where the part zero is located on the machine table. Select **5) WPC**.

This is the first or number 1 WPC. Type in a number **1** and press Enter.

We don't know where the workpiece is going to be located on the machine so we will have to input a zero for each of the axes. In the dialogue box you will see the prompt:

> **Workpiece Coordinate WPC - X?**
>
> Type in a **0** and press Enter.
>
> **Workpiece Coordinate WPC - Y?**
>
> Type in a **0** and press Enter.
>
> **Workpiece Coordinate WPC - θ?**
>
> This is for an angle of rotation of the workpiece or holding device. Type in a **0** and press Enter.
>
> **Workpiece Coordinate WPC - Z?**
>
> Type in a **0** and press Enter.

Unit 2 is the unit in which we begin to describe actual machining operations. Looking back at the part print in Figure 7–58, note that first machining operation we will want to perform is to facemill the top of the part. We will do this to cleanup and finish the top surface of the piece. From the initial Machining Unit menu Select **Face Machining** from the dialogue box.

Face Machining is used to machine the top of the part surface. Face machining can be used for a variety of different machining methods. Cutting steps, cutting pockets, and cutting slots are only a sample of the machining methods you will find in the Face Machining unit. These will be used later on in this exercise. For the part we are doing now we need to face the top of the part. Select **1) Face Mil** now.

Dist. WPC-Z=0 to Finish Surface? What this question in the dialogue box (Figure 7–59) refers to is the distance from the workpiece Z zero point to the finish surface of the part. We want the top finished surface of our workpiece to be equal to Z0.0. Type in a **0** and press Enter.

```
┌─────────────────────────────────────────┬──────────┐
│ Dist. WPC-Z=0 to Finish Surface?        │          │
└─────────────────────────────────────────┴──────────┘
```

Figure 7–59 *Dialogue box asking for the distance from the workpiece Z zero point to the finish surface of the part.*

Z-Axis Stock Removal? This is the amount of material the tool will be removing in the Z axis. This would equate to the depth of cut on the Z axis. When we set our WPC-Z we are going to touch off with the tool on the top of the workpiece and tell it to remove .150 inches when facemilling the top of the workpiece. This will allow us to use 3/4-inch thick material and facemill it to a final thickness of .600. Type in **.15** and press Enter.

Bottom Roughness? This question in the dialogue box is asking for the surface finish requirements for this particular machining unit. The higher the number, the better the surface finish. The machine will adjust the speed, feed, depth of cut, and the amount of material that should be left for finishing according to what is input for surface requirements. Select **2)** for bottom roughness.

Finish Allowance -Z? How much stock would you like to leave for finishing on the Z axis or the top or face of this part? The machine will recommend or "default" an amount based on the selection(s) you have made previously. Type in **.03** and press Enter.

The next area of this unit is called the SNO area or sequence area. In this are we will be describing the cutting tool, how it gets to a part, and the cutting conditions (speeds and feeds).

Which Type Of Tool? The dialogue box gives you three choices. The control is defaulting to Facemill because of the prior choices you have made. You can accept the default of Facemill by pressing Enter or you can select **Facemill** by using the arrow keys to highlight your choice, then press Enter to accept this choice.

Nominal Diameter? What size cutting tool will we be using? Type **3.0** and press Enter for a 3-inch facemill.

Tool I.D. Code? The machine will pick the first 3.0-inch facemill it sees in the tool carousel. If you want to use a certain type of cutter, e.g., a roughing endmill, you would attach an I.D. code to this tool. We will accept any 3-inch facemill. Select **1)** for no tool I.D. code.

Machining Priority Number? The Mazatrol control will allow you to prioritize how you want the part to machined. Typically, we write the program in the sequence we would like it to be machined; however, there are times when we would like to change the sequence of the machining units. This priority number allows you to prioritize the sequence of machining units. We want to machine the part in the sequence it was written. Select **1) No Priority**.

> ─? = Unknown──────────
> Approach Point X?

FIGURE **7–60** *Dialogue box asking for the approach point X.*

Approach Point X? Approach point X is the precise point on the X axis where the tool will rapid traverse to just prior to the start of the machining process. In the dialogue box you will see a question mark for unknown (Figure 7–60). If you would like the control to calculate for you the approach point type in a **?.** Otherwise, you can input the exact coordinate in the box and press Enter. The ? feature is one of the major advantages of this control. Type a **?** and press Enter.

Approach Point Y? Approach point Y is the precise point on the Y axis where the tool will rapid traverse to just prior to the start the machining process. In the dialogue box you will see a question mark for unknown. If you would like the control to calculate for you the approach point type in a ?. Otherwise, you can input the exact coordinate in the box and press Enter. The ? feature is one of the major advantages of this control. Type a **?** and press Enter.

Cutting Direction? The cutting direction refers to how you want the piece machined. You are given six choices on how you want the part machined. Figure 7–61 will help to explain what these choices mean.

X BI - DIR - In Figure 7–61 you notice how the cutter path relates to the name of the cut direction. X BI - DIR defined means that the X axis is the major axis of the cutter path and BI - DIR means that the cutter path travels in both directions along the X axis.

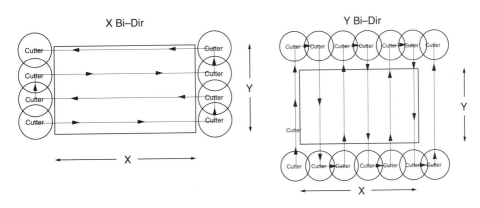

FIGURE **7–61** *Details of the cutting choices.*

Y BI - DIR - In Figure 7–61 notice how the cutter path relates to the name of the cut direction. Y BI - DIR defined means that the Y axis is the major axis of the cutter path and BI - DIR means that the cutter path travels in both directions along the Y axis.

X UNI - DIR - XBI - DIR defined means that the X axis is the major axis of the cutter path and UNI - DIR means that the cutter path travels in only one direction along the X axis. Once the cutter has reached the end of the cut in that axis it picks up and moves back to the start of the next cut.

Y UNI - DIR - YBI - DIR defined means that the Y axis is the major axis of the cutter path and UNI - DIR means that the cutter path travels in only one direction along the Y axis. Once the cutter has reached the end of the cut in that axis it picks up and moves back to the start of the next cut.

X SHORT BI - DIR - X SHORT BI - DIR defined means that the X axis is the major axis of the cutter path and BI - DIR means that the cutter path travels in both directions along the X axis. Note in Figure 7–61 that the tool, at the end of each pass, is completely past the end of the part. "Short" tells us that the tool will never completely travel past the end of the part.

Y SHORT BI - DIR - Y SHORT BI - DIR defined means that the Y axis is the major axis of the cutter path and BI - DIR means that the cutter path travels in both directions along the Y axis. Note in Figure 7–61 that the tool, at the end of each pass, is completely past the end of the part. "Short" tells us that the tool will never completely travel past the end of the part.

For the purposes of our face machining we will use the X axis short bi-directional cutting direction. Select **5)** and press Enter.

Depth Of Cut? The depth of cut is the amount of material we want to remove in one pass on the Z axis. We need to remove a total of .150 inches, but we will do it in .07 inch increments. Type in **.07** and press Enter.

Width Of Cut? The width of cut is the amount the tool is going to step over on each pass when facing off the part. The control will automatically calculate the amount to step over based on the tool diameter. The step-over distance is approximately 2/3 the diameter of the cutting tool. Accept the default by pressing Enter.

Cutting Speed? The operator needs to look up the cutting speed in surface feet per minute for the workpiece material and the cutting tool material. Type in **800** and press Enter.

Feed Rate? The operator needs to look up the feed rate in feed per tooth (chip load) for the workpiece material and the cutting tool. Type in **.025** and press Enter.

M-Code? This entry will allow us to add Miscellaneous functions to our machining unit. With Mazatrol programming you don't have to turn on the spindle or put in tool changes using M-codes; this is done already. What we

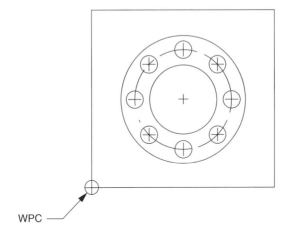

FIGURE 7–62 *Figure showing the WPC location for this part.*

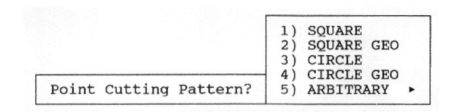

FIGURE 7–63 *Dialogue box showing point cutting pattern choices.*

will use the M-code entry for is to turn on the coolant. To turn on the coolant type in an **8** for flood coolant on and press Enter.

The next area of this machining unit is called the Figure area or FIG. In this area we will actually be describing the figure we want machined. Keep in mind that we are going to be using the lower left-hand corner of the part as our workpiece coordinate zero position (Figure 7–62).

Point Cutting Pattern? The point cutting pattern gives a general shape description of the portion of the part we want to be face machined in this unit. Your selection choices (Figure 7–63) are in the dialogue box.

Square will include any four-sided figure with parallel opposing sides. In other words, a rectangle is actually a square under these circumstances.

The shape of our face machining operation is a square. Select the choice of **Square**.

Corner 1 Coordinate X? This is the X coordinate of the first corner of the square Figure. A square can be described by the coordinates of the two corners: corner one and corner three.

Corner one is our workpiece X zero point. Type in **0.0** and press Enter.

Corner 1 Coordinate Y? This is the Y coordinate of corner one of the square figure. This is our workpiece Y zero point. Type in **0.0** and press Enter.

Corner 3 Coordinate X? This is the X coordinate of corner three of the square figure. Looking back at our part print (Figure 7–58), X axis corner three will be 3.50. Type in **3.5** and press Enter.

Corner 3 Coordinate Y? This is the Y coordinate of corner three of the square figure. Looking back at our part print (Figure 7–58), the Y axis corner three will be 3.25. Type in **3.25** and press Enter.

Corner 1 Chamfer? If the corner or corners of the part have chamfers or radii, they can input now. We have no corner chamfer or radii, so what we need to do is to skip the rest of the questions in this unit. Select the **down arrow**.

We are done with the face machining unit so select **Shape End**.

POCKET MILLING

Unit 2 is the unit which we will use to mill the 1.30 diameter pocket. Pocket milling is performed on the surface or face off the piece. From the Machining Unit menu select **Face Machining** from the dialogue box.

Face Machining is used to machine the top of the part surface. Face machining can be used for a variety of different machining methods (see Figure 7–64). Cutting steps, cutting pockets, and cutting slots are only a sample of the machining methods you will find in the Face Machining unit. For the part we are doing now we need to cut a circular pocket in the center of the part. Select **4) Pocket** now.

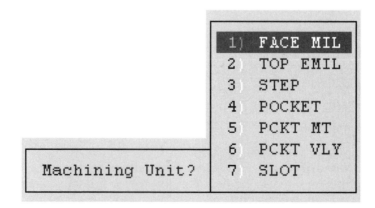

FIGURE 7–64 *Dialogue box listing machining options.*

Dist. WPC-Z = 0 to Finish Surface? What this question in the dialogue box refers to is the distance from the workpiece Z zero point to the finish surface of the pocket. We want the bottom surface of our pocket to .600 - .195. Type in **.405** and press Enter.

Z-Axis Stock Removal? This is the amount of material the tool will be removing in the Z axis. This would equate to the depth of the pocket on the Z axis. Type **.405** and press Enter.

Bottom Roughness? This question in the dialogue box is asking for the surface finish requirements for this particular machining unit. The higher the number, the better the surface finish. Select **3)** for bottom roughness.

Wall Roughness? This question in the dialogue box is asking for the surface finish requirements for the wall of the pocket. The higher the number, the better the surface finish. Select **3)** for wall roughness.

Finish Allowance -Z? How much stock would you like to leave for finishing on the Z axis? Type in a **.01** and press Enter

Finish Allowance -R? How much stock, in a radial value, would you like to leave for finishing on the wall of the pocket? We want to leave .005 inches per side. Type in a **.005** and press Enter.

Finish Allowance -R? Inter-R value? This is for multiple finish passes. Type in a **0** and press Enter.

Finish Allowance -R? Chamfer? There is no chamfer on the pocket. Type a **0** and press Enter.

The next area of this unit is called the SNO area or sequence area. In this are we will be describing the cutting tool, how it gets to a part, and the cutting conditions (speeds and feeds).

Which Type Of Tool? The dialogue box gives you two choices. The control is defaulting to endmill because of the prior choices you have made. Notice on the screen that this is the R1 endmill This is the first or roughing endmill. You can accept the default of endmill by pressing Enter or you can select **Endmill** by using the arrow keys to highlight your choice, then press Enter to accept this choice.

Nominal Diameter? What size cutting tool will we be using? Type **.50** and press Enter for a .50-inch endmill.

Tool I.D. Code? The machine will pick the first .50-inch endmill it sees in the tool carousel. If you want to use a certain type of cutter, e.g., a roughing endmill, you would attach an I.D. code to this tool. We will accept any .50-inch endmill. Select **1)** for no tool I.D. code.

Machining Priority Number? The Mazatrol control will allow you to prioritize how you want the part to machined. Typically, we write the program in the sequence we would like it to be machined; however, there are times when

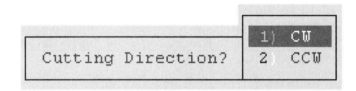

FIGURE 7–65 *Dialogue box listing cutting direction options.*

we would like to change the sequence of the machining units. This priority number allows you to prioritize the sequence of machining units. We want to machine the part in the sequence it was written. Select **1) No Priority**.

Approach Point X? Approach point X is the precise point on the X axis where the tool will rapid traverse to just prior to the start of the machining process. If you would like the control to calculate the approach point type in a ?. Otherwise, you can input the exact coordinate in the dialogue box and press Enter. The ? feature is one of the major advantages of this control. Type a **?** and press Enter.

Approach Point Y? Type a **?** and press Enter.

Cutting Direction? The cutting direction refers to how you want the piece machined (see Figure 7–65). CW or clockwise would be climb milling. CCW or counterclockwise would be conventional milling. Select **CW**.

Feed Rate – Z? You have three choices. Do you want to rapid traverse to depth or do you want a controlled feed down to depth? Other data does not apply to our particular situation. Select **G01** to feed down to depth from the approach point.

Depth Of Cut? The depth of cut should be defaulting to .395. This is the total depth minus the finish amount. We really should take more than one roughing cut to remove this amount of material. Type in **.150** and press Enter.

Width Of Cut? The width of cut is the amount the tool is going to step over on each pass when pocketing the part. The control will automatically calculate the amount to step over based on the tool diameter. The step-over distance is approximately 2/3 the diameter of the cutting tool. Accept the default by pressing Enter.

Cutting Speed? The operator needs to look up the cutting speed in surface feet per minute for the workpiece material and the cutting tool material. Type in **200** and press Enter.

Feed Rate? The operator needs to look up the feed rate in feed per tooth (chip load) for the workpiece material and the cutting tool. Type in **.01** and press Enter.

M-Code? This entry will allow us to add Miscellaneous functions to our machining unit. With Mazatrol programming you don't have to turn on the spindle or put in tool changes using M-codes; this is done already. What have already turned on the coolant in an earlier unit. Select the **down arrow** to continue.

Which Type Of Tool? This is the finishing endmill. You can accept the default of endmill by pressing Enter or you can select **Endmill** by using the arrow keys to highlight your choice, then press Enter to accept this choice.

Nominal Diameter? What size cutting tool will we be using for finishing? Type **.50** and press Enter for a .50-inch endmill.

Tool I.D. Code? Select **1)** for no tool I.D. code.

Machining Priority Number? Select **1) No Priority**.

Approach Point X? Type a **?** and press Enter.

Approach Point Y? Type a **?** and press Enter.

Cutting Direction? The cutting direction refers to how you want the piece machined. Select **CCW** for finish machining the pocket.

Feed Rate - Z? Select **G01** to feed down to depth from the approach point.

Depth Of Cut? The depth of cut has already been calculated by Mazatrol. No input is possible.

Width Of Cut? Accept the default by pressing Enter.

Cutting Speed? The surface speed should be increased for the finishing pass. Type in **300** and press Enter.

Feed Rate? The feed rate needs to be decreased for finishing. Type in **.006** and press Enter.

M-Code? Select the **down arrow** to continue.

The next area of this machining unit is called the Figure area or FIG (see Figure 7–66). In this area we will actually be describing the pocket we want

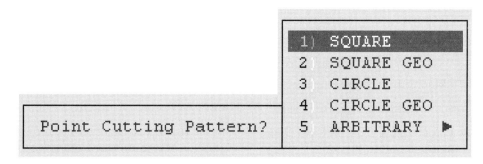

FIGURE **7–66** *Dialogue box listing point cutting pattern options.*

machined. Keep in mind that we are going to be using the lower left-hand corner of the part as our workpiece coordinate zero position.

Point Cutting Pattern? The point cutting pattern you need to select is dictated by the shape of the pocket you want to machine. Looking back at the part print (Figure 7–58), you can see that the shape of the pocket is circular. Select **Circle** from the point cutting pattern menu.

Circle Center X? What is the X axis coordinate of the centerpoint of the circular pocket figure? Type in **1.75** and press Enter.

Circle Center Y? What is the Y axis coordinate of the centerpoint of the circular pocket figure? Type in **1.625** and press Enter.

Circle Radius R? What is the radius of the circular pocket figure? Type in **.65** and press Enter.

This concludes the circular pocket milling routine. Select **Shape End** from the dialogue box.

Let's check to make sure that we have input everything correctly by using the **Shape Check** option. Use the mouse to move the cursor to the Menu Bar along the top of the screen. Select **Griffo**. Under the Griffo menu select **Shape Check**. Under the Shape Check menu select **Check Cont**. On the graphics screen you should now see two shapes, one inside the other. The outside box depicts the face machining and the inside circle depicts the pocket we just finished. If your screen doesn't show this, go back and check the numbers you input. To go back to the main programming screen select **End Check**.

POCKET MOUNTAIN MACHINING OPERATIONS

Machining Unit? Which machining unit would you like to initiate next? In unit 3 we are going to machine the 2.37-inch diameter circular step. In this step we are going to machine around the circular figure all the way out to the outside of the part. This type of machining in Mazatrol is known as a Pocket Mountain. The pocket is the area from the circular figure out to the square that makes up the outside profile of the part. The mountain is the circular feature that will end up being the raised feature of the part (see Figure 7–67).

Pocket Mountain is found under the Face Machining menu. Select **Face Machining** from the dialogue box.

Face Machining as you can see is used for more than just facemilling the part.

From the **Machining Unit?** select **5) Pocket MT** now (see Figure 7–68).

Dist. WPC-Z = 0 to Finish Surface? What this question in the dialogue box refers to is the distance from the workpiece Z zero point to the finish surface of the pocket. We want the bottom surface of our pocket to .600 - .26. Type in **.34** and press Enter.

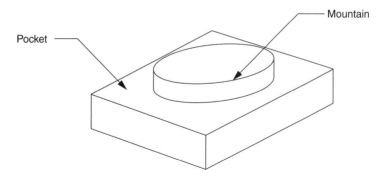

FIGURE 7–67 *A pocket mountain example.*

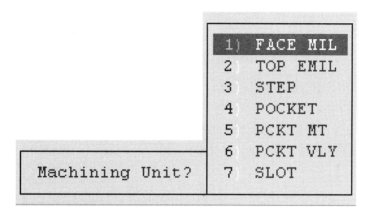

FIGURE 7–68 *Dialogue box showing machining unit options.*

Z-Axis Stock Removal? This is the amount of material the tool will be removing in the Z axis. This would equate to the depth of the pocket on the Z axis. Type **.34** and press Enter.

Bottom Roughness? This question in the dialogue box is asking for the surface finish requirements for this particular machining unit. The higher the number, the better the surface finish. Select **1)** for bottom roughness.

Wall Roughness? This question in the dialogue box is asking for the surface finish requirements for the wall of the mountain. The higher the number, the better the surface finish. Select **3)** for wall roughness.

Finish Allowance -Z? How much stock would you like to leave for finishing on the Z axis? Type in a **0** and press Enter.

Finish Allowance -R? How much stock, in a radial value, would you like to leave for finishing on the wall of the mountain? We want to leave .01 inches per side. Type in a **.01** and press Enter.

The next area of this unit is called the SNO area or sequence area. In this area we will be describing the cutting tool, how it gets to a part, and the cutting conditions (speeds and feeds).

Which Type Of Tool? The dialogue box gives you two choices. The control is defaulting to Endmill because of the prior choices you have made. Notice on the screen that this is the R1 endmill This is the first or roughing endmill. You can accept the default of Endmill by pressing Enter or you can select **Endmill** by using the arrow keys to highlight your choice, then press Enter to accept this choice.

Nominal Diameter? What size cutting tool will we be using? It is very important that you select an endmill size that will fit between the wall of the mountain figure and the programmed wall of the pocket. We realize that there really is no stock boundary for the outside of the pocket, but the computer doesn't realize that (see Figure 7–69). Type **.38** and press Enter for a 3/8-inch endmill.

Tool I.D. Code? The machine will pick the first 3/8-inch endmill it sees in the tool carousel. If you want to use a certain type of cutter, e.g., a roughing endmill, you would attach an I.D. code to this tool. We will accept any 3/8-inch endmill. Select **1)** for no tool I.D. code.

Machining Priority Number? The Mazatrol control will allow you to prioritize how you want the part to machined. This priority number allows you

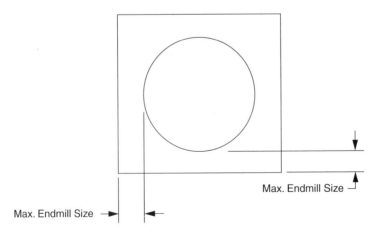

Max. Endmill Size

Max. Endmill Size

FIGURE 7–69 *Note that it is crucial that the endmill be smaller than the distance between the mountain and the pocket wall or the endmill will not fit. The program will not work if the endmill is larger than the gap between the mountain and the pocket wall.*

to prioritize the sequence of machining units. We want to machine the part in the sequence it was written. Select **1) No Priority**.

Approach Point X? Approach point X is the precise point on the X axis where the tool will rapid traverse to just prior to the start of the machining process. If you would like the control to calculate the approach point type in a ?. Otherwise you can input the exact coordinate in the dialogue box and press Enter. The ? feature is one of the major advantages of this control. Type a **?** and press Enter.

Approach Point Y? Type a **?** and press Enter.

Cutting Direction? The cutting direction refers to how you want the piece machined (see Figure 7–70). CW or clockwise would be climb milling. CCW or counterclockwise would be conventional milling. Select **CW**.

Feed Rate - Z? You have three choices. Do you want to rapid traverse to depth or do you want a controlled feed down to depth. Other data does not apply to our particular situation. Select **G01** to feed down to depth from the approach point.

Depth Of Cut? The depth of cut should be defaulting to .34. This is the total depth minus the finish amount. We really should take more than one roughing cut to remove this amount of material. Type in **.150** and press Enter.

Width Of Cut? The width of cut is the amount the tool is going to step over on each pass when pocketing the part. The control will automatically calculate the amount to step over based on the tool diameter. The step-over distance is approximately 2/3 the diameter of the cutting tool. Accept the default by pressing Enter.

Cutting Speed? The operator needs to look up the cutting speed in surface feet per minute for the workpiece material and the cutting tool material. Type in **225** and press Enter.

Feed Rate? The operator needs to look up the feed rate in feed per tooth (chip load) for the workpiece material and the cutting tool. Type in **.008** and press Enter.

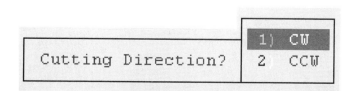

FIGURE **7–70** *Dialogue box showing cutting direction options.*

M-Code? This entry will allow us to add Miscellaneous functions to our machining unit. With Mazatrol programming you don't have to turn on the spindle or put in tool changes using M-codes; this is done already. We have already turned on the coolant in an earlier unit. Select the **down arrow** to continue.

Which Type Of Tool? This is the finishing endmill. You can accept the default of Endmill by pressing Enter or you can select **Endmill** by using the arrow keys to highlight your choice, then press Enter to accept this choice.

Nominal Diameter? What size cutting tool will we be using for finishing? Type **.38** and press Enter for a 3/8-inch endmill.

Tool I.D. Code? Select **1)** for no tool I.D. code.

Machining Priority Number? Select **1) No Priority**.

Approach Point X? Type a **?** and press Enter.

Approach Point Y? Type a **?** and press Enter.

Cutting Direction? Select **CW** to finish machining the pocket.

Feed Rate - Z? Select **G01** to feed down to depth from the approach point.

Depth Of Cut? The depth of cut has already been calculated by Mazatrol. No input is possible.

Width Of Cut? Accept the default by pressing Enter.

Cutting Speed? The surface speed should be increased for the finishing pass. Type in **300** and press Enter.

Feed Rate? The feed rate needs to be decreased for finishing. Type in **.006** and press Enter.

M-Code? Select the **down arrow** to continue.

The next area of this machining unit is called the Figure area or FIG (see Figure 7–71). In this area we will actually be describing the pocket we want

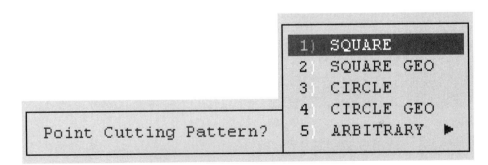

FIGURE **7–71** *Dialogue box showing point cutting pattern options.*

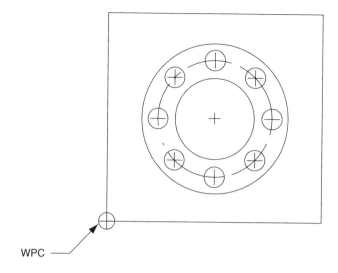

WPC

FIGURE 7–72 *The lower left-hand corner of the part serves as our coordinate zero position.*

machined. Keep in mind that we are going to be using the lower left-hand corner of the part as our workpiece coordinate zero position.

Point Cutting Pattern? The point cutting pattern you need to select is dictated by the shape of the pocket you want to machine. Looking back at the part print (Figure 7–58), you can see that the shape of the pocket is a square. Select **Square** from the point cutting pattern menu. Keep in mind that we are going to be using the lower left-hand corner of the part as our workpiece coordinate zero position (Figure 7–72).

Corner 1 Coordinate X? This is the X coordinate of the first corner of the square figure. A square can be described by the coordinates of the two corners: corner one and corner three. Because the square pocket, in reality, is not bounded by the outside walls we could make the square bigger than it actually is. This would give us the ability to use a bigger endmill. Remember if the endmill cannot fit between the pocket and the mountain Mazatrol will register an error. We used a 3/8-inch endmill. We can describe the pocket and the mountain in their true sizes. This will help to avoid confusion.

Corner one is our workpiece X zero point. Type in **0.0** and press Enter.

Corner 1 Coordinate Y? This is the Y coordinate of corner one of the square figure. This is our workpiece Y zero point. Type in **0.0** and press Enter.

Corner 3 Coordinate X? This is the X coordinate of corner three of the square figure. Looking back at our part print (Figure 7–58), X axis corner three will be 3.50. Type in **3.5** and press Enter.

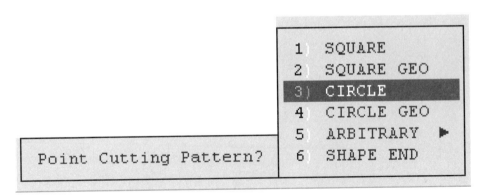

FIGURE 7–73 *Dialogue box showing point cutting pattern options.*

Corner 3 Coordinate Y? This is the Y coordinate of corner three of the square figure. Looking back at our part print (Figure 7–58) the Y axis corner 3 will be 3.25. Type in **3.25** and press Enter.

Corner 1 Chamfer? If the corner or corners of the part have chamfers or radii, they can input now. We have no corner chamfer or radii, so what we need to do is to skip the rest of the questions in this unit. Select the **down arrow.**

We are done with the pocket shape and now we need to describe the mountain shape. If the mountain were any shape other than a square or circle we would need to select Arbitrary shape (see Figure 7–73).

Because the shape of our mountain is circular, select **Circle** from the Point Cutting Pattern menu.

Circle Center X? What is the X axis coordinate of the centerpoint of the circular mountain figure? Type in **1.75** and press Enter.

Circle Center Y? What is the Y axis coordinate of the centerpoint of the circular mountain figure? Type in **1.625** and press Enter.

Circle Radius R? What is the radius of the circular mountain figure? Type in **1.185** and press Enter.

This concludes the pocket mountain milling routine. Select **Shape End** from the dialogue box.

Let's check to make sure that we have input everything correctly by using the **Shape Check** option. Use the mouse to move the cursor to the Menu Bar along the top of the screen. Select **Griffo**. Under the Griffo menu select **Shape Check**. Under the Shape Check menu select **Check Cont**. On the graphics screen you should now see a donut shape in the middle of a square. The outside box depicts the face machining and the square pocket the inside circles depict the pocket and the mountain figure we just finished. If your screen doesn't show this, go back and check the numbers you input. To go back to the main programming screen select **End Check**.

BOLT HOLE CIRCLE MACHINING

Machining Unit? Which machining unit would you like to initiate next? The final machining we need to perform is creating the .325-inch diameter holes in the 1.835-inch diameter bolt hole circle. Notice that the holes have an odd size diameter and they have a three place dimension. The proper way to machine these holes is with a circle milling routine. From the machining unit menu select **Point Machining**. Under point machining you will see a variety of hole machining operations (see Figure 7–74)

From the Machining Unit menu select **8) CIRC MIL**.
Hole Diameter? The hole is .325 inch diameter. Type **.325** and press Enter.

Hole Depth? The holes go through a .600 thick part. Type **.625** and press Enter. We allow .025 to assure the hole goes all the way through the part.

Chamfer Width? There is no chamfer. Type **0.0** and press Enter.

Bottom Roughness? Select **1** for bottom surface roughness.

Prepared Hole Diameter? We have not drilled any clearance holes for the circle mills. Type in **0** and press Enter. Because we have not previously prepared the holes we must make sure that we have an end cutting endmill.

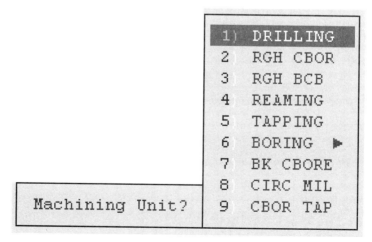

FIGURE 7–74 *Dialogue box showing machining unit options.*

Chamfer Width? The print does not call for a chamfer. Type **0** and press Enter.

Type of Tool? The control is defaulting to Endmill. Select **Endmill** or accept the default by pressing Enter.

Tool Diameter? We have to select an endmill which is smaller than **.325**. We will use a 1/4-inch endmill. Type **.25** and press Enter.

Tool I.D. Code? The machine will pick the first .25 endmill it sees in the tool carousel. If you want to use a certain type of cutter, e.g., a long endmill or an endmill that can plunge, you would attach an I.D. code to this tool. We will need an extended length 1/4-inch end cutting endmill. Select **2)** to attach an A code to the tool selection. When we setup the tools for machining we will have to notify the control with the same code letter identifier.

Tool I.D. Code? You can add additional priorities to the tool selection that will be made by the control by selecting weighted. We do not need a weighted selection. Select **Normal**.

Machining Priority Number? The Mazatrol control will allow you to prioritize how you want the part to machined. We want to machine the part in the sequence it was written. Select **1) No Priority**.

Hole Diameter? Accept the **.35** default by pressing Enter.

Hole Depth? Accept the **.625** default by pressing Enter.

Prepared Hole Diameter? Accept the **0** default by pressing Enter.

Bottom Roughness? Accept the **1** default by pressing Enter.

Width of Cut? The control uses 2/3 of the diameter as a step-over vector per pass. Accept the **.150** default by pressing Enter.

Cutting Speed? The surface speed should be decreased because you will be machining in a confined area. Type in **200** and press Enter.

Feed Rate? The feed rate needs to be decreased also. Type in **.006** and press Enter.

M-Code? Select the **down arrow** to continue.

Point Cutting Pattern? What circle milling pattern are we going to be programming. From Figure 7–75, you can get a little better idea what point cutting patterns are available. Again, this is the real strength of Mazatrol programming.

Looking back at the part print (Figure 7–58), you will note that the pattern of holes we need is a bolt hole circle. Select **5) Circle** from the dialogue box.

Z Value of The Work Surface? What is the Z coordinate value of the top of the surface where the drilled hole resides. It is the top of the part or Z0.0. Type **0.0** and press enter.

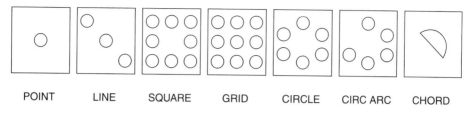

POINT   LINE   SQUARE   GRID   CIRCLE   CIRC ARC   CHORD

FIGURE 7–75 *Available point cutting patterns.*

Circle Center X? What is the X axis coordinate of the centerpoint of the bolt hole circle, not the individual circles? Type in **1.75** and press Enter.

Circle Center Y? What is the Y axis coordinate of the centerpoint of the bolt hole circle? Type in **1.625** and press Enter.

Angle of Start PT From X Axis? When you program a cutting pattern that has a circular hole pattern the control needs to know where the first hole is located. Looking back at the part print our hole pattern can start at 0, 45, or 90 degrees rotated from the X axis (see Figure 7–76). Type in **0** and press Enter.

Circle Radius R? What is the radius of the bolt hole circle? The print states the bolt hole as a 1.835-inch diameter. 1.835 divided in half is .9175. The radius is .9175. Type in **.9175** and press Enter.

Number of Holes? How many holes in this bolt hole circle? Type in **8** and press Enter.

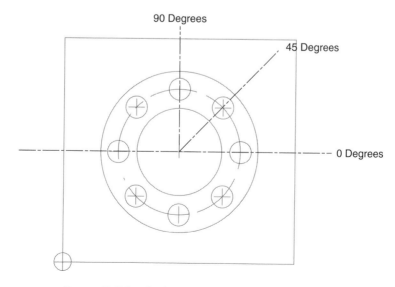

FIGURE 7-76 *The hole pattern can start at 0, 45, or 90 degrees from the X axis.*

Return Position <INIT:0, R:1>? After the control has circle milled the last hole in this process, where do you want the endmill to return to on the Z axis? Type in a **0** for initial point clearance and press Enter.

We have reached the end of the circle milling cycle. At the next question in the dialogue box select **9) END SHAPE**.

We have also reached the end of the program. At the next question in the dialogue box select **7) END**.

Continue < Y:1, N:0 >? No, we don't wish to continue. Type in a **0** and press enter.

Parts Counter < Y:1, N:0 >? No, we don't wish for parts counting. Type in a **0** and press enter.

If you are using Griffo Brothers off-line programming software we now need to save our program. If you are programming at the machine control, the program is automatically saved. In Griffo Brothers, save the program by using the mouse to move the cursor to the Menu Bar along the top of the screen. Select File. Under the File menu select **Save**.

Save Data File? Type a **Y** for yes or accept the default of yes by pressing Enter.

Let's check to make sure that we have input everything correctly by using the **Shape Check** option. Use the mouse to move the cursor back to the Menu Bar along the top of the screen. Select **Griffo**. Under the Griffo menu select **Shape Check**. Under the Shape Check menu select **Check Cont**. On the graphics screen you should now see two boxes, one inside the other and the drilled holes. The outside box depicts the face machining and the inside box depicts the step frame we just finished and the circles are the drilled holes. If your screen doesn't show this, go back and check the numbers you input. To go back to the main programming screen select **End Check**.

That concludes the tutorial portion of this chapter on Mazatrol programming. Try programming the additional exercises at the end of the chapter to sharpen your Mazatrol programming skills.

PROJECT 1

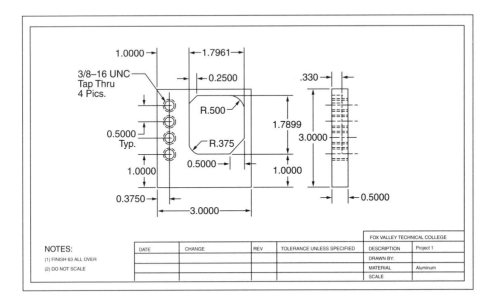

NOTES:

(1) FINISH 63 ALL OVER

(2) DO NOT SCALE

DATE	CHANGE	REV	TOLERANCE UNLESS SPECIFIED

FOX VALLEY TECHNICAL COLLEGE

DESCRIPTION	Project 1
DRAWN BY:	
MATERIAL	Aluminum
SCALE	

PROJECT 2

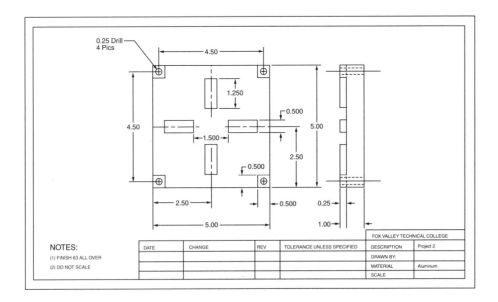

NOTES:

(1) FINISH 63 ALL OVER

(2) DO NOT SCALE

DATE	CHANGE	REV	TOLERANCE UNLESS SPECIFIED

FOX VALLEY TECHNICAL COLLEGE

DESCRIPTION	Project 2
DRAWN BY:	
MATERIAL	Aluminum
SCALE	

PROJECT 3

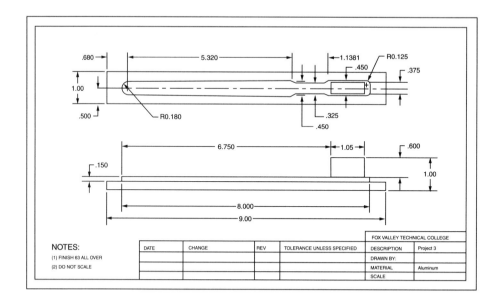

NOTES:

(1) FINISH 63 ALL OVER

(2) DO NOT SCALE

DATE	CHANGE	REV	TOLERANCE UNLESS SPECIFIED	FOX VALLEY TECHNICAL COLLEGE	
				DESCRIPTION	Project 3
				DRAWN BY:	
				MATERIAL	Aluminum
				SCALE	

Chapter 8

CNC TURNING MACHINES

INTRODUCTION

A CNC lathe or turning center is a numerically controlled lathe. Its main function is to create high-quality cylindrical parts in a minimum amount of time. Turning centers can machine internal and external surfaces. Other machine operations, such as drilling, tapping, boring, and threading, are also done on turning centers.

OBJECTIVES

Upon completion of this chapter, the reader will be able to:

- *Name the major components of the turning center.*
- *State the purpose of tool turrets.*
- *Describe the advantage of the slant-bed lathe.*
- *Correctly identify the major axes on turning centers.*
- *Describe the three major work-holding devices used on CNC lathes.*
- *Describe the common machining operations and the tools associated with them.*
- *Explain tool-wear offsets.*
- *Describe how geometry offsets or workpiece coordinates are set.*
- *Explain the use of the more common material-handling equipment.*
- *Correctly identify safe working habits associated with CNC lathes.*
- *Identify and explain common machine controls.*

INTRODUCTION TO TURNING CENTERS

The turning center is one of the most productive machine tools in the machine shop. It is capable of producing cylindrically shaped parts in great volumes and with incredible accuracy. The first numerically controlled turning machines were developed in the mid-1960s and were little more than an engine lathe retrofitted with a control and drive motors. Today's turning centers can be equipped with dual tool turrets, dual spindles, milling head attachments, and a variety of other specialized features to make them capable of machining even the most complex parts in one setup. This chapter looks at all of the concepts and features of the turning center, but pays particularly close attention to the fundamentals.

TYPES

The first numerically controlled turning machines were conventional lathes fitted with numerical controls and drive systems. These lathes were significantly more accurate and more productive than conventional lathes.

The standard flat-bed configuration is still evident on some CNC lathes; however, most turning machines today have a slant-bed configuration (see Figure 8–1). Slanting the bed on a CNC lathe allows the chips to fall away from the slideways and allows the operator easy access to load and unload parts.

COMPONENTS OF CNC LATHES

The main components of a CNC lathe or turning center are the headstock, tailstock, turret, bed, and carriage (see Figure 8–2).

FIGURE **8–1** *Slant-bed-style turning center.*
(Courtesy of Mori Seiki Co., Ltd.)

FIGURE **8–2** *Slant-bed lathe components.*
(Courtesy of Mori Seiki Co., Ltd.)

HEADSTOCK

The headstock contains the spindle and transmission gearing, which rotates the workpiece. The headstock spindle is driven by a variable speed motor. This motor, which is programmable in revolutions per minute, delivers the required horsepower and torque through a drive belt or series of drive belts (see Figure 8–3).

TAILSTOCK

The tailstock is used to support one end of the workpiece. The tailstock slides along its own set of slideways on some turning machines and on the same set of slideways as the carriage on conventional-style CNC lathes. The tailstock has a sliding spindle much like that of the tailstock on a manual lathe. Two types of tailstocks are available on CNC turning machines: manual and programmable. The manual tailstock is moved into position by the use of a switch or hand wheel. The programmable tailstock can be moved manually or can be programmed like the tool turret (see Figure 8–4).

Figure 8–3 *The gearless headstock configuration delivers maximum power and speed with very little noise or vibration. (Courtesy of Giddings & Lewis, Inc.)*

TOOL TURRETS

Tool turrets on turning machines come in all styles and sizes, but the basic function of the turret is to hold and quickly index the cutting tool. Each tool or tool position is numbered for identification. When the tool needs to be changed, the turret moves to a clearance position and indexes, bringing the new tool into the cutting position. Most turning center turrets can move bi-directionally to assure the fastest tool indexing time. Tool turrets can also be indexed by hand, using a button or switch located on the control panel (see Figure 8–5).

BED

The bed of the turning center supports and aligns the axis and cutting tool components of the machine. The bed is made of high-quality cast iron and will absorb the shock and vibration associated with metal cutting conditions. The bed of the turning center lies either flat or at a slant to accommodate chip removal. The slant of the bed is usually 30 to 45 degrees (see Figure 8–6).

FIGURE **8–4** *Programmable tailstock.*
(Courtesy of Mori Seiki Co., Ltd.)

CARRIAGE

The carriage slides along the bed and controls the movement of the tool (see Figure 8–2).

TURNING MACHINE AXES IDENTIFICATION

The basic CNC lathe has two major axes, the X and Z axes. We will concentrate on these axes in this text (see Figure 8–7).

The Z axis always lies in the same plane as the spindle, just as on machining centers. The X or cross-slide axis runs perpendicular to the Z axis. Negative Z axis (–Z) motion moves the tool turret closer to the headstock. Positive Z (+Z) motion moves the turret or tool away from the headstock or toward the tailstock. Negative X (–X) motion moves the tool or turret toward the centerline of the spindle and positive X (+X) motion moves the tool or turret away from the centerline. Some CNC turning centers are equipped with a programmable tailstock or third axis known as the W axis. More complex turning centers have four axes and

FIGURE 8–5 *Twelve-station tool turret. Each turret position is numbered for identification purposes. (Courtesy of Mori Seiki Co., Ltd.)*

have two opposing turrets. The turrets are independent of one another and can do different machining operations at the same time (see Figure 8–8).

WORK HOLDING

Work-holding devices are an integral part of your CNC lathe. As the demand for higher production increases, so does the need for an understanding of the different types of work-holding devices. The most common work-holding method used on turning machines is the chuck.

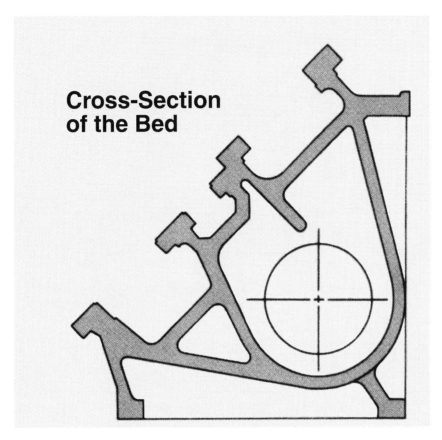

FIGURE **8–6** *The slant bed is designed for quick chip removal and easy operator access. (Courtesy of Mori Seiki Co., Ltd.)*

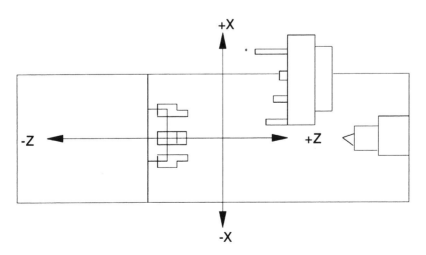

FIGURE **8–7** *Lathe axes of motion. Note that the Z axis is always in line with the spindle.*

FIGURE 8–8 *Two independent turrets allow two machining operations to be done simultaneously. (Courtesy of Mori Seiki Co., Ltd.)*

CHUCKS

There are many different types of chucks, but the most common type is the three-jaw, self-centering, hydraulic chuck (see Figure 8–9).

This type of chuck has three jaws, all of which move in unison under hydraulic power. The chucks are activated by a foot switch that opens and closes the chuck jaws.

There are two types of chuck jaws, hardened jaws and soft jaws. Hardened jaws are used where maximum holding power is needed on unfinished surfaces.

Soft jaws are used on parts that can have little runout or on finished surfaces that cannot be marred. Soft jaws are typically made from soft steel and are turned to fit each type of part. Soft jaws can be machined with ordinary carbide tooling.

FIGURE 8–9 *Three-jaw hydraulic chuck.*
(Courtesy of Mori Seiki Co., Ltd.)

COLLET CHUCKS

The collet chuck is an ideal work-holding device for small parts where accuracy is needed. The collet chuck assembly consists of a draw tube and a hollow cylinder with collet pads. Collets are available for holding hexagonal, square, and round stock. They are typically used in bar feeding systems and are covered in the material handling section of this chapter.

FIXTURES

Fixtures are work-holding devices used for odd-shaped or other hard to hold workpieces. Fixtures may be held in the chuck or can be bolted directly to the spindle. Fixtures are spun in the spindle, so they must be balanced. Unbalanced fixtures can cause severe damage to the machine and possible injury to the operator.

CUTTING TOOLS

Modern turning machines use tool holders with indexable inserts. The tool holders on CNC machines come in a variety of styles, each suited for a particular type of cutting operation. The machining operations discussed in this chapter include facing, turning, grooving, parting, boring, and threading.

FACING

Facing operations involve squaring the face or end of the stock. The tool needs to be fed into the stock in a direction that will push the insert toward the pocket of the holder (see Figure 8–10).

TURNING

Turning operations remove material from the outside diameter of the rotating stock. Rough turning removes the maximum amount of material from the workpiece and should be done with an insert with a large included angle.

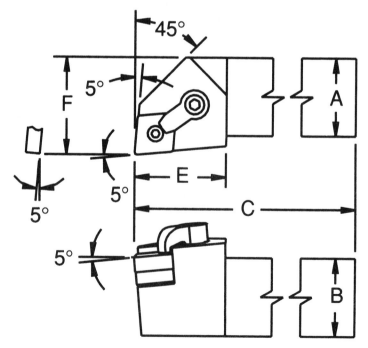

FIGURE 8–10 *Tool style L can be used for turning and facing using an 80-degree diamond insert. (Courtesy of Kennametal, Inc.)*

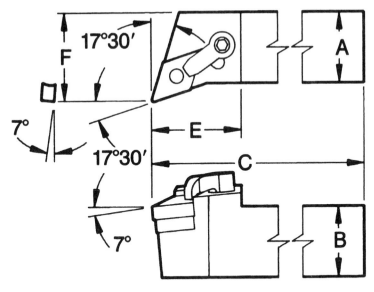

FIGURE 8–11 *Tool style Q is used for profile turning using a 55-degree diamond insert. (Courtesy of Kennametal, Inc.)*

The large included angle will insure that the tool has the proper strength to withstand the cutting forces being exerted. Profile turning uses an insert with a smaller included angle. If the finish profile warrants the use of a small, sharp-angled insert, a series of semi-finish passes are necessary to insure against tool breakage (see Figure 8–11).

GROOVING

For internal and external grooving the tool is fed straight into the workpiece at a right angle to its centerline. The cutting insert is located at the end of the tool. Grooving operations include thread relieving, shoulder relieving, snap-ring grooving, O-ring grooving, and oil reservoir grooving (see Figure 8–12).

PARTING

Parting is a machine operation that cuts the finished part off of the rough stock. This operation is similar to grooving. The tool is fed into the part at a right angle to the centerline of the workpiece and is fed down past the center-line of the work, thus separating it from the rough stock. The parting tool has a carbide insert located at the end of the tool and has a slight back taper along the insert for clearance (see Figure 8–13).

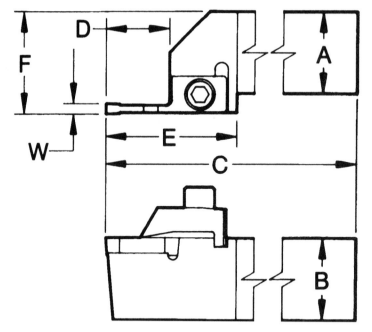

FIGURE **8–12** *Tool style NG for grooving.*
(Courtesy of Kennametal, Inc.)

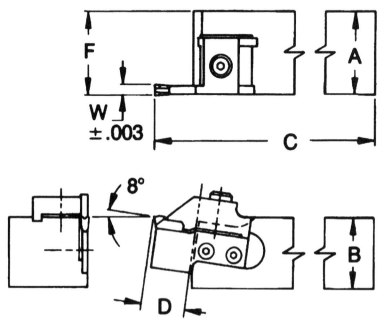

FIGURE **8–13** *Tool style KGSP for parting.*
(Courtesy of Kennametal, Inc.)

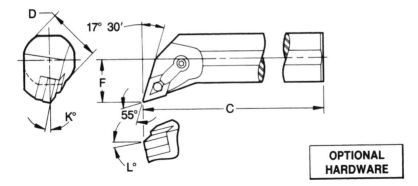

FIGURE 8–14 *LP-style boring bar using a standard 55-degree diamond insert. (Courtesy of Kennametal, Inc.)*

BORING

Boring is an internal turning operation that enlarges, trues, and contours previously drilled or existing holes. Boring is done with a boring bar (see Figure 8–14).

THREADING

Threading is the process of forming a helical groove on the outside or inside surface of a cylinder or cone. Threads can be cut in several different manners, but for this tooling section we will concentrate on single-point threading tools (see Figure 8–15). Single-point threading tools are typically 60-degree carbide inserts clamped in a tool holder. The threading tool is fed into the work and along the part at a feed rate equal to the pitch. (The pitch of a thread is the distance from the one thread to the next.) On single start threads, the pitch can be calculated by dividing 1 inch by the number of threads. For example, if you had eight threads per inch, the calculation would appear as 1/8 or .125.

PRESETTING TOOLS

Presetting tooling involves setting the cutting point of the tool in relation to a predetermined dimension. Modern tool presetting is done with a toolset arm (see Figure 8–16), which is equipped with sensors.

The operator simply moves each tool close to the sensor and touches off the tool on the X and Z sensor to set the dimension. When the tool tip comes in contact with the sensors, the offset dimension is recorded in an offset page. When all the tools are measured, the operator uses one premeasured tool and touches off on the end of the workpiece.

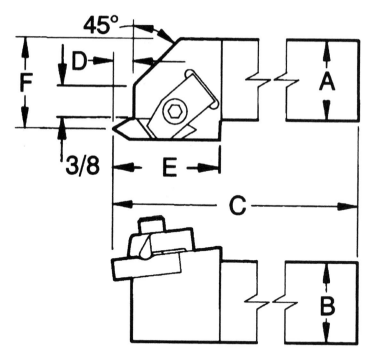

FIGURE 8–15 *NS-style threading tool. This style holder can be equipped with different angled threading inserts for different types of threads. (Courtesy of Kennametal, Inc.)*

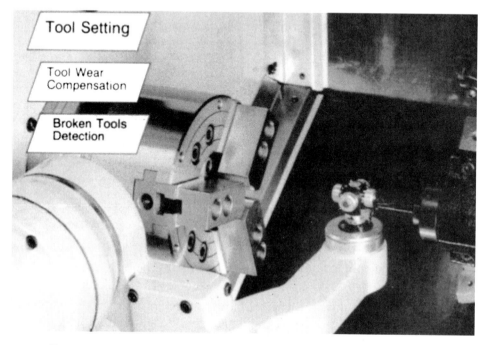

Tool Setting

Tool Wear Compensation

Broken Tools Detection

FIGURE 8–16 *The tool presetting arm greatly reduces the time required to measure tools. (Courtesy of Yamazaki Mazak Corporation.)*

The control then uses this information to determine where the workpiece is located and also calculates and adjusts for the different tool tip locations.

Some controls can use the tool measurement arm to check tools automatically between cutting operations to be sure tools have not been worn, damaged, or broken during cutting.

Machines that are not equipped with tool presetting arms use standard geometry offsets, discussed later in this chapter.

OFFSETS

There are two kinds of offsets used on CNC turning machines: tool offsets and geometry offsets.

TOOL OFFSETS

Tool offsets, also called tool wear offsets, are an electronic feature for adjusting the length and diameter of machined surfaces. CNC turning machines have offset tables in which the operator can input or change numbers to adjust part sizes without changing the program.

Adjusting the offsets is the biggest responsibility of the CNC machine operator. If the part sizes don't meet the part print requirements, the operator alters the offset, which corresponds to the tool in the tool wear offset table (see Figure 8-17).

GEOMETRY OFFSETS

Geometry offsets or workpiece coordinates are used to tell the control where the workpiece is located. The workpiece coordinate is the distance from the tool tip, at the home position, to the workpiece zero point. The workpiece zero point is normally located at the end and center of the workpiece or at the chuck face and center of the machine (see Figure 8–18).

Figure 8–18 Tool geometry offsets, also known as workpiece coordinate settings, are generally set as the distance from machine home to the end and center of the workpiece.

Geometry offsets, or workpiece coordinates, can be registered two different ways. The most common approach is to use a preparatory or G-code such as a G50 or a G92. The offset distance is determined by touching the tool tip off on the workpiece and recording the distance.

There are two kinds of offsets used on CNC turning machines: tool offsets and geometry offsets.

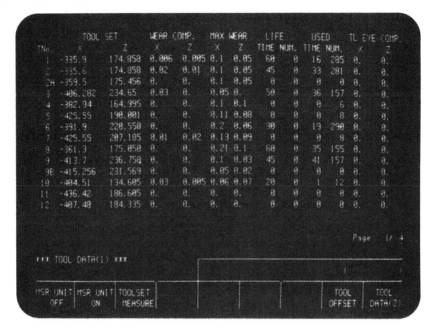

FIGURE **8–17** *Tool offset table. Notice that this is a wear offset table. (Courtesy of Yamazaki Mazak Corporation.)*

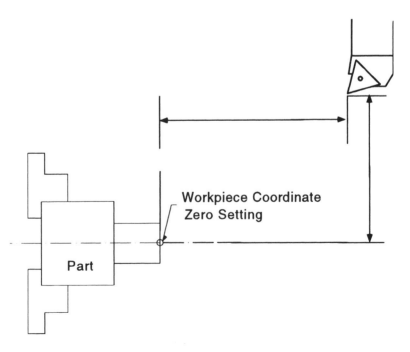

FIGURE **8–18** *Tool geometry offsets, also known as workpiece coordinate settings, are generally set as the distance from machine home to the end and center of the workpiece.*

TOOL OFFSET (Geometry)					N0000
No	X Axis	Z Axis	Radius	Tip	
01	00.000	00.000	00	00	**Machine Position (Relative)**
02	00.000	00.000	00	00	X00.0000
03	00.000	00.000	00	00	Z00.0000
04	00.000	00.000	00	00	
05	00.000	00.000	00	00	
06	00.000	00.000	00	00	
07	-12.346	-08.567	.032	3	
08	-6.5671	-6.987	.015	2	
09	-4.5672	-3.7865	.032	3	
10	-3.7890	-4.8923	.032	3	
11	00.000	00.000	00	00	
12	-10.500	-4.876	.031	1	
13	-9.5624	-5.8763	.015	3	
14	00.000	00.000	00	00	
15	00.000	00.000	00	00	
16	-8.5390	-7.9845	.015	3	
		(INCH)			

FIGURE 8–19 *Tool offset table.*

The position or distance can be determined accurately by using the position screen on the control. For example, if the distance from the home position to the face of the workpiece is 16.500 inches and the distance from the tool tip to the centerline of the workpiece is 8.500 inches, the G50 or G92 would be: G92 X8.500 Z16.500.

To find the centerline of the workpiece, the operator takes a skim cut off of the outside of the workpiece. The operator then measures the turned diameter and adds that dimension to the X-axis machine position. If a geometry offset is used, the X and Z values would be loaded into the geometry offset table under the tool offset number, and the G50 or G92 would not be needed. It is important to remember that every tool that is used in the program will need to be measured in this manner. Figure 8–19 shows a tool offset table.

MATERIAL HANDLING

Material handling devices increase production rates and reduce labor costs. The types of devices used are determined by part size, shape, and production levels.

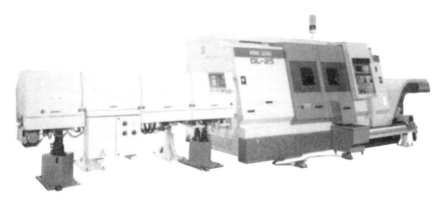

FIGURE **8–20** *Bar feeding mechanisms are capable of handling full-length bars of stock, which are sometimes 20 feet in length. (Courtesy of Mori Seiki Co., Ltd.)*

BAR FEEDERS

Bar feeders automatically load rough stock into the work-holding device. The raw stock is fed into the machine by use of pneumatic or hydraulic pressure. The stock is fed the same distance each time through the use of stock stops. When the stock reaches the stock stop, the work-holding device closes and clamps the workpiece in place. Bar feeders eliminate the need for the operator to manually load individual part blanks (see Figure 8–20).

PART LOADERS AND UNLOADERS

Individual part blanks can be automatically loaded using part loaders. Part loaders take up less space than bar feeders, but the parts have to be cut to length prior to loading. A part loader is an auxiliary arm that places the precut stock into the chuck or collet. The auxiliary arm can also unload parts after the necessary machining has been done.

ROBOTIC LOADING SYSTEMS

The use of robotic equipment represents a major trend in automated manufacturing. Robots can be used to load and unload parts, retrieve parts from pallets, and change chuck jaws.

Robots can communicate directly to the machine tools by using their own controllers, switches, and sensors. The robot controllers are designed to be easy to program and are already integrated to the machine control.

Special fixtures are sometimes needed to use robotic loading systems. Grippers need to be designed and built so that the robot can handle different part

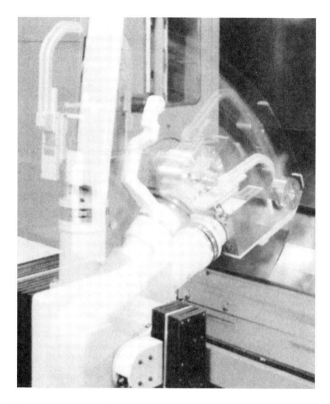

FIGURE **8–21** *Robotic handling systems lend flexibility and productivity to turning centers. (Courtesy of Yamazaki Mazak Corporation.)*

shapes. Some robots have automatic tool changing capability so that they can even change grippers automatically as required for different parts.

In the future, artificial intelligence and vision systems will undoubtedly reduce the need for special fixtures and give robots more flexibility (see Figure 8–21).

PARTS CATCHERS

Parts catchers are used on small-diameter workpieces. Parts catchers consist of a tray that, prior to the tool cutting off the part, tips forward and catches the completed part and delivers the part outside the machine.

CHIP CONVEYERS

Chip conveyers automatically remove chips from the bed of the machine. The chips produced by machining operations fall onto the conveyer track and are transported to scrap or recycling containers (see Figure 8–22).

FIGURE **8–22** *Chip conveyers provide unmanned chip disposal to provide a chip-free work environment. (Courtesy of Mori Seiki Co., Ltd.)*

MACHINE CONTROL OPERATION

SAFETY

Before you operate any machine, remember that no one has ever thought he/she was going to be injured. But it happens! It can happen in a split second when you are least expecting it, and an injury can affect you for the rest of your life.

You must be safety minded at all times. Please get to know your machine before operating any part of the machine control and please keep in mind these safety precautions.

1. *Wear safety glasses and side shields at all times.*

2. *Do not wear rings or jewelry that could get caught in a machine.*

3. *Do not wear long sleeves, ties, loose fitting clothes, or gloves when operating a machine. These can easily get caught in a moving spindle or chuck and cause severe injuries.*

4. *Keep long hair covered or tied back while operating a machine. Many severe accidents have occurred when long hair became entangled in moving tooling and machinery.*

5. *Keep hands away from moving machine parts.*

6. *Use caution when changing tools. Many cuts occur when a wrench slips.*

7. *Stop the spindle completely before doing any setup or piece loading and unloading.*

8. *Do not operate a machine unless all safety guards are in place.*

9. *Metal cutting produces very hot, rapidly moving chips that are very dangerous. Long chips are especially dangerous. You should be protected from chips by guards or shields. You must also always wear safety glasses with side shields to prevent chips from flying into your eyes. Shorts should not be worn because hot chips can easily burn your legs. Hot chips that land on the floor can easily burn though thin-soled shoes.*

10. *Many injuries occur during chip handling. Never remove chips from a moving tool. Never handle chips with your hands. Do not use air to remove chips. They are dangerous when blown around and can also be blown into areas of the machine where they can damage the machine.*

11. *Securely clamp all parts. Make sure your setup is adequate for the job.*

12. *Use proper methods to lift heavy materials. A back injury can ruin your career. It does not take an extremely heavy load to ruin your back; bad methods are enough.*

13. *Safety shoes with steel toes and oil-resistant soles should be worn to protect your feet from dropped objects.*

14. *Watch out for burrs on machined parts. They are very sharp.*

15. *Keep tools off of the machine and its moving parts.*

16. *Keep your area clean. Sweep up chips and clean up any oil or coolant that people could slip on.*

17. *Use proper speeds and feeds. Reduce feed and speed if you notice unusual vibration or noise.*

18. *Dull or damaged tools break easily and unexpectedly. Use sharp tools and keep tool overhang short.*

As you look at the control of the turning center, you see what seems to be an endless number of handles, buttons, and switches (see Figure 8–23). Although every manufacturer has its own style of control, they all have basically the same types of features, and if you have a good understanding of one machine, the next control will be that much easier to learn.

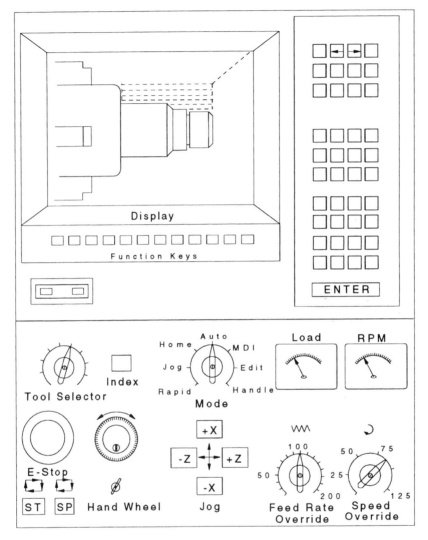

FIGURE 8–23 *Typical computer numerical control configuration.*

MANUAL CONTROL

Manual control features are buttons or switches that control machine movement (see Figure 8–24).

EMERGENCY STOP BUTTON

The emergency stop button is the most important component of the machine control. This button has saved more than one operator from disaster. The

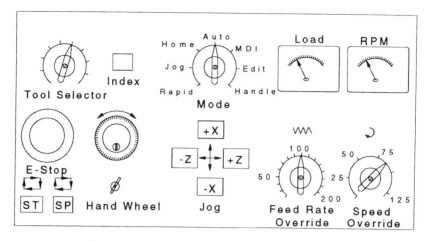

FIGURE **8–24** *Manual machine control features.*

emergency stop button, which shuts down all machine movement, is a big red button with the word *reset* on the front. Emergency stop buttons should be used when it is evident that a collision or tool breakage is going to occur. Emergency stop buttons are located in more than one area on the machine tool and should be located prior to doing any machine operations.

MOVING THE AXES OF THE MACHINE

Manual movement of the machine axes is done a number of different ways. Most controls are equipped with a pulse-generating hand wheel (see Figure 8–25).

The manual hand wheel feature gives the operator a great deal of control of the machine axes.

The hand wheel has an axis selection switch that allows the operator to choose which axis he/she wants to move. The handle sends a signal or electronic pulse to the motors, which move the carriage and the cross slide.

Hand wheel

FIGURE **8–25** *This hand wheel would be used to move an axis.*

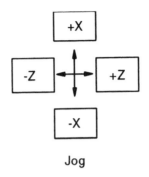

Jog

FIGURE 8–26 *Axes jog buttons.*

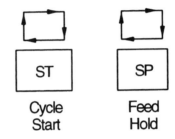

Cycle
Start

Feed
Hold

FIGURE 8–27 *Cycle start and feed hold buttons.*

Some machines are equipped with jog buttons (see Figure 8–26). When the jog button for a certain axis is pressed, the axis moves. The distance or speed at which the machines move is selected by the operator prior to the move.

The joystick, a popular feature on some types of machine tools, moves the machine axes in the direction that the joystick is moved. For example, if the joystick is moved to the down position, the cross slide moves down; if the joystick is moved to the left, the carriage moves toward the headstock, and so on. The distance or speed at which the machine moves is selected in much the same way as the jog buttons. The selection options include rapid traverse, selected feed rate, or incremental distance.

CYCLE START/FEED HOLD BUTTONS

The two most commonly used buttons on the control are the cycle start and feed hold buttons (see Figure 8–27). The cycle start button is used to start execution of the program. The feed hold will stop execution of the program without stopping the spindle or any other miscellaneous functions. By pushing cycle start, the operator can restart the execution of the program.

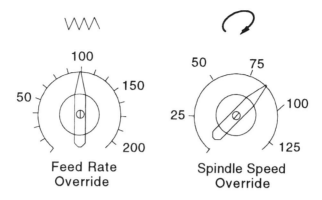

FIGURE 8–28 *Feed rate and spindle speed override controls.*

SPINDLE SPEED AND FEED RATE OVERRIDE SWITCHES

Spindle speed and feed rate overrides are used to speed up or slow down the feed and speed of the machine during cutting operations (see Figure 8–28). The override controls are typically used by the operator to adjust to changes in cutting conditions, such as hard spots in the material. Feed rates can be adjusted from 0 to 150 percent of the programmed feed rate. Spindle speeds can be adjusted from 0 to 200 percent of the programmed spindle speed.

Override switches allow the operator to adjust the spindle RPM and feed rate to changes in machining conditions.

SINGLE BLOCK OPERATION

The single block option on the control is used to advance through the program one block at a time. When the single block switch is on, the operator presses the button each time he/she wants to execute a program block. When the operator wants the program to run automatically, he/she can turn the single block off and press cycle start, and the program will run through without stopping. The single block switch allows the operator to watch each operation of the program carefully. It is primarily used on unproven programs.

OTHER CONTROL FEATURES

Some control features do not control the machine directly as do the manual controls. Many control modes are used as information input devices or information access features.

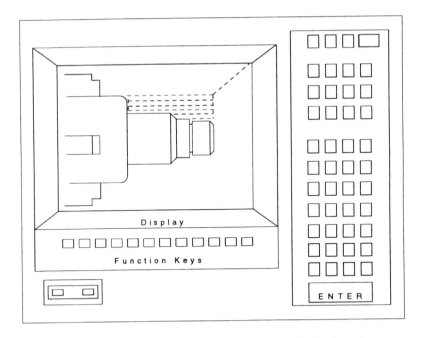

FIGURE 8–29 *A comprehensive alphanumeric keyboard, which contains a wide variety of keys, allows the operator to input information without having to continually switch screen pages.*

MANUAL DATA INPUT

Manual data input or MDI is an input method that can be used for making changes to a previously loaded program or as a means of inputting data for the machine to act on manually, especially for setup purposes. MDI is done through the alphanumeric keyboard located on the control (see Figure 8–29). The keyboard is made up of letters, numbers, and symbols. These keys allow the operator to input a series of commands or a whole program, although that could take a considerable amount of time.

PROGRAM EDITING

When a part program is loaded, it almost always needs some changes and these errors in programming usually show up on the shop floor. The operator or programmer can make changes at the machine control using the program edit mode. The programmer uses the display screen to locate the program errors and the keyboard to correct the errors.

CATHODE RAY TUBE

The cathode ray tube (CRT) is commonly referred to as the *screen*. The screen displays information such as the program or part graphics. In most cases the program is too long to fit on the screen and is separated into pages. The page or cursor button allows you to move through consecutive parts of the program. Graphics are also displayed on the screen if you have purchased graphics capabilities from your manufacturer. Graphics are a representation of the part and the tool path generated by the active program. Graphics are also used to aid the programmer when doing conversational programming.

DIAGNOSTICS

The diagnostics mode consists of several routines that detect errors in the machine system. An error number and message will be displayed on the screen. If any error is found in a CNC operation or servo system, the error message will prompt the operator or service technician to the cause of the problem.

CONVERSATIONAL PROGRAMMING

Conversational programming is a built-in feature that allows the programmer to respond to a set of questions displayed on the graphics screen. The questions guide the programmer through each phase of machining operations such as turning, threading, or grooving. After each response to a question, further questions are presented until the operation is complete. This type of manual data input is quicker than using machine code language. There is no standard conversational part programming language, and each system can be quite different; however, once the program is complete, most controls will transform the conversational language into standard EIA/ISO machine language. This is why we concentrate on EIA/ISO programming language in this textbook.

The programmer's function is to take information from the part print and convert it into data that the control will understand. Figure 8-30 shows an operator programming a CNC machine. The controls take the data that you have input and control the machine movements. Take this information and some time, and become familiar with the operation of your machine control so that you both work together effectively.

FIGURE 8–30 *An operator programs a CNC lathe.*
(Photo courtesy Kazuko Akashi.)

Chapter Questions

1. Name four of the main components that make up the CNC turning center.

2. What is the purpose of the tool turret?

3. State two advantages that the slant-bed-style CNC turning center has over the flat-bed-style CNC lathe.

4. Which of the two major axes associated with the turning center always lies in the same plane as the spindle?

5. What is the most common type of work-holding device used on the turning center?

6. Describe threading.

7. Which machining operation cuts the finished part off the rough stock?

8. When are tool wear offsets used?

9. What is a geometry offset?

10. How can an operator accurately judge the position of the tool?

11. From what location is the workpiece zero or geometry offset typically calculated?

12. What is a bar feeder?

13. How can chips be automatically removed from the bed of the turning center?

14. Name two manual control devices that allow the operator to move the axes of the machine.

15. What tasks do override switches perform?

Chapter 9

PROGRAMMING CNC TURNING MACHINES

INTRODUCTION

Chapter 2 covered the basics of programming with information to program basic moves on a CNC turning center. In this chapter we cover, in greater detail, the steps necessary to properly plan, set up, and program a turned part. The vast majority of turning centers in machine shops today have the capabilities to reduce part programming time and increase part quality through the use of canned cycles and tool-nose radius compensation. For this reason, we will pay particular attention to these programming techniques. You may need to rely on information from previous chapters.

OBJECTIVES

Upon completion of this chapter, the reader will be able to:

- *Identify the two main axes of movement associated with the turning center.*
- *Describe the recommended sequence of operations for the turning center.*
- *Define the term "tool-nose radius compensation."*
- *Explain the use of tool-nose direction vectors.*
- *Program a part using G70 and G71 canned cycles.*
- *Program a part that uses a G75 grooving cycle.*
- *Cut a thread using a G76 thread-cutting cycle.*

REVIEW OF TURNING CENTERS

Understanding how to program turning centers is generally easier than understanding how to program machining centers because the turning center uses two basic axes of movement and the machining center uses three. The two basic axes of movement that we will be concerned with on the turning center are the X axis, which controls the diameter of the part, and the Z axis, which controls the length.

The X axis is normally programmed with diameter rather than radius values, so the actual position of the tool would be the radial distance from the centerline. The spindle centerline would be an X0 position. When positioning the tool in the X axis of travel we will seldom position to a –X position, except in the case of a facing operation. The Z axis part origin or zero position can be either at the right end of the part or a position located near the spindle of the machine.

PLANNING THE PROGRAM

When planning a program for the turning center, we need to be aware of the tooling, the work-holding device, and the part print. The part drawing or part print gives the programmer detailed information on the part requirements. The shape of the part, part tolerances, material requirements, surface finishes, and the quantity of parts all have an impact on the program, as well as where the part zero or datum will be located.

WORK HOLDING

Work-holding devices on turning centers will typically be chuck jaws, collets, or centers. A typical CNC turning center will have a hydraulically actuated chuck. A foot pedal, located in front of the machine, will open and close the chuck jaws or collet. Chuck jaws are either soft or hardened. Hardened chuck jaws are used for maximum holding power, but they will sometimes mar the surface of the parts. Parts that cannot be marred or that must be concentric can be held in soft jaws, made of soft, low-carbon steel or aluminum and usually machined to fit the workpiece.

Chuck jaws need to be repositioned quite often for different size workpieces. When the jaws are repositioned, you must be sure that the jaws are tight and that they are all the same distance from the centerline of the chuck. This insures the part will run true. The maximum travel of the chuck jaws is usually no more than 3/8 of an inch, which is why some machine shops use collets. Collets are available in standard fractional sizes and can be changed quickly for various size parts. They are also used when greater part concentricity is needed.

Chucked parts tend to run out, especially parts held in hardened chuck jaws. All work-holding information needs to be included in the process plan or setup sheet to insure the reliability of the program.

TOOLING CONSIDERATIONS

The necessary tooling is determined by the configuration of the part. Typical tools used on turning centers were discussed in Chapter 4. The most common tools used on turning centers are 80-degree diamond inserts for roughing work, 35-degree diamond inserts for finishing, grooving inserts, and 60-degree threading inserts.

The type and style of tool holders used will be based upon the clearance conditions, which are dictated by the part configuration and the amount of stock to be removed. The tools and tool-holder styles needed will also have to be chosen and entered into the setup sheet or process plan.

THE PROCESS PLAN

Process planning involves deciding when certain turning operations will take place. Primary machining operations are those operations that will take place on the CNC machine. Although the part configuration will have a decided effect on the sequence of operations, there are some general rules to follow when deciding on the sequence of machining operations. The recommended procedures for turning, threading, and grooving are as follows:

1. *facing*
2. *rough turning of the profile of the part*
3. *finish turning of the profile of the part*
4. *drilling*
5. *rough boring*
6. *finish boring*
7. *grooving*
8. *threading*

Roughing operations should be performed when the maximum amount of holding material is still in place to insure that the part will not flex or deflect away from the cutting tool. When the part needs to be turned around to machine the back side of the part, special consideration in planning is required. This process planning is usually done by the operator and/or programmer in smaller job-shop settings. In large shops the process plan would come down from the engineering area and would include information for each step in the complete turning of the part. In a small job shop, the operator becomes the manufacturing engineer and determines the best, most economical way to produce the part. The operator determines the operating sequence, types of cutting tools, cutter path, work-holding devices, and machining conditions (cutting speed, feed rate, and depth of cut). This operator must have a good background in the machine tool field.

Part #_____		Written by_____
Machine_____		Date ___/___/____Sheet___/___

Notes:	
Part Datum X0 Z0	

Tool #	Tool Description	Operation

FIGURE 9–1 *The process plan outlines the*
machining steps to be done on the part.

Whether process planning is done by the manufacturing engineer or the operator, a plan for each setup needs to be developed. The process plan is done with paper and pencil on a process planning sheet in much the same manner as the one used for machining centers (see Figure 9–1). A well-organized planning sheet will be the programmer's ready reference. A well-planned job is half done!

THE SETUP SHEET

The NC turning center setup sheet is a detailed explanation of how the parts are to be set up, the type of work-holding device to be used, where the part datum is located, and the type of tools to be used (see Figure 9–2). The setup sheet communicates to the setup personnel exactly what the programmer had in mind while programming the part.

In small job shops the programmer and the operator are usually the same person. The setup sheet is useful to reference if the same or similar parts are

Part #1_____ Written by: Kelly Curran
Machine: Mazak Q10N Date 07/03/95 Sheet 1 of 1

Notes:

Part Datum
X0 __Center____
Z0 __Right end____

Tool #	Tool Description	Operation
1	80-degree diamond	Rough turn
2	35-degree diamond	Finish profile
3	.125 grooving tool	Groove
4	60-degree threading tool	Thread
5	1" diameter drill	1" hole

FIGURE 9–2 *The NC turning center setup sheet usually includes a sketch for clarity. Setup sheets help assure consistent quality and rapid setups. They help assure that whoever sets up a job and runs it does it in the same manner each time.*

programmed in the future. The setup sheet should contain all of the necessary information to prepare for the job, including sketches of parts and soft-jaw configurations, and can drastically reduce setup times.

QUICK REVIEW OF PROGRAMMING

This section reviews word address programming, linear programming, arc programming, and offsets. If the material is unclear, you may need to review Chapter 2 before continuing.

WORD ADDRESS PROGRAMMING

The word address programming format is a system of characters, typically letters, arranged into blocks of information. G-codes, or preparatory functions, and M-codes, or miscellaneous functions, are paramount in word address programming. Figures 9–3 and 9–4 list the more commonly used M- and G-codes found in industry. Use these or the codes for your turning center as a reference when completing the exercises in this chapter.

M00	Program stop
M01	Program stop
M03	Spindle start clockwise
M04	Spindle start counterclockwise
M05	Spindle stop
M08	Coolant on
M09	Coolant off
M30	End of program
M41	Low gear range
M42	Intermediate gear range
M43	High gear range

FIGURE 9–3 *Commonly used turning center M (miscellaneous) functions.*

G00	Rapid positioning	Modal
G01	Linear positioning at a feed rate	Modal
G02	Clockwise arc	Modal
G03	Counterclockwise arc	Modal
G28	Zero or home return	Non-modal
G40	Tool nose radius compensation-cancel	Modal
G41	Tool nose radius compensation-left	Modal
G42	Tool nose radius compensation-right	Modal
G50	Workpiece coordinate setting/maximum spindle RPM setting	Modal
G70	Inch programming	Modal
G75	Grooving cycle	
G76	Threading cycle	
G90	Absolute coordinate positioning	Modal
G91	Incremental positioning	Modal
G92	Workpiece coordinate setting	Modal
G96	Constant surface footage	
G97	RPM input	Modal
G98	Feed rate per minute	Modal
G99	Feed rate per revolution	Modal

FIGURE 9–4 *Commonly used turning center G-codes or preparatory functions.*

REVIEW OF PROGRAMMING PROCEDURES

Whether you are programming a machining center or a turning center, all CNC programs basically follow a common format.

The procedure is as follows:

1. *start-up procedures*
2. *tool call*
3. *workpiece location block*
4. *spindle speed control*
5. *tool motion blocks*
6. *home return*
7. *program end procedures*

STARTUP OR PRELIMINARY PROCEDURES

Startup procedures are those commands or functions that are necessary at the beginning of the program. Every manufacturer of machine tools has its own protocol for startup. Throughout the text we have been using a protocol that could be used on most machines. Some machines are different, however. A standard startup procedure usually involves cancellation of compensation, absolute or incremental programming, standard or metric, and the setting of the work-plane axis. Below is an example of typical startup procedure blocks.

N0001 G90; Absolute programming.

N0002 G20 G40; Inch units, tool nose radius compensation cancel.

TOOL CHANGE AND TOOL CALL BLOCK

An M06 tool change code is not used on the turning center for changing or selecting tools. It is usually used for clamping or unclamping the work-holding device. The T-code or tool code is sufficient to tell the control which tool turret position the tool is in. The tool call is also accompanied by the offset call, which is the last two digits in the tool call. These offsets are used for offsetting the tool path to accommodate for tool wear or exact sizing of the part. Because the tool turret can index at any position, return the axis to home or to a safe position with a G28 or G29 preparatory function before calling a new tool. Here is a look at a tool change block.

N0001 T0101; The first two numbers call for tool #1, the second two numbers call for offset number 1. Normally, we correspond the tool offset number and the tool number to eliminate mistakes in machining and damage to equipment.

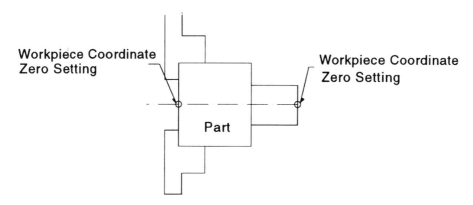

FIGURE 9–5 *Workpiece coordinate location.*

WORKPIECE COORDINATE SETTING

Workpiece coordinate setting is considerably easier on turning machines than on machining centers. The spindle or workpiece center is the location of the X0 position and the Z0 position is typically the right end of the workpiece or the chuck face (see Figure 9–5).

The workpiece location is set using either a G50 or a G92, or it may be called directly from the tool number. When using the G50 or G92, the part datum location will accompany the code. For example, G92 X8.4500 Z10.400; this would be the distance from the tool tip at the machine home position to the end and center of the workpiece (see Figure 9–6).

Because each tool differs in length and shape, every new tool used in the program must be accompanied by its own G50 or G92 workpiece coordinate setting.

SPINDLE START BLOCK

On the turning center, three codes control the spindle. A large majority of turning centers have the capability to speed up and slow down as the part diameter changes, called *constant surface speed control*. Constant surface speed control is important for efficient use of cutting tools, tool life, and proper surface finish.

Constant surface speed is controlled with a G96 preparatory code. Accompanying the G96 code should be the proper surface footage per minute (SFPM) setting for the cutting tool material and part material, set with an S.

When not using constant surface speed control, a G97 RPM input code is used. For thread cutting or drilling, a G97 code is used. Accompanying the G97 will be the properly calculated RPM, also set with an S. An M03 or M04

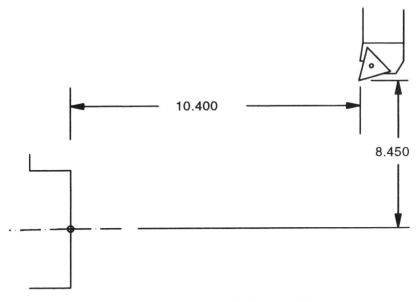

FIGURE 9–6 *Setting the workpiece coordinate.*

tells the spindle to start in a clockwise or counterclockwise direction. A typical spindle start block using constant surface speed control follows:

N0010 G96 S450 M03;

Block number 10 would set constant surface speed control to 450 SFPM and start the spindle in a clockwise direction.

TOOL MOTION BLOCKS

The tool motion blocks are the body of the program. The tool is positioned and the cutting takes place in these blocks.

HOME RETURN

As stated earlier, the tool needs to be returned to home whenever a tool change takes place. Most machine controls use a G28 command to rapid position the tool to home. When a home return is commanded, the path the tool takes to get there is crucial. If we use only a G28, the tool will take the straightest route to home.

We can accompany the G28 with directions on how to get home. If we wanted the tool to move up to a X2.00 position before returning home, we command G28 X2.00. This command would move the tool to X2.00 and then would automatically return home (see Figure 9–7). This can be especially

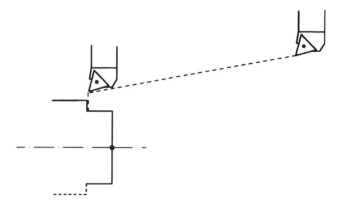

FIGURE 9–7 *G28 return to home position command.*

useful when grooving because we want the tool to clear the work-piece before going to the home position. If the tool were inside a bore, we would command the tool to come out of the hole before rapid positioning to home.

PROGRAM END BLOCKS

There are a number of different ways to end the program. Some controls require that you turn off the coolant and the spindle with individual miscellaneous function codes. Other controls will end the program, rewind the program, and turn off miscellaneous functions, all with an M30 code. No matter what type of control you have, it is always a good idea to cancel any offsets that may be active when ending your program.

Next, we examine a typical turning program that incorporates many of the elements that have been covered. The part is shown in Figure 9–8.

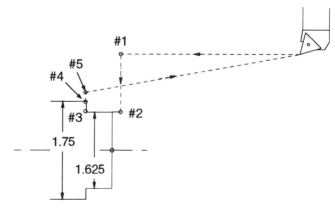

FIGURE 9–8 *The center of the workpiece is zero and the right end of the stock is Z0.0.*

The program for the part shown in Figure 9–8 is listed below.

N0001 G90 G20; (Absolute programming, inch programming)
N0005 G40; (Cancel tool-nose radius compensation)
N0010 T0101; (Call tool #1, offset #1)
N0015 G92 X5.800 Z10.250; (Workpiece coordinate setting)
N0020 G96 S400 M03; (Constant surface speed setting, spindle CW)
N0030 G00 Z.100; (Rapid position #1)
N0035 G00 X1.625; (Rapid position #2)
N0040 G01 Z-1.25 F0.01; (Feed to position #3)
N0045 G01 X1.75; (Feed to position #4)
N0050 G28 X2.00; (Rapid home through position #5)
N0055 T0100; (Cancel offset #1)
N0060 M30; (Stop and rewind program)

CIRCULAR INTERPOLATION

Only straight-line moves were used for the program to machine the part shown in Figure 9-8. As discussed in earlier chapters, one of the most important features of a CNC machine is its ability to do circular cutting motions. CNC turning centers are capable of cutting any arc of a specified radius value. Arc or radius cutting is known as *circular interpolation.*

Circular interpolation is carried out precisely the same way as on machining centers, with the use of G02 or G03 preparatory codes. To cut an arc, the programmer needs to follow a very specific procedure. To start cutting an arc, the tool needs to be positioned to the start point of the arc. Next, we need to tell the control the direction of the arc: clockwise or counterclockwise. The third piece of information is the end point of the arc. The last piece of information the control needs is the position of the arc center or, if you are using the radius method of circular interpolation, the radius value of the arc (see Figure 9–9).

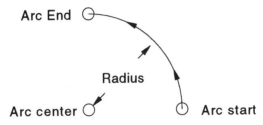

FIGURE 9–9 *The critical pieces of information needed to cut an arc are the arc start point, arc direction, arc endpoint, and arc centerpoint location.*

ARC START POINT

The arc start point is the coordinate location of the start point of the arc. The tool is moved to the arc start point in the line prior to the arc generation line. Simply stated, the start point of the arc is the point the tool is presently at when you want to generate an arc.

ARC DIRECTION (G02, G03)

Circular interpolation can be carried out in two directions: clockwise and counterclockwise. Two G-codes specify arc direction. The G02 code is used for circular interpolation in a clockwise direction, and the G03 code is used for circular interpolation in a counterclockwise direction. Both codes are modal. G02 and G03 codes are controlled by a feed rate (F) code, just like a G01.

ARC END POINT

The computer control requires the tool to be positioned at the start point of the arc prior to a G02 or G03 command. The current tool position becomes the arc starting point and the arc end point is the coordinate position for the end point of the arc. The arc start point and arc end point set up the tool path, which is generated according to the arc center position. Remember, we are turning a round piece of material that is separated by a centerline.

Most CNC turning center controllers are programmed using diameter, so if we move the tool up on the part 1/4 of an inch, we change the diameter by 1/2 of an inch. If we have an arc of a .25 radius, the diameter of the part can change by .50 of an inch (see Figure 9–10).

ARC CENTERPOINTS

To generate an arc path, the controller has to know where the center of the arc is. There are two methods of specifying arc centerpoints: the coordinate arc center point method and the radius method. When using the coordinate arc center method, a particular problem arises: How do we describe the position of the arc center? If we use the traditional X, Y, Z coordinate position words to describe the end point of the arc, how will the controller discriminate between the end points coordinates and the arc center coordinates? We discriminate by using different letters to describe the same axes. Secondary axes addresses are used to designate arc centerpoints. The secondary axes addresses for the primary axes are:

> *I = X axis coordinate of an arc centerpoint*
> *K = Z axis coordinate of an arc centerpoint*

When we cut an arc on the turning center, the X/Z axes are the primary axes, and the I/K letter addresses are used to describe the arc centerpoint. The type

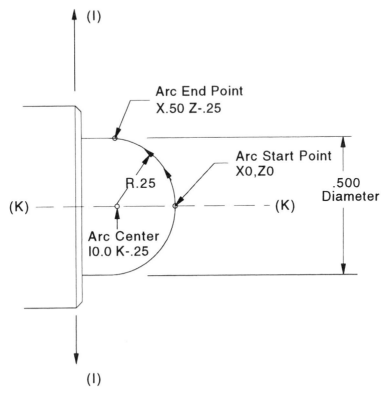

FIGURE 9–10 *G02 arc generation.*

of controller that you use determines how these secondary axes are located. With most controllers, such as the Fanuc controller, the arc centerpoint position is described as the incremental distance from the arc start point to the arc center. This is the most common method of specifying the arc center.

Some CNC controllers can calculate the centerpoint of the arc by merely stating the arc size and the end point of the arc; however, because these controls are a small minority, we will concentrate on the incremental method of arc center locating. Practice generating some arcs using this method. Keep in mind, we are locating the arc center using the incremental method. If the arc centerpoint is located down or to the left of the start point, a negative sign (–) must precede the coordinate dimension.

The next programming example (see Figure 9–11) incorporates arc generation. Pay particular attention to the locations of the start, end, and centerpoint locations.

N0001 G90 G20; (Absolute programming, inch programming)
N0005 G40; (Cancel tool nose radius compensation)
N0010 T0101; (Call tool #1, offset #1)

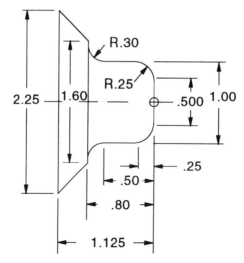

FIGURE 9–11 *Arc interpolation program example.*

N0015 G92 X5.800 Z10.250; (Workpiece coordinate setting)
N0020 G96 S400 M03; (Constant surface speed setting, spindle CW)
N0030 G00 Z.100; (Rapid position)
N0035 G00 X0.0; (Rapid position)
N0040 G01 Z0.0 F.008; (Feed move to end and center of workpiece)
N0045 G01 X.50; (Feed move to arc start point)
N0050 G03 X1.00 Z-.25 I0.0 K-.25; (CCW arc to end point)
N0055 G01 Z-.50; (Linear feed move to arc start point)
N0060 G02 X1.600 Z-.80 I.30 K0.0; (CW arc to end point)
N0065 G01 X2.25 Z-1.125; (Linear feed move to cut the taper)
N0070 G28 X3.00; (Return to home through X3.00)
N0075 T0100; (Cancel offset #1)
N0080 M30; (Stop and rewind program)

TOOL-NOSE RADIUS COMPENSATION

Tool-nose radius compensation is another type of offset used to control the shape of machined features. In Chapter 3 you learned to compensate for the diameter of the tool by offsetting the tool path by the radius of the tool diameter.

The turning center control can offset the path of the tool so we can program the part just as it appears on the part print. There is an error that becomes apparent when we use the tool edges to set our workpiece coordinate position

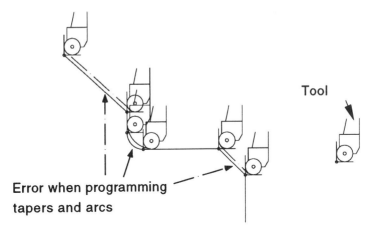

Tool

Error when programming
tapers and arcs

FIGURE **9–12** *Tool-nose radius compensation allows us to program the part, not the tool path. The mathematical calculations that are needed to program a part profile with angles and radii, without the aid of cutter diameter compensation, can be very involved.*

on a turning center. When we set the X and Z axes of the tool, we create a square point on the tool. Most of the tools we use for turning have radii. To compensate for the radii, we need to use TNR compensation, which saves us from having to mathematically calculate the cutter path (see Figure 9–12). TNR compensation also lets us use the same program for a variety of tool types. With TNR compensation capabilities, the insert radius size can be ignored and the part profile can be programmed. The exact size of the cutting tool to be used is entered into the offset file, and when the offset is called, the tool path will automatically be offset by the tool radius.

Tool-nose radius compensation can be to the right or left of the part profile. To determine which offset you need, imagine yourself walking behind the cutting tool. Do you want the tool to the left of the programmed path or to the right (see Figure 9–13)?

Compensation direction is controlled by a G-code. When a compensation to the left is desired, a G41 is used. When a compensation to the right is needed, a G42 is used. When using these cutter compensation codes, you need to specify how much the controller is to offset. The size of the radius is placed in the nose radius offset table, which is typically located adjacent to the tool file under the tool number being used. Tool-nose radius information can be determined from catalogs or the insert package.

Other information needed to insure proper compensation is the tool nose direction vector. The tool tip or imaginary tool tip of turning tools has a specific

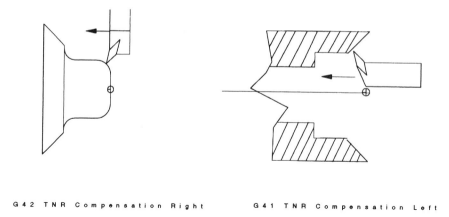

G 4 2   T N R   C o m p e n s a t i o n   R i g h t   G 4 1   T N R   C o m p e n s a t i o n   L e f t

FIGURE **9–13** *Tool-nose compensation.*

location or direction from the center of the tool-nose radius. The tool nose vector tells the control which direction it must compensate for individual types of tools. Standard tool-nose direction vectors are shown in Figure 9–14. The direction vector number is usually placed in the same tool offset table as the radius value.

To invoke compensation, the programmer will have to make a machine move (ramp on). This move allows the control to evaluate its present position and make the necessary adjustment from tool-edge positioning to tool-nose radius positioning. This adjustment move must be greater than the radius value of the tool. To cancel the cutter compensation and return to cutter-edge programming, the programmer must make a linear move (ramp off) to invoke a cutter compensation cancellation (G40).

Figure 9–15 illustrates a typical part that uses tool-nose radius compensation. The program for the part follows the figure.

N0001 G90 G20;

N0005 G40;

N0010 T0101;

N0015 G92 X5.800 Z10.250;

N0020 G96 S400 M03;

N0025 G00 G42 X1.30 Z.100; (Rapid position ramp on move and TNR compensation to the right)

N0035 G00 X.375; (Profile of the part)

N0045 G01 Z0.0 F.01; (Profile of the part)

N0050 G01 X.500 Z-.0625; (Profile of the part)

N0055 G01 Z-1.00; (Profile of the part)

N0060 G02 X1.00 Z-1.25 I.25 K0.0; (Profile of the part)

N0065 G01 X1.25; (Profile of the part)

Tool nose direction vector numbers

A= Point to which the control moves
the tool when commanded.

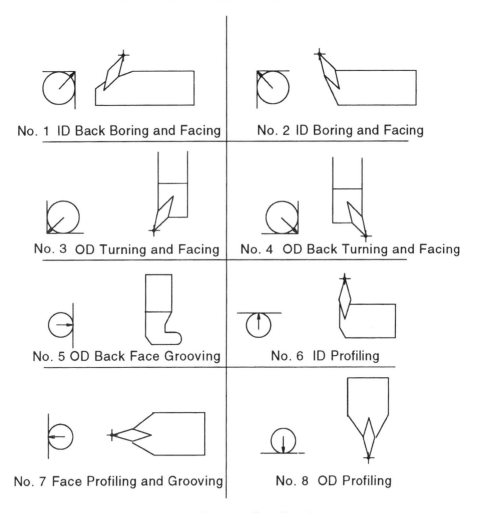

FIGURE 9–14 *Tool-nose radius direction vectors.*

N0070 G40 G00 X2.0 Z2.0; (TNR compensation cancel and rapid position
ramp off move)
N0075 G28;
N0080 T0100;
N0085 M30;

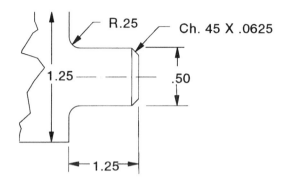

FIGURE 9–15 *Program that utilizes tool-nose radius compensation.*

CANNED CYCLES FOR TURNING CENTERS

Canned cycles (fixed) cycles are used to simplify the programming of repetitive turning operations, such as rough turning, threading, and grooving. Canned cycles are sets of preprogrammed instructions that eliminate the need for many lines of programming. Programming a simple part without the use of a canned cycle can take up to four or five times the number of lines needed for a part programmed with canned cycles. Think of the lines that are needed to produce a thread: (1) position the X and Z axes to the proper coordinates with a rapid traverse move (G00), (2) position the tool for the proper lead angle, (3) feed the tool across, (4) rapid position the tool back to the clearance plan, (5) feed the tool across. That is only two threading passes. With a canned threading cycle, a thread can be done with one line of programming. Standard canned cycles, or fixed cycles, are common to most CNC machines. See Figure 9–4 for a general list of the most commonly used canned cycles for turning centers.

ROUGHING OR TURNING CYCLE (G71)

The G71 automatically takes roughing passes to turn down a workpiece to a specific diameter at a specified depth of cut. The G71 cycle reads a specified number of blocks to determine the part profile and determines each pass, the depth of cut for each pass, and the number of repeat passes for the cycle. Cutting is accomplished through parallel moves of the tool in the Z axis direction.

A certain procedure needs to be followed when using canned cycles. In the first procedure, the tool needs to be positioned to the rough stock boundaries. This procedure has a two-fold purpose: it tells the control how big the stock is, and it creates a Z clearance position that the tool rapids back to on each pass. The G71 uses letters to give the controller information on the part profile, the amount of stock we are going to leave for finishing, the depth of cut, and the feed rate. A G71 roughing cycle command follows.

N0010 G71 P40 Q85 U.03 W.010 D750 F.012;

G71 is the roughing cycle call.

P40 is the block or line number that designates the start of the part profile.

Q85 is the block or line number that designates the end of the part profile.

U.03 tells the controller that we want to leave .03 of an inch stock on the X axis of the profile for finishing.

W.01 tells the controller that we want to leave .01 of an inch stock on the Z axis of the profile for finishing.

D750 tells the controller we want to take .0750 of an inch per pass per side depth of cut. Notice that the decimal point was left off. No decimal point input is possible for depth of cut. When the D is commanded the controller reads from the right and decides what the depth of cut will be. Each number in each decimal position gets a value. If we wanted to take .0500 depth of cut per side, we would write it as D500. The first zero from the right has 0 tenths of a thousandths value. The next zero from the right has 0 thousandths of an inch value. The 5 in the third position from the right has 5 ten-thousandths of an inch value or 50 thousandths of an inch.

F.012 is the feed rate of the roughing passes.

Next we take the example program from Figure 9–15 and convert it to a program that utilizes a roughing canned cycle (see Figure 9–16). We will eliminate the TNR compensation just for ease of understanding.

```
N0001 G90 G20;
N0005 G40;
N0010 T0101;
N0015 G92 X5.800 Z10.250;
N0020 G96 S400 M03;
N0025 G00 X1.30 Z.100; (Rapid position that indicates to the controller
our stock size and Z clearance point)
N0030 G71 P35 Q65 U.03 W.01 D600 F.010; (Canned roughing cycle call)
N0035 G00 X.375; (Profile of the part)
N0045 G01 Z0.0 F.01; (Profile of the part)
N0050 G01 X.500 Z-.0625; (Profile of the part)
N0055 G01 Z-1.00; (Profile of the part)
N0060 G02 X1.00 Z-1.25 I.25 K0.0; (Profile of the part)
N0065 G01 X1.25; (Profile of the part)
N0070 G00 X2.0 Z2.0;
N0075 G28;
N0080 T0100;
N0085 M30;
```

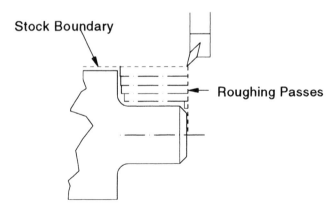

FIGURE 9–16 *Roughing cycle example*

FINISHING CYCLE (G70)

The G70 command calculates the finish part profile, then executes a finish pass on the part. The finishing cycle is called with a G70, followed by a letter address P for the start line of the finish part profile and the letter address Q for the end line of the part profile. A finishing feed rate can also be included in this block. When the finish cycle is commanded, it reads the program blocks designated by the P and Q and formulates a finishing cycle. As in the roughing cycle, the tool needs to be positioned to a Z clearance plane or stock boundary prior to the calling of the G70 finishing cycle. Next we will finish the program shown in Figure 9–15.

N0001 G90 G20;

N0005 G40;

N0010 T0101;

N0015 G92 X5.800 Z10.250;

N0020 G96 S400 M03;

N0025 G00 X1.30 Z.100; (Rapid position that indicates to the controller our stock size and Z clearance point)

N0030 G71 P35 Q65 U.03 W.01 D600 F.010; (Canned roughing cycle call)

N0035 G00 X.375; (Profile of the part)

N0045 G01 Z0.0 F.01; (Profile of the part)

N0050 G01 X.500 Z-.0625; (Profile of the part)

N0055 G01 Z-1.00; (Profile of the part)

N0060 G02 X1.00 Z-1.25 I.25 K0.0; (Profile of the part)

N0065 G01 X1.25; (Profile of the part)

N0070 G00 X2.0 Z2.0;

N0075 G28;

N0080 T0202; (Calls for finishing tool, with offset #2)

N0080 G00 X1.30 Z.100; (Rapid position that indicates to the controller our stock size and Z clearance point)

N0085 G70 P35 Q65 F.008; (Canned finishing cycle call)

N0090 G28; (Return to home from position on line N0080)

N0100 T0200; (Tool offset cancel)

N0110 M30;

PECK DRILLING CYCLE (G74)

The G74 peck drilling cycle will peck drill holes with automatic retract and incremental depth of cut. The G74 command relays to the controller the incremental depth of cut, the full depth of the hole, and the feed rate through the command variables K, Z, and F. The next command shows the proper format for peck drilling.

N0010 G74 X0.0 Z-1.25 K.125 F.010;

G74 is the peck drill cycle call.

X0.0 is the center of the workpiece (X is always zero).

Z-1.25 is the full depth of the drilled hole.

K.125 is the depth of each peck.

F.01 is the drilling feed rate.

The drill must be positioned to a clearance plane in the Z axis and also to X0.0 prior to the calling of the G74 peck drilling cycle. The spindle should also be reprogrammed for direct RPM input using a G97 when drilling. Examine the sample peck drilling cycle in Figure 9–17.

N0001 G90 G20;

N0005 G40;

N0010 T0606; (Tool #6, .500 diameter drill)

N0015 G92 X5.800 Z10.250;

N0020 G97 S800 M03; (RPM 800)

N0025 G00 X0.0 Z.200; (Rapid position the drill to the center of the stock and .200 of an inch in front of the work face)

N0030 G74 X0.0 Z-1.25 F0.01 K.125; (Peck drilling cycle call)

N0035 G28 Z1.00; (Return to home through Z1.00)

N0040 T0600;

N0045 M30;

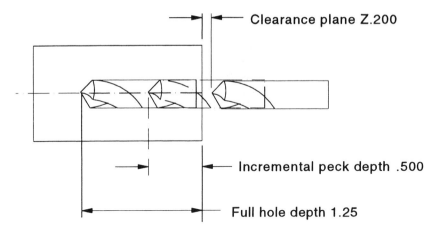

Clearance plane Z.200

Incremental peck depth .500

Full hole depth 1.25

FIGURE 9–17 *Peck drilling cycle.*

GROOVING CYCLE (G75)

The grooving cycle is used to cut grooves of varying widths automatically. To use the grooving cycle, the tool must be positioned to the start of the groove prior to calling the grooving cycle. Through a series of letter addresses, the controller can be commanded to cut a groove of varying width and depth. The next example shows the proper format for the grooving cycle.

N0010 G75 X.750 Z-.50 F0.125 D0 I.125 K.125;

G75 is the grooving cycle call.

X.750 is the diameter at the bottom of the groove.

Z-.50 is the end position of the groove.

F0.125 is the incremental retract of the grooving tool.

I is the depth of cut on the X axis.

K is the depth of cut on the Z axis.

The controller looks at the position of the tool prior to the calling of the grooving cycle and uses that information to establish the groove width and depth positions. The programmer, when positioning the grooving tool, will have to take into consideration which corner of the tool is the leading edge. She/he may have to make adjustments for the width of the tool. Note that the last feed rate that was active prior to the grooving cycle will be the grooving feed rate. It is not possible to express a feed rate within a grooving cycle. Take a look at a sample grooving cycle in Figure 9–18.

N0001 G90 G20;
N0005 G40;

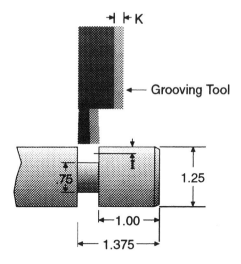

FIGURE **9–18** *This figure shows a part that would be appropriate for a G75 grooving cycle.*

N0010 T0505; (Tool #5, .125 wide grooving tool)

N0015 G92 X5.800 Z10.250;

N0020 G96 S200 M03;

N0025 G00 X1.250 Z-1.125; (Rapid position that indicates to the controller our stock diameter at the groove location and the Z position of the start of the groove)

N0030 G75 X.750 Z-1.375 F0.125 I.125 K.125; (Grooving cycle call)

N0035 G28 X2.00 (Return to home through X2.00)

N0040 T0500;

N0045 M30;

The grooving cycle is primarily used to cut a groove that is wider than the tool. If the groove is the same width as the tool it may be just as easy to program the groove in linear blocks.

THREAD-CUTTING CYCLE (G76)

The G76 thread-cutting cycle can cut multi-pass threads with one block of information. By using several letter address parameters, the control will automatically calculate the correct number of cut passes, depth of cut for each pass, and the starting point for each pass. To use the G76 thread-cutting canned cycle, the following commands need to be programmed:

N00010 G76 X.712 Z-1.125 I0.0 K.076 D0.012 F.0147 A60;

X.712 is the minor diameter of the thread.

Z-1.125 is the absolute Z position of the end of the thread.

I0.0 is the radial difference between the thread starting point and the thread ending point. The I is used for cutting tapered threads. For cutting straight threads a zero should be programmed.

K.076 is the thread height expressed as a radius value (i.e., [major diameter – minor diameter] divided by 2).

D0.012 is the depth of cut for the first pass (in a radius value). Note: Every pass after the first pass will be decreasing in depth.

F is the thread lead (i.e., 1 divided by the number of threads per inch). Note: The F can sometimes be substituted with E.

A is the included angle of the thread.

Prior to calling the G76 thread-cutting cycle, the tool must be positioned to the major diameter of the thread. The tool should also be positioned in front of the thread Z axis start position by a distance of at least double the thread lead. This insures that the proper lead will be cut throughout the length of the thread. The spindle should be running in direct RPM (G97), not constant surface footage control. Consider the sample G76 thread-cutting cycle in Figure 9–19.

    N0001 G90 G20;

    N0005 G40;

    N0010 T0505; (Tool #5, 60-degree threading tool)

    N0015 G92 X5.800 Z10.250;

    N0020 G97 S300 M03; (Direct RPM input)

    N0030 G00 Z.200; (Z axis rapid position in front of the thread start point).

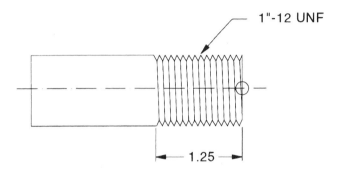

FIGURE 9–19 *This part would be appropriate for a G76 thread-cutting cycle.*

N0040 G00 X1.00; (X axis rapid position to major thread diameter)
N0050 G76 X.897 Z-1.25 K.051 D120 F.0833 A60; (Threading cycle)
N0060 G28; (Return home)
N0070 T0500;
N0080 M30;

Now that we have an understanding of canned cycles and how they are used, we need to put our knowledge to work. After the chapter questions there are some part prints. Use the part prints and the tool table shown in Figure 9–20 to program these parts. The programs should include canned cycles and tool-nose radius compensation where appropriate.

Part #1_____ Written by: Kelly Curran		
Machine: Mazak Q10N Date 07/03/95 Sheet 1 of 1		
Notes:		
Part Datum X0 __Center____ Z0 __Right end____		
Tool #	**Tool Description**	**Operation**
1	80-degree diamond	Rough turn
2	35-degree diamond	Finish profile
3	.125 grooving tool	Groove
4	60-degree threading tool	Thread
5	1" diameter drill	1" hole

FIGURE 9–20 *Tool table.*

CHAPTER QUESTIONS
....

1. Which turning center axis controls the diameter of the part?

2. What are the potential drawbacks of using hardened chuck jaws?

3. Define constant surface footage control.

4. What secondary axes addresses are used to define the centerpoint position when cutting radii on a turning center?

5. Explain why tool-nose radius compensation is needed when cutting tapers or radii.

6. Which letter address controls the depth of cut when using canned cycles?

7. What is the letter address command to take .100 of an inch off the diameter of the part per pass when using a roughing cycle?

8. Describe when a grooving cycle would be used.

9. What letter address controls the pitch or lead of the thread when using a G76 thread cutting cycle?

10. What must be done prior to the calling of a G76 thread cutting cycle?

11. Program the part shown in Figure 9–21. Use canned cycles where appropriate and the tooling shown in Figure 9–20.

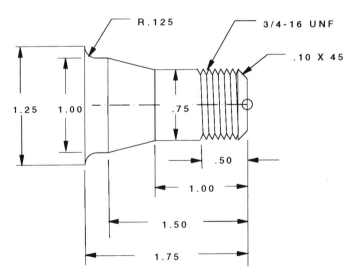

FIGURE 9–21 *Use with question 11.*

12. Program the part shown in Figure 9–22.

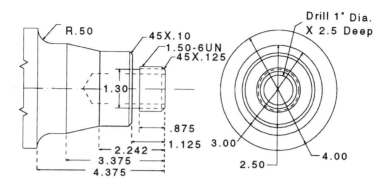

FIGURE 9–22 *Use with question 12.*

13. Program the part shown in Figure 9–23.

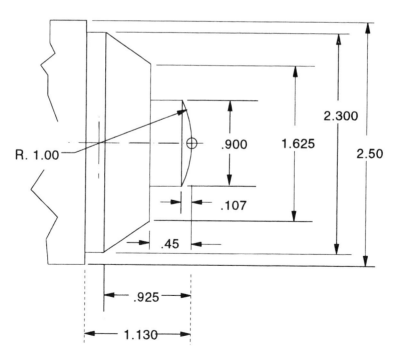

FIGURE 9–23 *Use with question 13.*

14. Program the part shown in Figure 9–24.

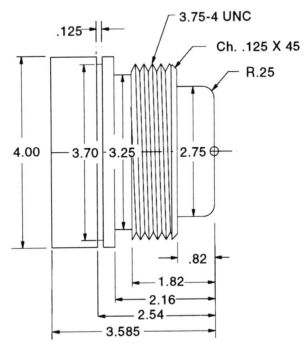

FIGURE **9–24** *Use with question 14.*

15. Program the part shown in Figure 9–25.

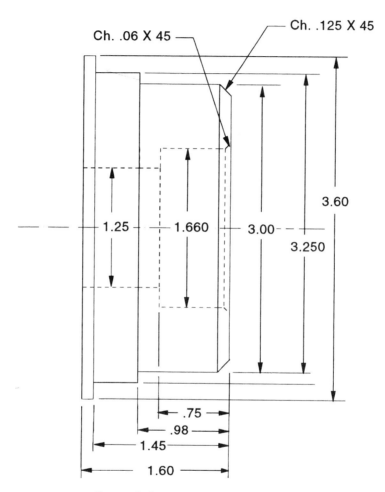

FIGURE 9–25 *Use with question 15.*

Chapter 10

··

INTRODUCTION TO BRIDGEPORT EZPATH PROGRAMMING

INTRODUCTION

Bridgeport's Ezpath conversational programming language offers a straightforward approach to computer numerical control programming. The software uses menus that help guide the programmer. The menus will prompt the programmer to input variables into Ezpath cutting cycles. The Ezpath cutting cycles work in practically the same manner as the cycles we used in EIA/ISO programming.

OBJECTIVES

Upon completion of this chapter the reader will be able to:

- *Describe each function on the Basic Operation screen.*
- *Enter a new program into Ezpath.*
- *Program miscellaneous preliminary functions, such as tool changes, and spindle functions.*
- *Describe the part using the StartPath command.*
- *Apply a Rough cycle to the part path.*
- *Apply a Profile cycle to the part path.*
- *Apply a Groove cycle where applicable*
- *Apply a Thread cycle where applicable.*
- *End the program.*
- *Save the program.*
- *View the tool path geometry*

STARTING A NEW PROGRAM

Programming with the Ezpath software is best taught through tutorials. A Bridgeport Expath control is shown in Figure 10–1. The tutorials we will use are simple turning examples that we will take you through step by step. As

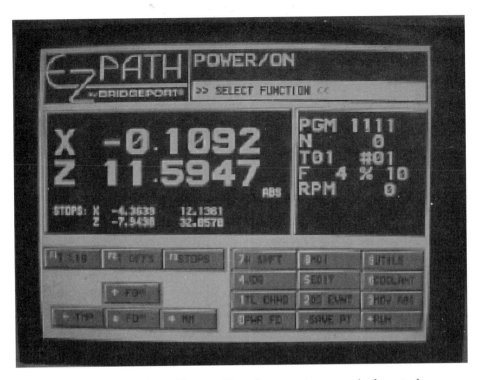

FIGURE **10–1** *The Bridgeport Ezpath computer numerical control lathe can be used both as a manual lathe as well as totally computer operated turning machine.*

you go through the examples, please refer to the part print. This will help you to understand the programming format.

When the Ezpath software is accessed, the Basic Operations screen is displayed (see Figure 10–2).

As you can see from Figure 10–2, each function on the Basic Operations screen is associated with a number. These key functions can be accessed by selecting these numbers on the keyboard of the computer or the numeric key pad on the machine control.

The key functions shown in the Basic Operations screen are as follows:

0 EXIT   This allows you to exit the program mode.

2 EDIT TXT   This command will enter a Code editor. In this mode you will be able to modify or enter standard EIA/ISO programs. This is typically available only on the Personal Computer version of this software not the machine-based software.

3 LIST TXT   This command will display the last program created or edited.

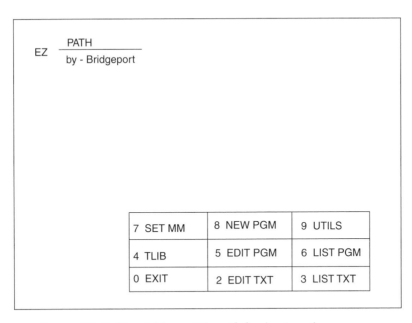

FIGURE 10–2 *The Bridgeport Ezpath basic operations screen.*

4 T LIB   The 4 key calls the tool library screen, which allows the operator to view and enter tool information.

5 EDIT PGM   This key allows the programmer to edit any previously created program.

6 LIST PGM   The 6 key will display the contents of the last active program.

7 SET MM (SET INCH)   The 7 key allows you to toggle between metric and standard values for each part program.

8 NEW PGM   It is from this operation that a new program is started.

9 UTILS   The utilities key in the Basic Operations mode temporarily exits the programming mode to allow the programmed to run the disk utilities. The disk utilities area allows the programmer to copy files, delete files, and otherwise manage files.

STARTING A NEW PROGRAM

Figure 10–3 is our first tutorial programming example. We will step through this programming example with the programming instructions. Please refer back to the part print, Figure 10–3 as we go through the example. Again, this will help you to understand what we are attempting to do.

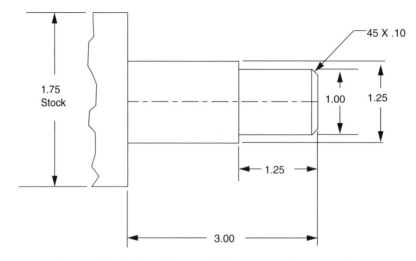

FIGURE 10–3 *Our first tutorial programming example.*

The tool we used to create this part is an 80-degree outside diameter turning tool. All of the tools are described in the tool library. Select the number **4** to view your tool library. Take note of the tool types and the tool numbers that correspond to them. The suggested material is 1-3/4 inch cold rolled steel.

EXERCISE 1

To create the new program, enter the NEWPGM mode from the Basic Operations screen (see Figure 10–4). Type a number **8** from the keyboard.

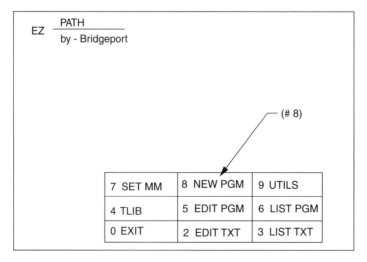

FIGURE 10–4 *Basic operation screen.*

The NEWPGM screen will appear and a number 0000 will be created. DO NOT attempt to delete this line.

The first line of the program is a rapid position move to the absolute coordinates X2.0, Z3.0.

This will give us a safe position to do a tool change.

Press the [**1**] **POS** key. Enter the following:

DIA ABS 2.0 <enter>
Z ABS 3.0 <enter>

The program line will appear:

0010 RAPID ABS X2.0000 Z3.0000

The second line of the program should be a Tool Change. This will create a screen Prompt so that the operator can tell the control which tool will be loaded.

Press the [**7**] **TLCHG** key. Enter the correct tool ID number, tool number, and offset number.

We will be using an 80-degree outside diameter turning tool.

In this case it is: tool ID number **1**, tool number **1**, and offset number **1**.

Input these values now.

Note: Your individual tools may have been pre-loaded into the tool library in a different order than ours. If your tool library differs from ours, input your tool numbers accordingly.

The new program line will appear:

0010 TLCHG I1 T01 01

The next line tells the operator to change the spindle RPM.

Press the [**6**] **SETRPM** key. Enter the following:

SET RPM 750 <enter>

The program line will appear:

0030 SETRPM S750

STARTPATH

The next line of the program is the **STARTPATH** command. This command tells the machine to create a path based on the following commands. If you remember back to EIA/ISO programming, this is the equivalent of the first line of the G71 roughing command (i.e., P0040).

Press the F1 **PATH** key. Enter the following:

Note: In the path mode only a few functions are available. They are

1. POS, 2. LINE , 3. ARC , —UNDO, 4. BL LINE , 5. BL ARC, 6. PTHSTP

    PATH NAME 1 <enter>

The program line will appear:

    0040 STARTPATH 1

The next program line is a RAPID command. It tells the machine to move to a specific point. The first line of a STARTPATH command must start with Position command.

Press the 1 **POS** key. Enter the following:

    DIA ABS 0.0 <enter>
    Z ABS 0.05 <enter>

The program line will appear:

    0050 RAPID ABS X00.000 Z00.050

Press the 2 **LINE** key. Enter the following:

    DIA ABS 0.0 <enter>
    Z ABS 0.0 <enter>
    F0.005

The feed rate selection at this point really doesn't matter because the feed rate will be controlled in the Roughing cycle, which is coming up later in this tutorial.

Press the 4 **BL LINE** key.

This BL LINE creates a chamfer, fillet, or radius on the end of the part. The end point of the blend line is the end diameter of the fillet or radius on the part. The control will back figure the start and end point of the chamfer or radius (see Figure 10–5).

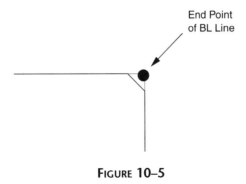

End Point
of BL Line

FIGURE 10–5

Enter the following:

DIA ABS 1.0 <enter>
Z ABS 0.0 <enter>
CHAMF/R BLEND 0.10 <enter>
CHAMF/CW/CCW 1 <enter>

Note: A number 1 is for a chamfer. A number 2 is for a clockwise arc. A number 3 is for a counterclockwise arc.

The program line will appear:

0070 CHAMFER ABS X1.000 Z0.000 P0.1000 P0.1000 F0.0050

Note: While creating a path the feed rate is automatically input with a default value.

Press the 2 **LINE** key. Enter the following:

DIA ABS 1.0 <enter>
Z ABS -1.25 <enter>

This line creates the 1.00-inch diameter turned end.

Press the 2 **LINE** key. Enter the following:

DIA ABS 1.25 <enter>
Z ABS -1.25 <enter>

Press the 2 **LINE** key. Enter the following:

DIA ABS 1.25 <enter>
Z ABS -3.00 <enter>

Press the ⎡2⎤ **LINE** key. Enter the following:

> **DIA ABS 1.75 <enter>**
> **Z ABS -3.00 <enter>**

The diameter programmed in the last line of the path creates the rapid clearance point for the rough stock. The last diameter *must* be greater than or equal to the rough stock diameter.

Press the ⎡6⎤ **PTHSTP** key.

The last line of the PATH is the PATHSTOP. This is necessary to let the machine know that the Profile of the part is complete.

Enter the following:

> **0 <enter>**

The program line will appear:

> 0120 PATHSTOP

ROUGHING

The next command tells the machine how to rough out the profile of the part.

Press the ⎡F2⎤ **ROUGH** key. Enter the following:

> **PATH ID 1 <enter>**
> **1-OD/2-ID/3-FACE 1 <enter>**
> **X FIN ALLOW 0.02 <enter>**
> **Z FIN ALLOW 0.01 <enter>**
> **ENGAGE FEED 0.01 <enter>**
> **CROSS FEED 0.01 <enter>**
> **RETRACT FEED 0.01 <enter>**
> **CUT STEP 0.06 <enter>**
> **CLEARANCE 0.1 <enter>**
> **WITHDRAW ANG 45.0 <enter>**
> **WITHDRAW LEN 0.05 <enter>**
> **CUT DIR 1-POS/2-NEG 2 <enter>**

UNDERCUT 1-ON/2-OFF 1 <enter>
AUTO ROUND 1-ON/2-OFF 1 <enter>

The program line will appear:

0130 ROUGH 1 I1 X0.02 Z0.01 F0.01 0.01 0.01 S0.06 C0.01
W45.00 W0.05 D2 U1 A1

Press the **1** **POS** key. Enter the following:

DIA ABS 2.0 <enter>
Z ABS 3.0 <enter>

This command instruct the tool to move away from the part.

The program line will appear:

0140 RAPID ABS X2.0000 Z3.0000

Press the **7** **TLCHG** key. Enter the correct tool ID number, tool

number, and offset number. In this case it is tool number **2**, and offset number **2**. This is the 55-degree finishing tool.

The next line tells the operator to change the spindle RPM for the finish cut.

Press the **6** **SETRPM** key. Enter the following:

SET RPM 900 <enter>

The program line will appear:

0160 SETRPM S900

PROFILING

The next command is a PROFILE command. This is the finishing pass.

Press the **F3** **PROFILE** key. Enter the following:

PATH ID 1 <enter>
TOOL POS 1-LEFT/2-RIGHT 1 <enter>
X FIN ALLOW 0.0 <enter>
Z FIN ALLOW 0.0 <enter>

FEED RATE 0.008 <enter>
CLEARANCE 0.10 <enter>
ENGAGE ANG 45.0 <enter>
WITHDRAW ANG 90.00 <enter>
UNDERCUT 1-ON/2-OFF 1 <enter>
AUTO ROUND 1-ON/2-OFF 1 <enter>

The program line will appear:

0170 PROFIL 1 X0.0000 Z0.0000 F0.0800 E 45.0 W90.0 U1 A1

This last position move brings the tool to a clearance position and will allow us to end the program.

Press the [**1**] **POS** key. Enter the following:

DIA ABS 2.0 <enter>
Z ABS 3.0 <enter>

ENDING THE PROGRAM

The Auxiliary function M30 will END the program and REWIND the program back to the start.

Press the [**·**] **AUXFUN** key. Enter the following:

PRGM 30 <enter>

The program line will appear:

0190 AUXFUN M30

The program is now completed.

SAVING THE PROGRAM

To SAVE the program, press the [**0**] **EXIT** key.

Enter a name for your program and then select the [**/**] **SAVE VIEW** key.

The control will now process the program and check for errors. If an error occurs the control will prompt you with a message in the upper left corner of the screen. If the control processes the program correctly the Preview Screen will appear (see Figure 10–6).

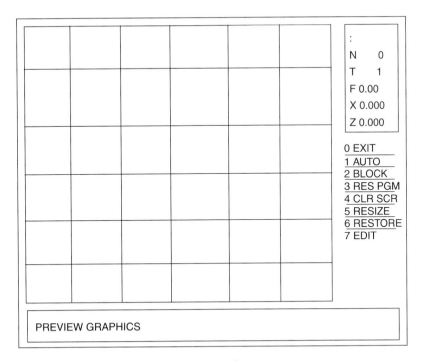

FIGURE 10–6 *Preview screen.*

USING PREVIEW

After you pressed the [/] SAVE/VIEW key, the control processed the program and automatically scales the screen to match the program. Press the **1 AUTO** key to begin simulation of the program. The screen will prompt you at tool changes and program stops. You will be instructed on the bottom of the screen on how to restart the graphics mode.

In this case, press the "**+**" key to continue. Cutting or linear feed moves will appear as solid lines. Rapid moves will appear as dotted lines. You can use the **2 BLOCK** key to run the program one line at a time. If the graphic looks correct, have an instructor check it. If the program doesn't look correct, use the copy of program 1 to look for errors.

PREVIEW GRAPHIC KEY FUNCTIONS

0 EXIT   The **EXIT** key leaves the Preview mode and returns to the
`   MDI screen.

1 AUTO   The AUTO command sets the loaded program to be run in continuous operation. The graphics will stop only for tool changes or program stops. The graphics will restart after pressing the **1 AUTO** key again.

2 BLOCK   The BLOCK command runs the program in a single step mode. The **2 BLOCK** must be pressed for each line of the program.

3 RES PGM   The RES PGM command resets the program back to the beginning.

4 CLR SCR   The CLR SCR command clears the screen.

5 RESIZE   The RESIZE command is used to set the preview window and select other viewing operations.

6 RESTORE   The RESTORE command returns the preview screen to the original size.

7 EDIT   The EDIT command returns the operator back to the program and allows for editing of the active program.

PROGRAM 1

```
0000 EZPATHISX 1 MODEIINCH
0010 RAPID ABS X2.0000 Z3.0000
0020 TLCHG I1 T01 01
0030 SETRPM S750
0040 STARTPATH 1
0050 RAPID ABS X0.0000 Z0.0500
0060 LINE ABS X0.0000 Z0.0000 F0.0050
0070 CHAMFER ABS X1.0000 Z0.0000 P0.1000 P0.1000 F0.0050
0080 LINE ABS X1.0000 Z-1.2500 F0.0050
0090 LINE ABS X1.2500 Z-1.2500 F0.0050
0100 LINE ABS X1.2500 Z-3.0000 F0.0050
0110 LINE ABS X1.7500 Z-3.0000 F0.0050
0120 PATHSTOP
0130 ROUGH 1 I1 X0.0200 Z0.0100 F0.0100 0.0100 0.0100 S0.0600
C0.1000W45.0000W0.0500D2U1A1
0140 RAPID ABS X2.0000 Z3.0000
```

0150 TLCHG I2 T02 02

0160 SETRPM S900

0170 PROFIL 1 1 X0.0000 Z0.0000 F0.0080 C0.1000 E45.0000 W90.0000 U1 A1

0180 RAPID ABS X2.0000 Z3.0000

0190 AUXFUN M30

EXAMPLE PROGRAM 2

As you go through the second example, please refer to the part print shown in Figure 10–7. Again, this will help you to understand the programming format. Every attempt will be made to explain each section, but if you don't under-stand a particular section please see your instructor for help.

Figure 10–7 is our second programming example. The tools used to create this part are an 80-degree outside diameter turning tool, a 35-degree finish profiling tool, and a .125-inch wide grooving tool. The suggested material is 1-3/4 inch cold rolled steel.

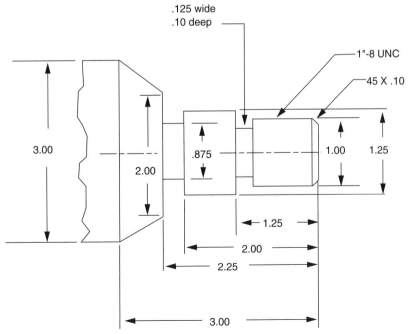

FIGURE **10–7** *Part print for example.*

EXERCISE 2

To create the new program, enter the NEWPGM mode from the Basic Operations screen (see Figure 10–8). Type a number **8** from the keyboard.

The NEWPGM screen will appear and a number 0000 will be created. DO NOT attempt to delete this line.

The first line of the program is a rapid position move to the absolute coordinates X4.0, Z3.0.

Press the **1** **POS** key. Enter the following:

   DIA ABS 4.0 <enter>
   Z ABS 3.0 <enter>

The program line will appear:

   0010 RAPID ABS X4.0000 Z3.0000

The second line of the program should be a Tool Change. This will create a screen prompt so that the operator can tell the control which tool will be loaded.

Press the **7** **TLCHG** key. Enter the correct tool ID number, tool number, and offset number.

In this case it is tool ID number **1**, tool number **1**, and offset number **1**.

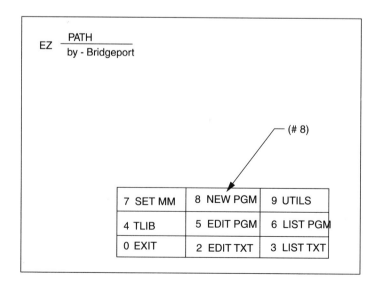

FIGURE **10–8** *Basic Operation Screen*

Enter these numbers now.

The program line will appear:

0020 TLCHG I1 T01 01

The next line tells the operator to change the spindle RPM.

Press the [**6**] **SET RPM** key. Enter the following:

SET RPM 650 <enter>

The program line will appear:

0030 SETRPM S650

STARTPATH

The next line of the program is the STARTPATH command. This command tells the machine to create a path or profile of the part.

Press the [**F1**] **PATH** key. Enter the following:

PATH NAME 1 <enter>

The program line will appear:

0040 STARTPATH 1

The next program line is a rapid position command. It tells the machine to move to a specific point.

Press the [**1**] **POS** key. Enter the following:

DIA ABS 0.0 <enter>
Z ABS 0.05 <enter>

The program line will appear:

0050 RAPID ABS X00.000 Z00.050

Press the [**2**] **LINE** key. Enter the following:

DIA ABS 0.0 <enter>
Z ABS 0.0 <enter>
F0.008

Remember: the feed within the STARTPATH defaults to the last preset feed rate. It cannot be changed. The feed rate is actually set in the roughing and profiling cycles.

Press the [4] **BL LINE** key. Enter the following:

> **DIA ABS 1.0 <enter>**
> **Z ABS 0.0 <enter>**
> **CHAMF/R BLEND 0.10 <enter>**
> **CHAMF/CW/CCW 1 <enter>**

The program line will appear:

> 0070 CHAMFER ABS X1.000 Z0.000 P0.1000 P0.1000 F0.0050

Note: While creating a path the feed rate is automatically input with a default value.

Press the [2] **LINE** key. Enter the following:

> **DIA ABS 1.0 <enter>**
> **Z ABS -1.25 <enter>**

Press the [2] **LINE** key. Enter the following:

> **DIA ABS 1.25 <enter>**
> **Z ABS -1.25 <enter>**

Press the [2] **LINE** key. Enter the following:

> **DIA ABS 1.25 <enter>**
> **Z ABS -2.25 <enter>**

Press the [2] **LINE** key. Enter the following:

> **DIA ABS 2.00 <enter>**
> **Z ABS -2.25 <enter>**

Press the [2] **LINE** key. Enter the following:

> **DIA ABS 3.00 <enter>**
> **Z ABS -3.00 <enter>**

Remember that the last diameter position in the final line of the path creates the clearance point for the rough stock. The last diameter should be greater than or equal to the rough stock diameter.

Press the [**6**] **PTHSTP** key. Enter the following:

The last line of the PATH is the PATHSTOP. This is necessary to let the machine know that the profile of the part is complete.

> **0 <enter>**

The program line will appear:

> 0130 PATHSTOP

ROUGHING

Press the [**F2**] ROUGH key. Enter the following:

> **PATH ID 1 <enter>**
> **1-OD/2-ID/3-FACE 1 <enter>**
> **X FIN ALLOW 0.02 <enter>**
> **Z FIN ALLOW 0.01 <enter>**
> **ENGAGE FEED 0.01 <enter>**
> **CROSS FEED 0.01 <enter>**
> **RETRACT FEED 0.01 <enter>**
> **CUT STEP 0.06 <enter>**
> **CLEARANCE 0.1 <enter>**
> **WITHDRAW ANG 45.0 <enter>**
> **WITHDRAW LEN 0.05 <enter>**
> **CUT DIR 1-POS/2-NEG 2 <enter>**
> **UNDERCUT 1-ON/2-OFF 1 <enter>**
> **AUTO ROUND 1-ON/2-OFF 1 <enter>**

The program line will appear:

> 0140 ROUGH 1 I1 X0.02 Z0.01 F0.01 0.01 0.01 S0.06 C0.01 W45.00 W0.05 D2 U1 A1

Press the [**1**] **POS** key. Enter the following:

> **DIA ABS 4.0 <enter>**
> **Z ABS 3.0 <enter>**

The program line will appear:

    0150 RAPID ABS X4.0000 Z3.0000

Press the **7** **TLCHG** key. Enter the correct tool ID number, tool number, and offset number.

In this case it is tool ID number **2**, tool number **2**, and offset number **2**. This is the Finishing tool. Enter these numbers now.

The next line tells the operator to change the spindle RPM for the finish cut.

Press the **6** **SET RPM** key. Enter the following:

SET RPM 850 <enter>

The program line will appear:

The next command is a PROFILE command. This is the finishing pass.

    0170 SETRPM S850

PROFILING

Press the **F3** **PROFILE** Key. Enter the following:

PATH ID 1 <enter>
TOOL POS 1-LEFT/2-RIGHT 1 <enter>
X FIN ALLOW 0.0 <enter>
Z FIN ALLOW 0.0 <enter>
FEED RATE 0.008 <enter>
CLEARANCE 0.10 <enter>
ENGAGE ANG 45.0 <enter>
WITHDRAW ANG 90.00 <enter>
UNDERCUT 1-ON/2-OFF 1 <enter>
AUTO ROUND 1-ON/2-OFF 1 <enter>

The program line will appear:

    0180 PROFIL 1 X0.0000 Z0.0000 F0.0800 E 45.0 W90.0 U1 A1

Press the **1** **POS** key. Enter the following:

DIA ABS 4.0 <enter>
Z ABS 3.0 <enter>

Press the **7** **TLCHG** key. Enter the correct tool ID number, tool number, and offset number.

In this case it is tool ID number **3**, tool number **3**, and offset number **3**. This is the grooving tool. Enter these number now.

The next line tells the operator to change the spindle RPM for grooving.

Press the **6** **SET RPM** key. Enter the following:

SET RPM 350 <enter>

The program line will appear:

0210 SETRPM S350

The next two commands tell the machine to move into position to cut the groove.

Press the **1** **POS** key. Enter the following:

DIA ABS 4.0 <enter>
Z ABS -1.25 <enter>

Press the **1** **POS** key. Enter the following:

DIA ABS 1.30 <enter>
Z ABS -1.25 <enter>

Press the **2** **LINE** key. Enter the following:

DIA ABS .90 <enter>
Z ABS -1.25 <enter>
F0.002

This next position line tells the machine to retract the tool from the groove.

Press the **1** **POS** key. Enter the following:

DIA ABS 4.0 <enter>
Z ABS -1.25 <enter>

On this line we will need to position for the next groove. Because this groove is greater than the .125 grooving tool we will need to do a grooving procedure.

Press the ⌷1⌷ **POS** key. Enter the following:

DIA ABS 4.0 <enter>
Z ABS -2.25 <enter>

GROOVING

The next line of the program is the STARTPATH command. This command tells the machine to create a groove path based on the following profile commands.

Press the ⌷F1⌷ **PATH** key. Enter the following:

This path name must be different from the previous path name, which was used for the part profile.

PATH NAME 2 <enter>

The program line will appear:

0260 STARTPATH 2

The next program line is a Position command. It tells the machine to move to a specific point.

Press the ⌷1⌷ **POS** key. Enter the following:

DIA ABS 2.0 <enter>
Z ABS -2.25 <enter>

The program line will appear:

0270 RAPID ABS X2.000 Z-2.250

In this path we need to describe the profile of the groove (see Figure 10–9).

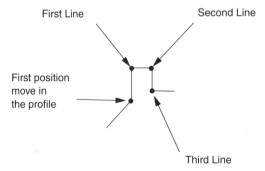

FIGURE 10–9 *Lines that make up the profile of the groove.*

Press the ☐ 2 ☐ **LINE** key. Enter the following:

> **DIA ABS 0.875 <enter>**
> **Z ABS -2.25 <enter>**
> **F0.002**

Press the ☐ 2 ☐ **LINE** key. Enter the following:

> **DIA ABS 0.875 <enter>**
> **Z ABS -2.00 <enter>**
> **F0.002**

Press the ☐ 2 ☐ **LINE** key. Enter the following:

> **DIA ABS 2.00 <enter>**
> **Z ABS -2.00 <enter>**
> **F0.002**

Press the ☐ 6 ☐ **PATHSTOP**. Enter the following:

> **0 <enter>**

Now that we have described the profile of the groove we must tell the machine to perform a grooving cycle.

Press the ☐ F4 ☐ **GROOVE** key. Enter the following:

> **PATH ID 2 <enter>**
> **1-ROUGH/2-FIN/3-BOTH 3 <enter>**
> **FIN ALLOW 0.005 <enter>**
> **PLUNGE FEED 0.002 <enter>**
> **RETRACT FEED 0.01 <enter>**
> **PLUNGE STEP 0.000 <enter>**
> **CLEARANCE 0.10 <enter>**
> **STEP OVER % 80.00 <enter>**
> **LIFT OFF 0.000 <enter>**
> **DWELL 0.000 <enter>**

The program line will appear:

> 0320 GROOVE 2 3 A0.005 F0.002 R0.01 P00.00 C0.100 O80.00 L0.000
> D0.000

Now that we have finished inputting the necessary values for the grooving cycle, let's answer some questions you might have about the grooving cycle.

1-ROUGH/2-FIN/3-BOTH	The grooving command can be used to rough, finish or both operations.
FIN ALLOW	This sets the amount of material left on the part for finishing.
PLUNGE FEED	This is the plunge cut feed rate.
RETRACT FEED	This sets the speed at which the tool retracts from the groove.
PLUNGE STEP	This sets the depth of each plunge or peck cut depth.
CLEARANCE	This sets how far from the part the tool begins its plunge move.
STEP OVER %	This sets how far the tool moves over for each cut. The percentage is a percentage of the tool width.
LIFT OFF	This value sets the distance the tool moves back after each plunge cut. It is similar to a peck retract on a drill cycle.
DWELL 0.000	The sets the time, in seconds, for the tool to dwell at the groove depth.

This next position move insures that the tool clears the part before going to a home position.

Press the ⬚**1**⬚ **POS** key. Enter the following:

> **DIA ABS 4.0 <enter>**
>
> **Z ABS -2.25 <enter>**

This last position move brings the tool to a clearance position and will allow us to end the program.

Press the ⬚**1**⬚ **POS** key. Enter the following:

> **DIA ABS 4.0 <enter>**
>
> **Z ABS 3.00 <enter>**

ENDING THE PROGRAM

Press the [·] **AUXFUN** key. Enter the following:

PRGM 30 <enter>

The program line will appear:

0350 AUXFUN M30

The program is now completed.

SAVING THE PROGRAM

To SAVE the program, press the [0] **EXIT** key.

Enter the name of the program and then select the [/] **SAVE VIEW** key.

The control will now process the program and check for errors. If an error occurs the control will prompt you with a message in the upper left corner of the screen. If the control processes the program correctly the Preview Screen will appear .

PROGRAM 2

```
0000 EZPATHISX 1 MODEIINCH
0010 RAPID ABS X4.0000 Z3.0000
0020 TLCHG I1 T01 01
0030 SETRPM S650
0040 STARTPATH 1
0050 RAPID ABS X0.0000 Z0.0500
0060 LINE ABS X0.0000 Z0.0000 F0.0050
0070 CHAMFER ABS X1.0000 Z0.0000 P0.1000 P0.1000 F0.0050
0080 LINE ABS X1.0000 Z-1.2500 F0.0050
0090 LINE ABS X1.2500 Z-1.2500 F0.0050
0100 LINE ABS X1.2500 Z-2.2500 F0.0050
0110 LINE ABS X2.0000 Z-2.2500 F0.0050
0120 LINE ABS X3.0000 Z-3.0000 F0.0050
```

0130 PATHSTOP

0140 ROUGH 1 I1 X0.0200 Z0.0100 F0.0100 0.0100 0.0100 S0.0600
C0.1000W45.0000W0.0500D2U1A1

0150 RAPID ABS X4.0000 Z3.0000

0160 TLCHG I2 T02 02

0170 SETRPM S850

0180 PROFIL 1 1 X0.0000 Z0.0000 F0.0080 C0.1000 E45.0000 W90.0000
U1 A1

0190 RAPID ABS X4.0000 Z3.0000

0200 TLCHG I3 T03 03

0210 SETRPM S350

0220 RAPID ABS X1.3000 Z-1.2500

0230 LINE ABS X0.9000 Z-1.2500 F0.0020

0240 RAPID ABS X4.0000 Z-1.2500

0250 RAPID ABS X4.0000 Z-2.2500

0260 STARTPATH 2

0270 RAPID ABS X2.0000 Z-2.2500

0280 LINE ABS X0.8750 Z-2.2500 F0.0020

0290 LINE ABS X0.8750 Z-2.0000 F0.0020

0300 LINE ABS X2.0000 Z-2.0000 F0.0020

0310 PATHSTOP

0320 GROOVE 2 3 A0.0050 F0.0020 R0.0100 P0.0000 C0.1000 O80.0000
L0.0000 D0.0000

0330 RAPID ABS X4.0000 Z-2.2500

0340 RAPID ABS X4.0000 Z3.0000

0350 AUXFUN M30

EXAMPLE PROGRAM 3

The third and final example is the tutorial in Figure 10–10. In this tutorial we will cover again everything that we have covered to this point, plus threading.

The tools used to create this part are an 80-degree outside diameter turning tool, a 35-degree finish profiling tool, and a .125-inch wide grooving tool, and a 60-degree threading tool. In this example we are preparing you for the projects that are to follow. In this exercise we will face the part off first. Please pay close attention, because this will be how we will want you to program the example projects at the end of this chapter.

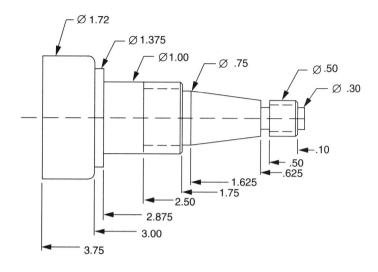

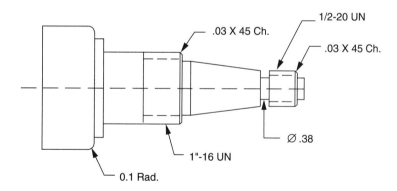

FIGURE 10–10 *Part print for exercise 3.*

EXERCISE 3

To create this new program, enter the NEWPGM mode from the Basic Operations screen. Type a number **8** from the keyboard.

The NEWPGM screen will appear and a number 0000 will be created.

The first line of the program is a rapid position move to the absolute coordinates X3.0, Z3.0.

Press the ⬚**1** **POS** key. Enter the following:

DIA ABS 3.0 <enter>
Z ABS 3.0 <enter>

The program line will appear:

0010 RAPID ABS X3.0000 Z3.0000

The second line of the program should be a Tool Change.

Press the ⬚**7** **TLCHG** key. Enter the correct tool ID number, tool number, and offset number.

In this case it is tool ID number **1**, tool number **1**, and offset number **1**.

Enter these values now.

The program line will appear:

0020 TLCHG I1 T01 01

The next line tells the operator to change the spindle RPM.

Press the ⬚**6** **SET RPM** key. Enter the following:

SET RPM 850 <enter>

The program line will appear:

0030 SETRPM S850

Press the ⬚**1** **POS** key. Enter the following:

DIA ABS 2.0 <enter>
Z ABS 0.00 <enter>

This position and line move set up for the face off of the part. This is different than how we did it in the previous examples.

The program line will appear:

0040 RAPID ABS X02.000 Z00.000

Press the ⬚**2** **LINE** key. Enter the following:

DIA ABS 0.0 <enter>

Z ABS 0.0 <enter>
F0.008

Press the ⎡1⎤ **POS** key. Enter the following:

DIA ABS 0.0 <enter>
Z ABS .10 <enter>

The program line will appear:

0060 RAPID ABS X0.0000 Z0.1000

STARTPATH

The next line of the program is the STARTPATH command. This command tells the machine to create a path based on the following commands.

Press the ⎡F1⎤ **PATH** key. Enter the following:

PATH NAME 1 <enter>

The program line will appear:

0070 STARTPATH 1

The next program line is a rapid position command.

Press the ⎡1⎤ **POS** key. Enter the following:

DIA ABS 0.3 <enter>
Z ABS 0.00 <enter>

The program line will appear:

0080 RAPID ABS X00.300 Z00.000

Press the ⎡2⎤ **LINE** key. Enter the following:

DIA ABS 0.3 <enter>
Z ABS -.10 <enter>
F0.008

Press the ⎡4⎤ **BL LINE** key. Enter the following:

DIA ABS .50 <enter>
Z ABS -.10 <enter>

CHAMF/R BLEND 0.03 <enter>
CHAMF/CW/CCW 1 <enter>

The program line will appear:

0100 CHAMFER ABS X.500 Z-.100 P0.03 P0.030 F0.0080

Note: While creating a path the feed rate is automatically input with a default value.

Press the | 2 | **LINE** key. Enter the following:

DIA ABS .50 <enter>
Z ABS -.625 <enter>

Press the | 2 | **LINE** key. Enter the following:

DIA ABS .75 <enter>
Z ABS -1.625 <enter>

Press the | 2 | **LINE** key. Enter the following:

DIA ABS .75 <enter>
Z ABS -1.75 <enter>

Press the | 4 | **BL LINE** key. Enter the following:

DIA ABS 1.0 <enter>
Z ABS -1.75 <enter>
CHAMF/R BLEND 0.03 <enter>
CHAMF/CW/CCW 1 <enter>

Press the | 2 | **LINE** key. Enter the following:

DIA ABS 1.0 <enter>
Z ABS -2.875 <enter>

Press the | 4 | **BL LINE** key. Enter the following:

DIA ABS 1.375 <enter>
Z ABS -2.875 <enter>
CHAMF/R BLEND 0.03 <enter>
CHAMF/CW/CCW 1 <enter>

Press the [2] **LINE** key. Enter the following:

DIA ABS 1.375 <enter>
Z ABS -3.00 <enter>

Press the [4] **BL LINE** key. Enter the following:

DIA ABS 1.72 <enter>
Z ABS -3.00 <enter>
CHAMF/R BLEND 0.10 <enter>
CHAMF/CW/CCW 2 <enter>

Press the [2] **LINE** key. Enter the following:

DIA ABS 1.72 <enter>
Z ABS -3.75 <enter>

Press the [2] **LINE** key. Enter the following:

DIA ABS 2.0 <enter>
Z ABS -3.75 <enter>

Press the [6] **PTHSTP** key. Enter the following:

0 <enter>

The program line will appear:

0210 PATHSTOP

ROUGHING

Press the [F2] **ROUGH** key. Enter the following:

PATH ID 1 <enter>
1-OD/2-ID/3-FACE 1 <enter>
X FIN ALLOW 0.02 <enter>
Z FIN ALLOW 0.01 <enter>
ENGAGE FEED 0.01 <enter>
CROSS FEED 0.01 <enter>
RETRACT FEED 0.01 <enter>

CUT STEP 0.06 <enter>
CLEARANCE 0.1 <enter>
WITHDRAW ANG 45.0 <enter>
WITHDRAW LEN 0.05 <enter>
CUT DIR 1-POS/2-NEG 2 <enter>
UNDERCUT 1-ON/2-OFF 1 <enter>
AUTO ROUND 1-ON/2-OFF 1 <enter>

The program line will appear:

0220 ROUGH 1 I1 X0.02 Z0.01 F0.01 0.01 0.01 S0.06 C0.01 W45.00
W0.05 D2 U1 A1

Press the   **1**   **POS** key. Enter the following:

DIA ABS 2.0 <enter>
Z ABS 3.0 <enter>

The program line will appear:

0230 RAPID ABS X2.0000 Z3.0000

Press the   **7**   **TLCHG** key. Enter the correct tool ID number, tool number, and offset number.

In this case it is tool ID number **2**, tool number **2**, and offset number **2**. This is the Finishing tool. Enter these values now.

The next line tells the operator to change the spindle RPM for the finish cut.

Press the   **6**   **SET RPM** key. Enter the following:

SET RPM 950 <enter>

The program line will appear:

0240 SETRPM S950

PROFILING

Press the   **F3**   **PROFILE** Key. Enter the following:

PATH ID 1 <enter>
TOOL POS 1-LEFT/2-RIGHT 1 <enter>

X FIN ALLOW 0.0 <enter>
Z FIN ALLOW 0.0 <enter>
FEED RATE 0.008 <enter>
CLEARANCE 0.10 <enter>
ENGAGE ANG 45.0 <enter>
WITHDRAW ANG 90.00 <enter>
UNDERCUT 1-ON/2-OFF 1 <enter>
AUTO ROUND 1-ON/2-OFF 1 <enter>

The program line will appear:

0260 PROFIL 1 X0.0000 Z0.0000 F0.0800 E 45.0 W90.0 U1 A1

Press the [**1**] **POS** key. Enter the following:

DIA ABS 2.0 <enter>
Z ABS 3.0 <enter>

Press the [**7**] **TLCHG** key. Enter the correct tool ID number, tool number, and offset number.

In this case it is tool ID number **3**, tool number **3**, and offset number **3**. This is the grooving tool. Enter these values now.

The next line tells the operator to change the spindle RPM for the grooving.

Press the [**6**] **SET RPM** key. Enter the following:

SET RPM 350 <enter>

The program line will appear:

0290 SETRPM S350

CUTTING SINGLE-PASS GROOVES

The next two commands tell the machine to move into position to cut each single-pass groove.

Press the [**1**] **POS** key. Enter the following:

DIA ABS 3.0 <enter>
Z ABS -.625 <enter>

Press the ⬜**1** **POS** key. Enter the following:

> **DIA ABS .60 <enter>**
> **Z ABS -.625 <enter>**

Press the ⬜**2** **LINE** key. Enter the following:

> **DIA ABS .38 <enter>**
> **Z ABS -.625 <enter>**
> **F0.002**

Press the ⬜**1** **POS** key. Enter the following:

> **DIA ABS 3.0 <enter>**
> **Z ABS -.625 <enter>**

Press the ⬜**1** **POS** key. Enter the following:

> **DIA ABS 3.0 <enter>**
> **Z ABS 3.0 <enter>**

Press the ⬜**7** **TLCHG** key. Enter the correct tool ID number, tool number, and offset number.

THREADING

In this case it is tool ID number **4**, tool number **4**, and offset number **4**. This is the threading tool. Enter these values now.

The next line tells the operator to change the spindle RPM for threading.

Press the ⬜**6** **SET RPM** key. Enter the following:

> **SET RPM 500 <enter>**

The program line will appear:

> 0360 SETRPM S500

Press the ⬜**F5** **THREAD** key. Enter the following:

> **1=OD/2=ID 1 <enter>**

Note: The thread height is automatically calculated after entering the lead.

> **LEAD 0.0769 <enter>**
> **THREAD HEIGHT 0.0472 <enter>**
> **STEP 1 0.005 <enter>**
> **STEP 2 0.003 <enter>**
> **MIN STEP 0.002 <enter>**
> **# SPRING PASSES 2 <enter>**
> **WITHDRAW CLEARANCE 0.1 <enter>**
> **ENGAGE ANGLE 30 <enter>**
> **START Z 0.15 <enter>**
> **END Z -.55 <enter>**
> **START DIA 0.5 <enter>**
> **END DIA 0.5 <enter>**

The program line will appear:

> 0370 THREAD 1 L0.0769 H0.0472 S0.0050 0.0030 0.0020
> #2 C0.100 Z0.1538 -0.5500 D0.500 0.500

Now that we have finished inputting the necessary values for the threading cycle, let's answer some questions you might have about the this cycle.

1=OD/2=ID	The command tells the control whether we are cutting an external or internal thread.
LEAD	The lead is the distance from one thread to the same position on the next thread. This is calculated by dividing the number of thread per inch by 1.
THREAD HEIGHT	This is the distance from the crest of the thread to the root of the thread. This value is automatically calculated when the lead is input.
STEP 1	This is the depth of the first pass. The first two cut passes dictate how many passes the threading tool will take. Each pass after he first pass decreases in depth.
STEP 2	This is the depth of the second pass.
MIN STEP	This sets the minimum cut pass depth
# SPRING PASSES	Spring passes are cuts taken, at the end of the cycle, in which there is no additional

in feed. Spring cuts are designed to clear the thread of any tool deflection.

WITHDRAW CLEARANCE This is the amount the tool backs away from the part on the X axis.

ENGAGE ANGLE This should be set to half of the included angle of the thread. Engage angle works in the same manner as the compound infeed on the manual lathe.

START Z This is the Z axis position to start the tool prior to cutting the thread.

END Z This is the Z axis end point of the thread.

START DIA This is the major diameter of the thread.

END DIA This is the diameter at the end point of the thread. Unless the thread is tapered, the end diameter of the thread will be the same as the start diameter.

Press the [1] **POS** key. Enter the following:

DIA ABS 1.0 <enter>
Z ABS 0.05 <enter>

Press the [1] **POS** key. Enter the following:

DIA ABS 1.0 <enter>
Z ABS -1.65 <enter>

Press the [F5] **THREAD** key. Enter the following:

Note: The thread height is automatically calculated.

1=OD/2=ID 1 <enter>LEAD 0.0625 <enter>
THREAD HEIGHT 0.0383 <enter>
STEP 1 0.005 <enter>
STEP 2 0.003 <enter>
MIN STEP 0.002 <enter>
 # SPRING PASSES 2 <enter>
WITHDRAW CLEARANCE 0.1 <enter>
ENGAGE ANGLE 30 <enter>

START Z -1.65 <enter>

END Z -2.50 <enter>

START DIA 1.0<enter>

END DIA 1.0 <enter>

The program line will appear:

0390 THREAD 1 L0.0625 H0.0383 S0.0050 0.0030 0.0020

#2 C0.100 Z-1.650 -2.500 D1.00 01.00

Press the ⎡ **1** ⎤ **POS** key. Enter the following:

DIA ABS 3.0 <enter>

Z ABS -1.65 <enter>

Press the ⎡ **1** ⎤ **POS** key. Enter the following:

DIA ABS 3.0 <enter>

Z ABS 3.0 <enter>

ENDING THE PROGRAM

Press the ⎡ **·** ⎤ **AUXFUN** key. Enter the following:

PRGM 30 <enter>

The program line will appear:

0430 AUXFUN M30

The program is now completed.

SAVING THE PROGRAM

To SAVE the program, press the ⎡ **0** ⎤ **EXIT** key.

Enter the name of the program and then select the ⎡ **/** ⎤ **SAVE VIEW** key.

The control will now process the program and check for errors. If an error occurs the control will prompt you with a message in the upper left corner of the screen. If the control processes the program correctly the Preview Screen will appear.

PROGRAM 3

```
0000 EZPATHISX 1 MODEIINCH
0010 RAPID ABS X3.0000 Z3.0000
0020 TLCHG I1 T01 01
0030 SETRPM S850
0040 RAPID ABS X2.0000 Z0.0000
0050 LINE ABS X0.0000 Z0.0000 F0.0080
0060 RAPID ABS X0.0000 Z0.1000
0070 STARTPATH 1
0080 RAPID ABS X0.3000 Z0.0000
0090 LINE ABS X0.3000 Z-0.1000 F0.0080
0100 CHAMFER ABS X0.5000 Z-0.1000 P0.0300 P0.0300 F0.0080
0110 LINE ABS X0.5000 Z-0.6250 F0.0080
0120 LINE ABS X0.7500 Z-1.6250 F0.0080
0130 LINE ABS X0.7500 Z-1.7500 F0.0080
0140 CHAMFER ABS X1.0000 Z-1.7500 P0.0300 P0.0300 F0.0080
0150 LINE ABS X1.0000 Z-2.8750 F0.0080
0160 CHAMFER ABS X1.3750 Z-2.8750 P0.0300 P0.0300 F0.0080
0170 LINE ABS X1.3750 Z-3.0000 F0.0080
0180 BLENDILN ABS X1.7200 Z-3.0000 R0.1000 CW F0.0080
0190 LINE ABS X1.7200 Z-3.7500 F0.0080
0200 LINE ABS X2.0000 Z-3.7500 F0.0080
0210 PATHSTOP
0220 ROUGH 1 I1 X0.0200 Z0.0100 F0.0100 0.0100 0.0100 S0.0600
C0.1000W45.0000W0.0500D2U1A1
0230 RAPID ABS X2.0000 Z3.0000
0240 TLCHG I2 T02 02
0250 SETRPM S950
0260 PROFIL 1 1 X0.0000 Z0.0000 F0.0080 C0.1000 E45.0000 W90.0000
U1 A1
0270 RAPID ABS X2.0000 Z3.0000
0280 TLCHG I3 T03 03
0290 SETRPM S350
0300 RAPID ABS X3.0000 Z-0.6250
0310 RAPID ABS X0.6000 Z-0.6250
0320 LINE ABS X0.3800 Z-0.6250 F0.0020
0330 RAPID ABS X3.0000 Z-0.6250
```

0340 RAPID ABS X3.0000 Z3.0000

0350 TLCHG I4 T04 04

0360 SETRPM S500

0370 THREAD 1 L0.0769 H0.0472 S0.0050 0.0030 0.0020 #2 C0.1000
Z0.1500 -0.5500D0.5000 0.5000 29.0000

0380 RAPID ABS X1.0000 Z0.0500

0390 RAPID ABS X1.0000 Z-1.6500

0400 THREAD 1 L0.0625 H0.0383 S0.0050 0.0030 0.0020 #2 C0.1000 Z-
1.6500 -2.5000D1.0000 1.0000 30.0000

0410 RAPID ABS X3.0000 Z-1.6500

0420 RAPID ABS X3.0000 Z3.0000

0430 AUXFUN M30

LOADING THE PROGRAM INTO THE MACHINE

If you have been doing your programming on the personal computer you will
need to know how to get your program loaded into the machine. You need to
move your program from the personal computer hard drive to your disk. This
is done in UTILITIES. From the main menu type the number **9 UTILITIES**
key. You will notice that you have a list of commands that are available to
you. We want to copy a file from EZPATH to a FLOPPY disk. Type the number
2 key. You should now be seeing a list of utilities that are available to you.
Again, we want to copy a files from EZPATH to a FLOPPY disk. Type the
number **7** key. Insert your disk into drive A:. At the prompt type in a name for
your program. Press **ENTER**. Now hit **any** key to continue. Press **ENTER** to
exit the utilities mode. Press the **Esc** key to quit.

BRIDGEPORT EZ PATH
PROGRAMMING PROJECT 1

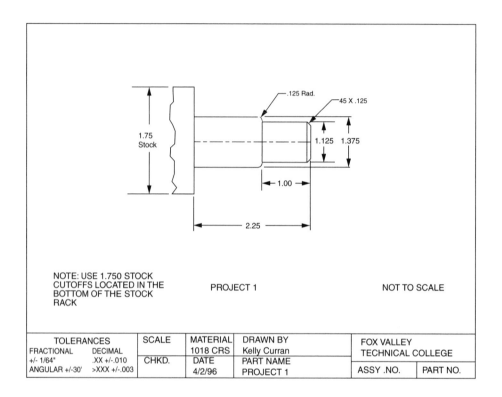

.125 Rad.

45 X .125

1.75
Stock

1.125   1.375

1.00

2.25

NOTE: USE 1.750 STOCK
CUTOFFS LOCATED IN THE
BOTTOM OF THE STOCK
RACK

PROJECT 1

NOT TO SCALE

TOLERANCES		SCALE	MATERIAL	DRAWN BY	FOX VALLEY
FRACTIONAL	DECIMAL		1018 CRS	Kelly Curran	TECHNICAL COLLEGE
+/- 1/64"	.XX +/-.010	CHKD.	DATE	PART NAME	ASSY .NO. / PART NO.
ANGULAR +/-30'	>XXX +/-.003		4/2/96	PROJECT 1	

BRIDGEPORT EZ PATH
PROJECT 2

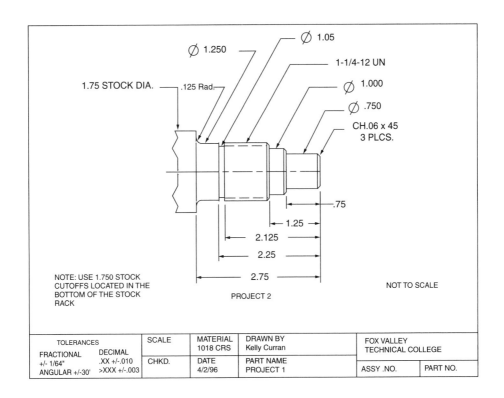

⌀ 1.05

⌀ 1.250

1-1/4-12 UN

1.75 STOCK DIA.   .125 Rad.

⌀ 1.000

⌀ .750

CH.06 x 45
3 PLCS.

.75

1.25

2.125

2.25

2.75

NOT TO SCALE

NOTE: USE 1.750 STOCK
CUTOFFS LOCATED IN THE
BOTTOM OF THE STOCK
RACK

PROJECT 2

TOLERANCES		SCALE	MATERIAL 1018 CRS	DRAWN BY Kelly Curran		FOX VALLEY TECHNICAL COLLEGE	
FRACTIONAL +/- 1/64" ANGULAR +/-30'	DECIMAL .XX +/-.010 >XXX +/-.003	CHKD.	DATE 4/2/96	PART NAME PROJECT 1		ASSY .NO.	PART NO.

BRIDGEPORT EZ PATH
PROJECT 3

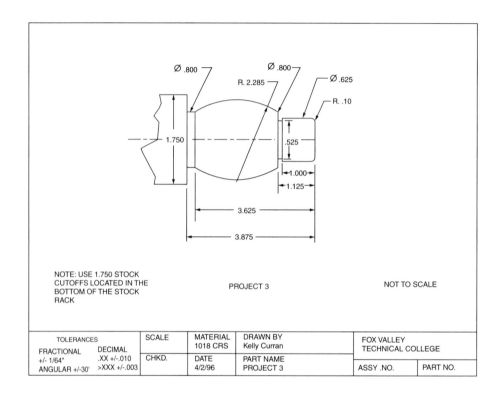

NOTE: USE 1.750 STOCK
CUTOFFS LOCATED IN THE
BOTTOM OF THE STOCK
RACK

PROJECT 3

NOT TO SCALE

TOLERANCES		SCALE	MATERIAL 1018 CRS	DRAWN BY Kelly Curran	FOX VALLEY TECHNICAL COLLEGE	
FRACTIONAL +/- 1/64" ANGULAR +/-30'	DECIMAL .XX +/-.010 >XXX +/-.003	CHKD.	DATE 4/2/96	PART NAME PROJECT 3	ASSY .NO.	PART NO.

Chapter 11

INTRODUCTION TO FANUC FAPT PROGRAMMING

INTRODUCTION

The FAPT programming format, which resides in the Fanuc control, is a very powerful machine programming language. The acronym FAPT stands for Fanuc Automatic Program Tool. Note that the format is conversational and that the control uses symbols on the control buttons to aide the programmer.

OBJECTIVES

Upon completion of this chapter the reader will be able to:

- *Initialize the FAPT programming side of the Fanuc control.*
- *Describe the blank and part.*
- *Describe the part figure.*
- *Define the home and tool index position.*
- *Describe and define the sequence of the machining operations.*
- *Prepare numerical control data.*

STARTING A NEW PROGRAM

Figure 11–1 shows a typical Fanuc CNC control. We will begin the first programming tutorial with Exercise 1 in Figure 11–2. Exercise 1 is a simple turning exercise, which includes rough turning and finishing. As we begin programming in FAPT, recognize that Fanuc uses defaults, which are probable answers to questions. The probable answers are based on how the control and the tool files are setup. The default answers to the questions will appear in the answer spaces before you have completed the questions. Use these defaults as a tool to enhance the programming capabilities of the control.

Upon initial start up of the machine, the machine control is on the FAPT programming side of the control. This is evident on the screen. If the machine is

FIGURE **11–1** *The Fanuc control is a very powerful and very popular computer numerical control.*

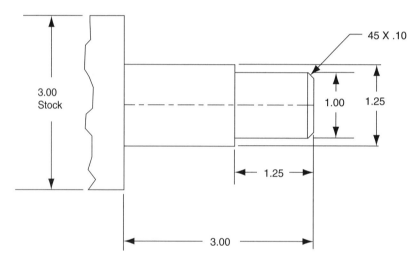

FIGURE **11–2** *Simple turning exercise that includes rough turning and finishing.*

on the NC/PC or numerical control side press the FAPT button located above the alphanumeric keyboard (see Figure 11–3).

You are now on the FAPT side of the control. You will notice that across the bottom of the screen you are given four choices (see Figure 11–4).

Select FAPT EXEC to enter the FAPT programming mode. Lets program the sample part in Figure 11–2.

BLANK & PART

Blank & Part deals with the raw piece of stock or billet that we are going to be using. In these two sections (1 & 2) you will be answering questions that deal ONLY with the raw stock.

NC/PC — ⌐ FAPT

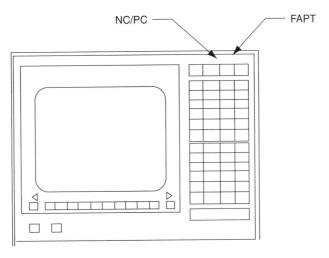

FIGURE **11–3** *Typical Fanuc control and
NC/PC and FAPT button locations.*

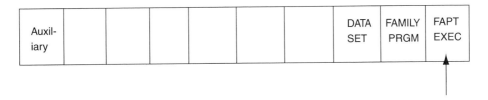

FIGURE **11–4** *Location of FAPT EXEC key.*

On the screen are the five steps to programming the part (see Figure 11–5).

These steps are in sequential order, this means we have to do them in the order that they appear on the screen. Type a **1** and press the **Exec** key located under the screen (see Figure 11–6).

MATERIAL TYPE

The first question that needs to be answered about the raw stock is the material type. On the screen you will see the possible material selection that have been previously programmed into the parameters of the machine (see Figure 11–7).

Sample material number 1 is 1018 plain carbon steel. If you look back on the print you will note that this is the type of steel we will be using. Type a **1** for 1018 steel and press the **INPUT** key located on the numeric key pad.

=*=FAPT=*=

1. Blank & Part (Drawing & Blank)
2. Blank & Part (Part Figure)
3. Home Position & Index Position
4. Definition of Machining
5. NC Data Preperation

0. End

NO. = ____

FIGURE 11–5 *Five steps to programming the part.*

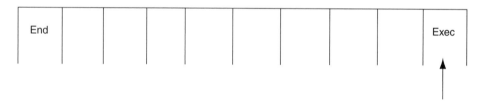

| End | | | | | | | | | | Exec |

FIGURE 11–6 *Location of Exec key.*

SURFACE ROUGHNESS DEFAULTS

The next blank and part question deals with the standard surface roughness. Surface roughness is controlled by the feed rate (see Figure 11–8).

Notice that the control is defaulting to number **3**. Number three is a standard surface finish of approximately 100 RMS. To accomplish this surface roughness it will use a finish feed rate default of 0.008 feed per revolution. This is acceptable for our part so press **INPUT** to accept this surface roughness default.

DRAWING FORMAT

The drawing format selection tells the control where your work piece coordinate is located and how the machining will take place, i.e., if the tools will be

```
*** Blank & Part ****    MATERIAL


No.    Sample of Material
1         1018
2         SCM
3         FC
4         AL
5         SUS







Material No..........MN= ____
```

FIGURE 11–7 *Material selection screen.*

```
*** Blank & Part ***        STANDARD SURFACE ROUGNESS


Surface Roughness  & Cutting Feed Rate

N0.    Kind of Surface Roughness        Cutting Feed Rate
1      Number of Triangle One ................ F1 = 0.012
2      Number of Triangle Two .................F2 = 0.010
3      Number of Triangle Three ..............F3 = 0.008
4      Number of Triangle Four ................F4 = 0.006






Standard Surface  Roughness.....NR  = 3
```

FIGURE 11–8 *Surface roughness screen.*

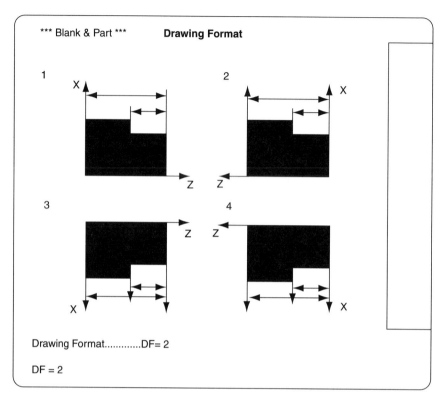

FIGURE 11–9 *The operator must choose the format that matches how the machining will actually be performed.*

cutting in a down/left direction, down/right direction, up/left direction or a up/right direction (see Figure 11–9). This may seem like an awkward question, but you have to realize that this control is put on many different types of turning machines, and some companies use the left end and center of the spindle as the part origin. Typically, the right end and center of the part are used as the part origin.

The control should now be defaulting to the drawing format number **2**. This is the one we want, so press **INPUT**.

BLANK FIGURE & BASE LINE

In this section we tell the control what the rough stock looks like, it's physical size, and whether we are going to face any of the end of the stock off (see Figure 11–10).

The first question that must be answered is whether the stock is a number 1 solid cylinder, a number 2 hollow cylinder (i.e., pipe), or a number 3 special

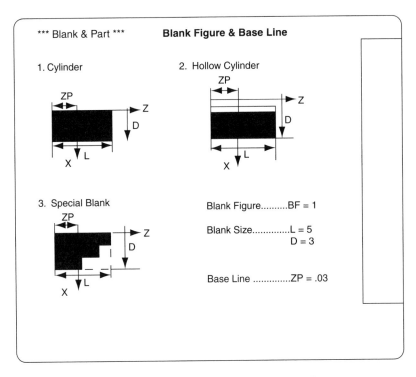

FIGURE 11–10 *This screen is used to define
the rough blank shape and size.*

blank (i.e., forging or casting). The stock we are using for this simple part is a solid cylinder. Type a **1** in for the BF __ and press **INPUT**.

The next question that needs to be answered deals with the physical size of the blank. The L stands for length. Estimate the length of the rough stock for our part. Make sure to add on what is going to be held in the chuck. Type in **5.0** and press **INPUT**. Next is the D. What is the diameter of the rough stock we will be using? Type in a **3.0** and press **INPUT**.

The ZP stands for Z axis positive. It establishes the Z base line or how much are we going to face off of the right end of the stock. Lets face off 0.030 inches. Type in **0.030** and press **INPUT**.

PART FIGURE

It is at this point in FAPT programming where we start to program the actual part shape. It is important to remember that we are not programming the part as it will be machined, but are just describing the part shape. You will notice from the screen that the plot axes are established by the solid lines that we described in the Drawing Format. The dotted lines depict the blank or rough stock dimensions that were described in Blank Figure and Base Line (see Figure 11–11).

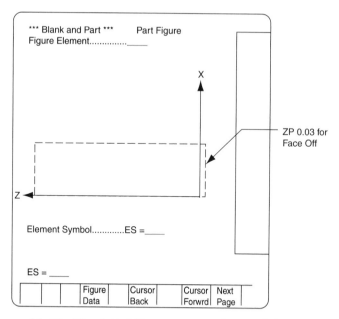

FIGURE 11–11 *The dotted lines represent the rough blank shape.*

To describe the part shape we need to use element symbols. The element symbols are located on the alphanumeric keypad (see Figure 11–12).You will notice in Figure 11-13 that these keys are used for auxiliary operations such as threading and grooving. Notice in Figure 11–14 that these keys are used to represent linear and arc figure lines. Looking back at the part print in Figure 11–2, note that the part origin begins at X0.0 Z0.0. Starting at the origin, the element symbol we need to press is the Up ARROW ↑ key or number **8** (see Figure 11–15). Select this key now and press **INPUT**. The control will now ask for the surface roughness requirement. The surface roughness is defaulting to **3** on the screen. This is allowable so press **INPUT**. The starting point in X is 0.0. Type in a **0.0** for the SDX. Press **INPUT**. SZ is the Z value of the start point. Type in 0.0 and press **INPUT**. The element question the control will prompt you for is PE, Part Exist. The part exists to the left of the origin. The control should be defaulting to **1**. Press **INPUT**. The control will now ask us for the X End Point of the line or "DX" (see Figure 11–16). Our first turned diameter is 1 inch. Type in **1.00** for the 1-inch diameter and press **INPUT**. Remember we don't have to consider the chamfer yet. We will let the control figure that out on the next figure element. The right end of the part should now appear on the screen in red. The control now prompts us for the next element. Our next element will be the Chamfer. Press the **Chamfer** key and press **INPUT**. The control now wants to know how big the chamfer is. Type in **.100** and press **INPUT**. Notice that the chamfer doesn't appear yet. The chamfer will appear upon the input of the next element. The next element

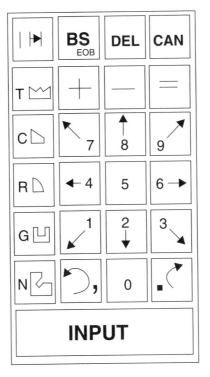

FIGURE 11–12 *Alphanumeric keypad.*

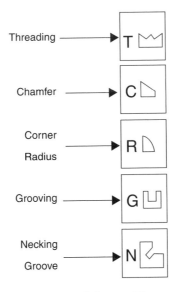

FIGURE 11–13 *Keys used for auxiliary operations.*

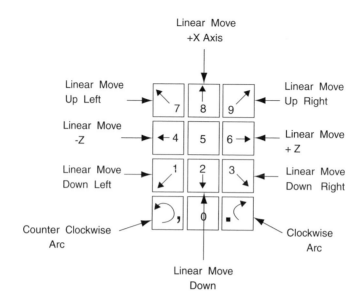

FIGURE 11–14 *Keys used to represent linear and arc figure lines.*

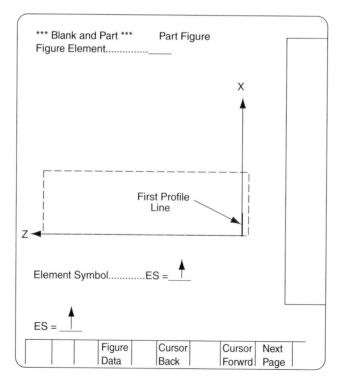

FIGURE 11–15 *The up arrow was chosen
to show the cutting direction.*

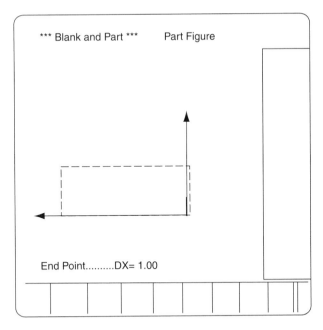

FIGURE 11–16 *An X point of 1.00 has been entered.*

will be a linear element to the left ←. Select the **4** key now and press **INPUT**. If at any time you select the wrong key, you can erase your selection by pressing the **CAN** or **Cancel** key, which is located on the keypad. Again, the surface roughness is defaulting to **3** on the screen. This is allowable so press **INPUT**. The control is now asking for the END POINT Z. The end point of this element is 1.25. Type **1.25** and press **INPUT**. You will notice that we are **not** inputting -Z numbers because the control realizes what direction the part lies in based on the drawing format. The chamfer should now appear on the screen. Note that the first linear element has not been trimmed back yet (see Figure 11–17). We can update the screen by pressing the **FIGURE DATA** button. Press it now. You will notice that this screen tells you what elements and what data you have input up to this point. If you press the **END** key you will now return to the graphics screen and your figure will be updated. Note that the chamfer is now trimmed.

Next element we need to press is the Up Arrow ↑ key **8.** Now press **INPUT**. The surface roughness is defaulting to **3** on the screen. This is allowable so press **INPUT**. The control will now ask us for the X End Point or DX. Type in **1.25** for the 1.25-inch diameter, press **INPUT**. The screen should now show our new element (see Figure 11–18).

The next element will be a linear element to the left ←. Select this key now. Press **INPUT**. Remember, the surface roughness is defaulting to **3** on the screen. This is allowable so press **INPUT**. The control is now asking for the

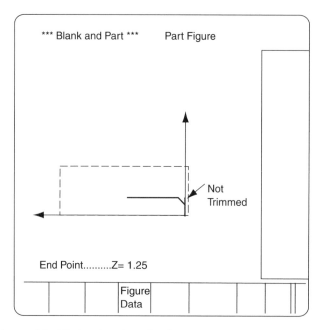

FIGURE **11–17** *Notice that the first element was not trimmed.*

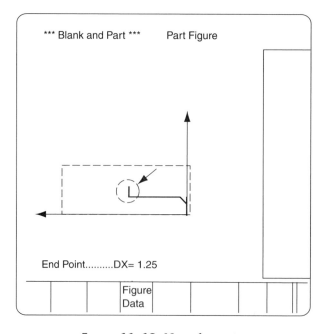

FIGURE **11–18** *New elements.*

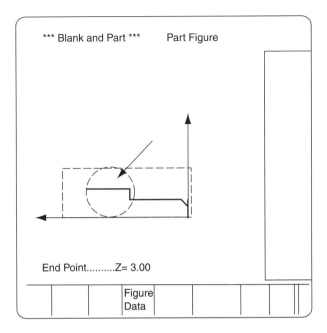

*** Blank and Part ***   Part Figure

End Point..........Z= 3.00

Figure Data

FIGURE **11–19** *New element.*

END POINT Z. The end point of this element is 3.00. Type **3.00** and press **INPUT**. The screen should now show our new element (see Figure 11–19).

Next element we need to press is the Up Arrow ↑ key **8**. Now press **INPUT**. The surface roughness is defaulting to **3** on the screen. This is allowable so press **INPUT**. The control will now ask us for the X End Point or DX. Type in **3.00** for the 3.00-inch diameter, press **INPUT**. The screen should now show our new element (see Figure 11–20).

We have now completed the Part Figure, step 2 in Blank & Part (Part Figure). We will now move on to step 3 by pressing the **NEXT PAGE** key (see Figure 11–21) located at the bottom of the screen. The shape already input on the screen will appear again and the following menu will be displayed in the lower portion of the screen.

FIGURE

1. NEW

2. CORRECT

3. CORRECT (ERASE GRAPHIC)

Because there are no corrections needed press the **NEXT PAGE** button at the bottom of the screen (see Figure 11–21).

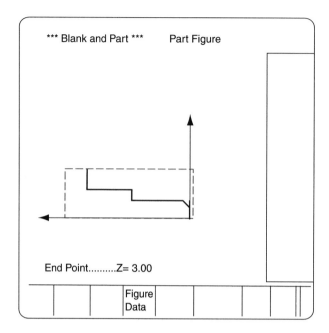

FIGURE 11–20 *New element.*

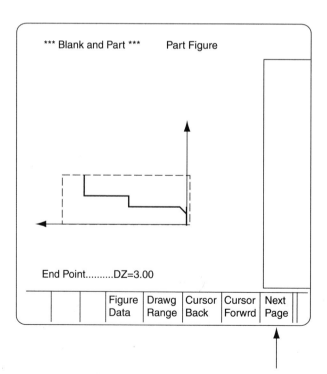

FIGURE 11–21 *Location of the Next Page key.*

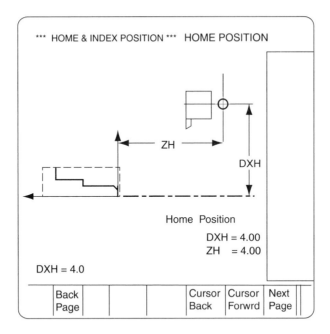

FIGURE 11–22 *Home position of machine.*

HOME POSITION & INDEX POSITION

The system is now asking you for the home position of the machine (see Figure 11–22). The home position may or may not be the actual home position of the machine, but it must always be a safe place to bring the turret back to for parts loading and part measurement.

HOME POSITION

DXH = Diameter value from the spindle center

ZH = Distance from the Z origin (workpiece face)

In most cases we will use 4 inches in the DXH, and 4 inches in the ZH. **UNDER NO CIRCUMSTANCES WILL THESE NUMBERS EVER BE NEGATIVE**.

Type in **4.00** for the DXH and press **INPUT**. Type in **4.00** for the ZH and press **INPUT**.

INDEX POSITION

The index position is the actual position where the tool turret will index during tool changes (see Figure 11–23). As you can imagine we will have to make sure this position is well away from the part and work holding device.

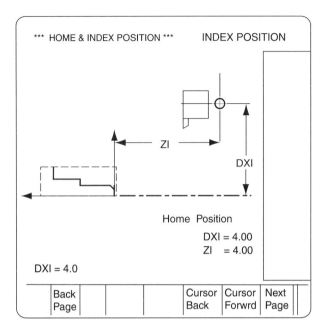

FIGURE 11–23 *Index position.*

The length of the tools will have a great deal to do with this position (i.e., a long drill or boring bar can be DEADLY if we don't go back far enough). We will use the same numbers for the INDEX POSITION as we did for the home position.

Type in **4.00** for the DXI and press **INPUT**. Type in **4.00** for the ZI and press **INPUT**.

DEFINITION OF MACHINING

The screen will change to the menu that you will use to define how you want the part to be machined (see Figure 11–24).

The system is asking you what machining operation you want to perform first. We want to face off the part and rough turn the outer figure. We need to answer the question: PROC. 01 = ___, with a number **3** for ROUGHING OF OUTER FIGURE. Press **INPUT**. The system now needs to know the tool data information. The screen should now be prompting you for TOOL DATA (see Figure 11–25).

The TOOL I.D. number is the number of the tool in the tool file. The tool file is a library of pre-described tools. The descriptions of the tools have been preloaded into the computer. The descriptions consist of nose-radius size, cutting-edge angle, nose angle, and virtual tool position. This information is of great importance because the control uses this information to decide if it can

*** MACHINING DEFINITION ***   KINDS OF MACHINING

PROC. 01   Roughing of Outer Figure   T0101   X4.000 Z4.00

KINDS OF MACHINING............PROC.01= <u>3</u>

1. CENTER DRILLING   6. SEMI FINISHING OF INNER FIG.
2. DRILLING   7. FINISHING OF OUTER FIGURE
3. ROUGHING OF OUTER FIGURE   8. FINISHING OF INNER FIGURE
4. ROUGHING OF INNER FIGURE   9. GROOVING OR NECKING
5. SEMI - FINISHING OF OUTER F.   10. THREADING

PROC. 01 = <u>3</u>

| | | | | | | | CURSOR BACK | CURSOR FORWRD | NEXT PAGE |

FIGURE 11–24 *Machining definition menu.*

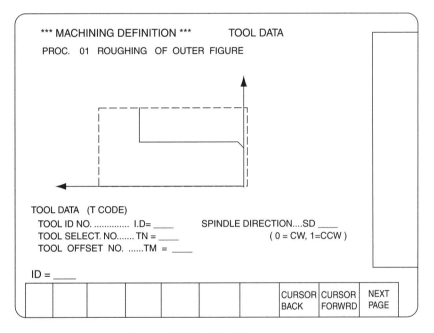

*** MACHINING DEFINITION ***   TOOL DATA
PROC. 01 ROUGHING OF OUTER FIGURE

TOOL DATA (T CODE)
  TOOL ID NO. .... I.D= ____   SPINDLE DIRECTION....SD ____
  TOOL SELECT. NO....... TN = ____   (0 = CW, 1=CCW)
  TOOL OFFSET NO. ......TM = ____

ID = ____

| | | | | | | | CURSOR BACK | CURSOR FORWRD | NEXT PAGE |

FIGURE 11–25 *Tool Data screen.*

cut certain profiles of the part. If you would like to see the tool file information in greater detail, look in the Tool File book that came with the machine. We have certain tools that will remain consistent with during this turning center example. They are the following:

Tool I.D. number 11, which is an 80-degree diamond roughing tool.

Tool I.D. number 114, which is a 35-degree diamond finishing tool.

Tool I.D. number 690, which is a .125-inch parting/grooving tool.

Tool I.D. number 720, which is a threading tool.

The tool I.D. number we will use for roughing the outer figure is Tool I.D. No. 11. Type an **11** and press **INPUT**. The TOOL SELECT is the tool turret position that the roughing tool is in. Type in the tool turret number of the roughing tool and press **INPUT**. The TOOL OFFSET NO. will be the same as the TOOL SELECT number. Example: Tool 11, offset 11. Type in the offset number and press **INPUT**. The spindle direction is based upon how the tool is set in the turret. When the spindle is turning toward you this is the CW or clockwise direction. This is "typically" the direction that the spindle will need to rotate. **PLEASE VERIFY THIS FOR EVERY TOOL**. Type in a **0** for the SD or spindle direction and press **INPUT**. When the tool identification numbers have been input, the screen will display the tool data information that was preset in the tool file (see Figure 11–26).

This tool data can be changed by moving the cursor to the area you wish to change. The **CURSOR FORWARD/CURSOR BACK** keys are located at the bottom of the screen. All of this information will be accepted by pressing the **NEXT PAGE** key located at the bottom right of the screen. Press this key now.

MACHINING START POSITION

When all of the tool data is input and complete, the system asks you for the Machining Start Position. The machining start position is the position or point that the tool will go through on its way to and from the home position. As you can imagine, it is important that this point lies well above and to the right of the part or rough stock. It is typical to use the same position numbers for the machine start position as you did for the Index Position. Type a **4.00** for the X-axis coordinate value of the Machine Start position and press **INPUT**. Type **4.00** for the Z-axis coordinate value of the Machine Start position (see Figure 11–27) and press **INPUT**.

CUTTING CONDITIONS

The system now asks you for the cutting conditions required for the machining process. The cutting conditions consist of information that the control will need to properly machine our part. The information is based on the material we

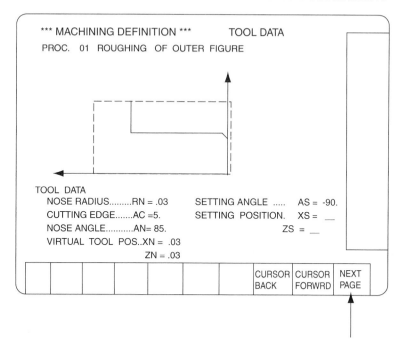

FIGURE 11–26 *Tool data in tool file.*

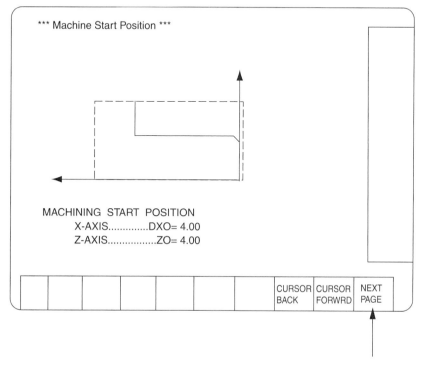

FIGURE 11–27 *Machine start position.*

have selected. The controller already has the parameters loaded so the cutting conditions will be defaulting on the screen. Carefully check this information. Change only the desired items. We are going to change the depth of cut. Press the **CURSOR FORWARD** key until the depth of cut is highlighted. Key in **.05** and press **INPUT** for the depth of cut. We will now need to input the MAX. RPM. Cursor Foreword to MAX. RPM. Key in **3000** and press **INPUT**.

CUTTING DIRECTION

The control is asking you for the cutting direction. We are rough facing the part. We rough face the part in an X negative or ↓ down direction. Press the **2** ↓ or down area key located on the keyboard. Press **INPUT**. The cutting direction or CD is very self explanatory. If the cutting direction is toward the head stock use the **4** ← key, and so on (see Figure 11–28).

CUTTING AREA DEFINITION

The cutting area definition defines the area that you want to machine in this particular operation. The area is defined by the cursors (■) as displayed along the part figure. The cursor is moved using the Cursor Foreword/Cursor Back keys located at the bottom of the screen. The cutting area START POINT cursor needs to be located at the point where the face of the part meets the chamfer start (see Figure 11–29).

Press the **CURSOR FORWARD** button until the blinking cursor is in the start point position as seen in the Figure 11–29. The control needs to know in which

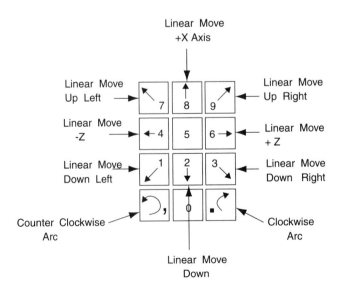

FIGURE 11–28 *Cutting direction selection keys.*

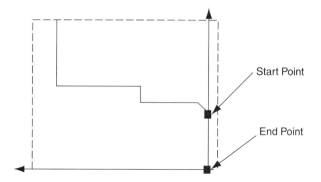

FIGURE 11–29 *Cutting area definition.*

direction the rough material is located from the blinking cursor. The control calls this the DS or DIVIDING DIRECTION. This concept may be difficult to understand. It is for this reason that the programmer will accept the control default practically 100 percent of the time. You will note that with the cursor at the start point, the control DIVIDING DIRECTION is defaulting ↑. This is correct, press **INPUT** to accept this. The control now prompt you for the END POINT. If the blinking cursor is not already in the end point position shown in Figure 11–29, move it there using the **CURSOR FORWARD** key. The control DIVIDING DIRECTION is defaulting →. This is correct, so press **INPUT** to accept this. Press **NEXT PAGE**. It is important to note that if the tool figure data was not input correctly in the tool file, the appropriate machining area cannot be cut. After the cutting area has been designated, the designated cutting area is faced off. This can be confirmed by looking at the graphic display. The blue dotted line represents the material that still needs to be cut.

The control now wants to know if you are going to use the same tool to cut another area. Remember we only rough faced the part. We need to rough the rest of the figure. Key in a **1** for yes, and press **INPUT**.

CUTTING CONDITIONS

The system asks you again for the cutting conditions required for the machining process. The cutting conditions are the same as we had for the rough facing operation. You can accept all of this information by simply pressing the **NEXT PAGE** key. Press this key now.

CUTTING DIRECTION

The control is asking you for the cutting direction for rough turning. We are turning in the Z negative direction. We faced the part with ↓ down direction now we will select the **4** ← from the keyboard for right hand turning. Press **INPUT**.

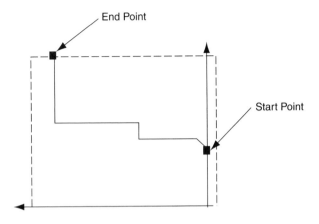

FIGURE 11–30 *Location of start point.*

CUTTING AREA DEFINITION

The cutting area definition will define the area that you want to machine in the rough turning operation. The area is defined by the cursors (■) as displayed along the part figure. The cursor is moved using the **CURSOR FORWARD/CURSOR BACK** key located at the bottom of the screen. The cutting area START POINT cursor needs to be located at the point where the face of the part meets the start of the chamfer (see Figure 11–30). Move the cursor there now.

The control now wants to know which direction the rough material is located from the blinking cursor. You will note that with the cursor at the start point, the control DIVIDING DIRECTION is defaulting →. This is correct so press **INPUT** to accept this. The control now prompt you for the END POINT. If the blinking cursor is not already in the end point position shown in Figure 11–30, move it there using the **CURSOR FORWARD** key. The control DIVIDING DIRECTION is defaulting ↑. This is correct, so press **INPUT** to accept this. Press the **NEXT PAGE** button. After the cutting area has been designated, the selected cutting area is rough turned. This can be confirmed by looking at the blue dotted line on the graphics screen. The blue dotted line represents the material that still needs to be cut. Notice that the blue dotted line closely follows the profile of the part, leaving just a small gap for finishing (see Figure 11–31). The control now wants to know if you are going to use the same tool to cut another area. Let's finish turn the part with a finishing tool. The control should be defaulting to "0." If not, key in a "**0**" for no, and press **INPUT**.

DEFINITION OF MACHINING

The screen will now change to the menu which you will use to define your next operation. The next operation will be FINISHING OF OUTER FIGURE, number 7. Type a **7** and press **INPUT** (see Figure 11–32).

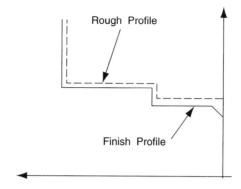

Rough Profile

Finish Profile

FIGURE 11–31 *The screen shows the rough profile with a blue dotted line.*

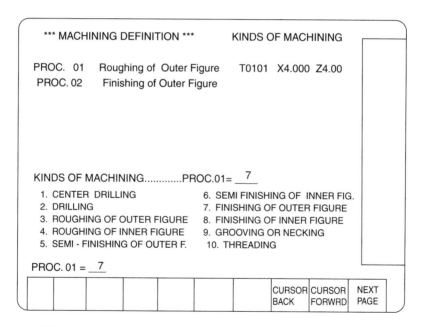

```
*** MACHINING DEFINITION ***        KINDS OF MACHINING

PROC. 01   Roughing of Outer Figure    T0101   X4.000  Z4.00
PROC. 02   Finishing of Outer Figure




KINDS OF MACHINING.............PROC.01= __7__
  1. CENTER DRILLING              6. SEMI FINISHING OF INNER FIG.
  2. DRILLING                     7. FINISHING OF OUTER FIGURE
  3. ROUGHING OF OUTER FIGURE     8. FINISHING OF INNER FIGURE
  4. ROUGHING OF INNER FIGURE     9. GROOVING OR NECKING
  5. SEMI - FINISHING OF OUTER F.  10. THREADING

PROC. 01 = __7__
```

| | | | | | | | CURSOR BACK | CURSOR FORWRD | NEXT PAGE |

FIGURE 11–32 *The menu used to define your next operation.*

The system now needs to know some information about the finishing tool. The screen should now be prompting you for TOOL DATA

TOOL DATA

The tool I.D. number we will use for finishing the outer figure is TOOL I.D. NO. 114. Type a **114** and press **INPUT**. Tool 114 is a 35-degree diamond finishing tool. The TOOL SELECT is the tool turret position that the finishing tool is in. Type in the tool turret number of the finishing tool and press INPUT.

The TOOL OFFSET NO. will be the same as the TOOL SELECT number. Example: Tool 5, offset 5. Type in the offset number and press **INPUT**. The spindle direction is based upon how the tool is set in the turret. When the spindle is turning toward you this is the CW or clockwise direction. Type in a **0** for the SD or spindle direction and press **INPUT**. When all of the tool numbers and spindle direction has been input the screen will display the tool data information that was preset in the tool file. This information can be changed by moving the cursor to the area you wish to change. The CURSOR FORWARD/CURSOR BACK keys are located at the bottom of the screen. All of this information can be accepted by pressing the **NEXT PAGE** key located at the bottom right of the screen. Press this key now.

MACHINING START POSITION

When all of the tool identification numbers are input, the system will ask you for the Machining Start Position. Remember, the machining start position is the position or point that the tool will go through on its way to and from the home position. We will use the same position numbers for the machine start position as you did for the start position of the roughing tool. Type in a **4.00** for the X axis coordinate value of the Machine Start position, press **INPUT**. Type in **4.00** for the Z axis coordinate value of the Machine Start position, press **INPUT**.

CUTTING CONDITIONS

The system asks you for the cutting conditions required for the finish machining process. The cutting conditions consist of information that the control will need to properly finish machine our part. The controller already has the parameters loaded so the cutting conditions will be defaulting on the screen. Use the **CURSOR FORWARD** key to MAX. RPM. Key in **3000** and press **INPUT**.

CUTTING DIRECTION

The control is asking you for the cutting direction. We are finish facing the part. We will finish face the part in an X negative or ↓ down direction. Press the **2** ↓ or down area key located on the keyboard. Press **INPUT**.

CUTTING AREA DEFINITION

The cutting area START POINT cursor needs to be located at the point where the face of the part meets the start of the chamfer (see Figure 11–33).

Press the **CURSOR FORWARD** button until the blinking cursor is in the start point position as seen in Figure 11–33. The control now wants to know in

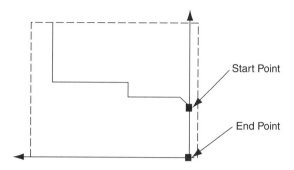

FIGURE 11–33 *The start point needs to be at the point where the face of the the part meets the start of the chamfer.*

which direction the rough material is located from the blinking cursor. You will note that with the cursor at the start point, the control DIVIDING DIRECTION is defaulting ↑. This is correct so press **INPUT** to accept this. The control now prompts you for the END POINT.

If the blinking cursor is not already in the end point position shown in Figure 11–33, move it there using the **CURSOR FORWARD** key. The control DIVIDING DIRECTION is defaulting →. This is correct, so press **INPUT**. Now Press **NEXT PAGE**. The control now needs to know if you are going to use the same tool to cut another area. Remember we only FINISH FACED the part, we now need to finish the rest of the figure. Key in a **1** for yes, and press **INPUT**.

CUTTING CONDITIONS

The system asks you again for the cutting conditions required for the machining process. The cutting conditions are the same as we had for the finish facing operation. You can accept all of this information by simply pressing the **NEXT PAGE** key. Press this key now.

CUTTING DIRECTION

The control is asking you for the cutting direction for turning. We are finish turning in the Z negative direction. We faced the part with ↓ down direction now we will press the **4 ←** from the keyboard for right-hand turning. Press **INPUT**.

CUTTING DIRECTION

The control is asking you for the cutting direction for turning. We will finish turn in the Z negative direction. We rough turned in **4 ←** direction and we

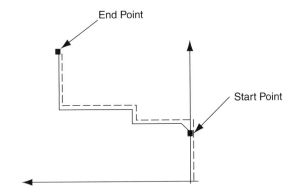

FIGURE 11–34 *The start point needs to be where the face
of the part meets the start of chamfer.*

will finish turn in the same direction. Press the 4 ← from the keyboard for
right-hand turning. Press **INPUT**.

CUTTING AREA DEFINITION

The cutting area START POINT cursor needs to be located at the point where
the face of the part meets the start of chamfer (see Figure 11–34). You will remember this is exactly the same position we used in rough turning.

The control DIVIDING DIRECTION is defaulting →. This is correct so press
INPUT to accept this. The control now prompts you for the END POINT. If
the blinking cursor is not already in the end point position shown in Figure
11–34, move it there. The control DIVIDING DIRECTION is defaulting ↑. This
is correct, so press **INPUT**. Press the **NEXT PAGE** key. After the cutting area
has been designated, the designated cutting area is finish turned. This can be
confirmed by looking at the blue dotted line on the screen. Note the blue
dotted line is the same as the profile of the part. The control now wants to
know if you are going to use the same tool to cut another area. The control
should be defaulting to 0. If not, key in a **0** for no, and press **INPUT**.

We have now completed the definition of machining for our part. The control
is now back to the Definition of Machining screen. Because we have completed this portion press the **NEXT PAGE** key.

NC DATA PREPARATION

We will now be executing the final step in FAPT programming, NC Data
Preparation. The system asks you for a Program Number. Type in a program
number and press **INPUT**. The NC or EIA/ISO code is generated according to

the machine process list, and a tool path is drawn on the graphics display. Press the **START** key located at the bottom of the screen. Carefully watch the tool path. After the tool path ends, press the **END** key. If the tool path looks correct, the operator must **REGISTER** the program to save it. Key in a **5** for NC data preparation to re-run and register your program. Press the **REGISTER** key before pressing the **START** key. After pressing the START key the control will now re-start the NC data preparation and save the program under the operator's program number. The control will automatically register a PROGRAM ERROR in the lower left-hand part of the screen if the operator has used a program number that has been previously registered. If this occurs re-do the NC data preparation using a different program number. Carefully check the tool path.

Our second attempt at programming will be Figure 11–35. As you follow along with this example, please refer to the part print. This will help you to understand what we are attempting to do.

If the machine is on the NC/PC or numerical control side press the FAPT button located above the alphanumeric keyboard (see Figure 11–36).

Your are now on the FAPT side of the control. You will notice that across the bottom of the screen you are given four choices (see Figure 11–37).

Select **FAPT EXEC** to enter the FAPT programming mode. Let's program the sample part in Figure 11–35.

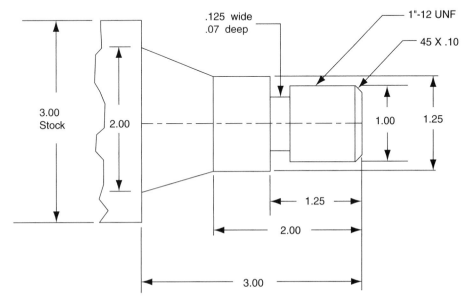

FIGURE 11–35 *Part print for exercise 2.*

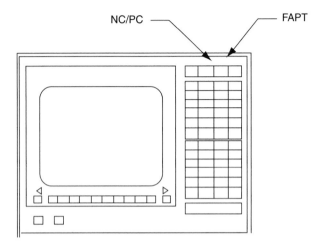

Figure 11–36 *Location of the FAPT button.*

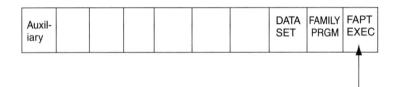

Figure 11–37 *These four choices appear across the bottom of the screen.*

BLANK & PART

Blank & Part deals with the raw piece of stock or billet that we are going to be using. In these two sections (1 & 2) you will be answering questions that deal ONLY with the raw stock.

On the screen are the five steps to programming the part (see Figure 11–38).

These steps are in sequential order, this means we have to do them in the order that they appear on the screen. Type a **1** and press the **Exec** key located under the screen (see Figure 11–39).

MATERIAL TYPE

Sample material number 1 is 1018 plain carbon steel (Figure 11–40). If you look back on the print you will note that this is the type of steel we will be using. Type a **1** for 1018 steel and press the **INPUT** key located on the numeric key pad.

```
=*=FAPT=*=


1. Blank & Part (Drawing & Blank)
2. Blank & Part (Part Figure)
3. Home Position & Index Position
4. Definition of Machining
5. NC Data Preperation

0. End



NO. = ____
```

FIGURE 11–38 *Five steps to programming the part.*

FIGURE 11–39 *Location of the Exec key.*

SURFACE ROUGHNESS DEFAULTS

The next blank and part question deals with the standard surface roughness. Surface roughness is controlled by the feed rate (see Figure 11–41).

Notice that the control is defaulting to number 3. Number three is a standard surface finish of approximately 100 RMS. To accomplish this surface roughness it will use a finish feed rate default of 0.008 feed per revolution. This is acceptable for our part so press **INPUT** to accept this surface roughness default.

DRAWING FORMAT

The drawing format selection tells the control where your workpiece coordinate is located and how the machining will take place, i.e., if the tools will be

```
*** Blank & Part ****    MATERIAL

No.    Sample of Material
1          1018
2          SCM
3          FC
4          AL
5          SUS




Material No.........MN= ____
```

FIGURE **11–40** *Material screen.*

```
*** Blank & Part ***        STANDARD SURFACE ROUGNESS


Surface Roughness  & Cutting Feed Rate

N0.    Kind of Surface Roughness      Cutting Feed Rate
1      Number of Triangle One ................. F1 = 0.012
2      Number of Triangle Two .................F2 = 0.010
3      Number of Triangle Three ..............F3 = 0.008
4      Number of Triangle Four ................F4 = 0.006




Standard Surface  Roughness.....NR  = 3
```

FIGURE **11–41** *Surface roughness screen.*

cutting in a down/left direction, down/right direction, up/left direction, or a up/right direction (see Figure 11–42). The machine is defaulting to the drawing format number 2. This is the one we want, so press **INPUT**.

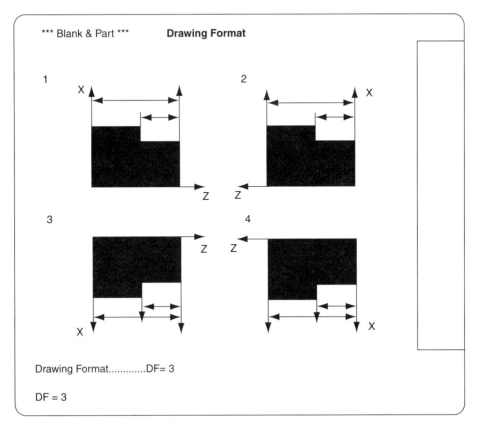

*** Blank & Part ***   **Drawing Format**

Drawing Format.............DF= 3

DF = 3

Figure 11–42 *Drawing format screen.*

BLANK FIGURE & BASE LINE

In this section we tell the control what the rough stock looks like, its physical size, and whether we are going to face off the end of the stock (see Figure 11–43).

The first question that must be answered is whether the stock is a number 1 solid cylinder, a number 2 hollow cylinder (i.e., pipe), or a number 3 special blank (i.e., forging or casting).

The stock we are using for this simple part is a solid cylinder. Type a **1** in for the BF __ and press **INPUT**.

The next question that needs to be answered deals with the physical size of the blank. The L stands for length. Estimate the length of the rough stock for our part. Make sure to add on what is going to be held in the chuck. Type in **5.0** and press **INPUT**. Next is the D. What is the diameter of the rough stock we will be using? Type in a **3** and press **INPUT**.

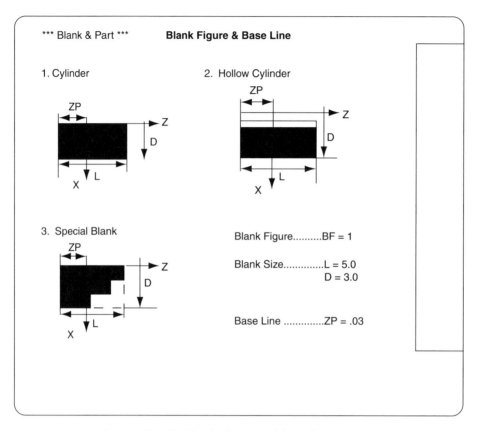

FIGURE 11–43 *Blank figure and base line screen.*

The ZP stands for Z axis positive. It establishes the Z base line or how much are we going to face off of the right end of the stock. Let's face off 0.030 inches. Type in **0.030** and press **INPUT**.

PART FIGURE

At this point in FAPT programming we start to program the actual part shape. Remember that we are not programming the part as it will be machined, but are just describing the part shape. The dotted lines depict the blank or rough stock dimensions that were described in Blank Figure and Base Line (see Figure 11–44).

Again, to describe the part shape we need to use Element Symbols. The element symbols are located on the alphanumeric keypad (see Figure 11–45).

Notice in Figure 11–46 that these keys are used for auxiliary operations such as threading and grooving. We will be using these in exercise 2.

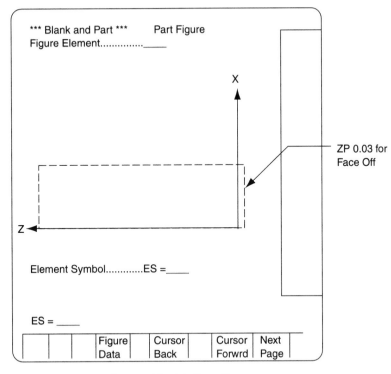

FIGURE **11–44** *Part figure screen.*

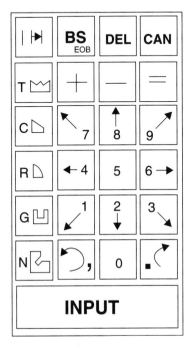

FIGURE **11–45** *Alphanumeric keypad.*

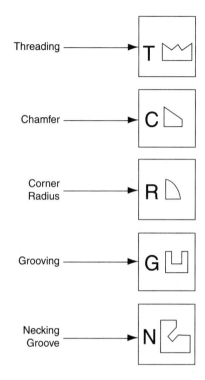

Figure 11–46 *Auxiliary operations keys.*

DESCRIBING THE PART FIGURE

Looking back at the simple part print, we need to describe our part starting at the part origin, X0.0 Z0.0. Starting at the origin, we need to press the **Up 8 Arrow** ↑ key in number **8** (see Figure 11–47) and press **INPUT**. The control will now ask for surface roughness. The surface roughness shows a default value on the screen. This is allowable so press **INPUT**. The starting point in X is 0.0. Type in a **0.0** for the SDX. Press **INPUT**. SZ is the Z value of the start point. Type in **0.0** and press **INPUT**. The part exists to the left of the Z0.0. The control should be defaulting to 1. Press **INPUT.** The control will now ask us for the X End Point or DX (see Figure 11–48). Type in **1.00** for the 1-inch diameter and press **INPUT**. The right end of the part should now appear on the screen in red. The control now prompts us for the next element. Our next element will be the Chamfer. Press the **C** ◁ **chamfer** key and **INPUT**. The control now wants to know how big the chamfer is. Type in **.100** and press **INPUT**. The next element will be the thread. Press the auxiliary function **thread** key ⊤ ⌣. Press **INPUT**. The next page brings a series of questions. The first question the control asks is ON WHICH ELEMENT? We have not described the element where the thread will reside yet. We will describe that as the next element. Answer the question by typing a **1** for NEXT ELEMENT and

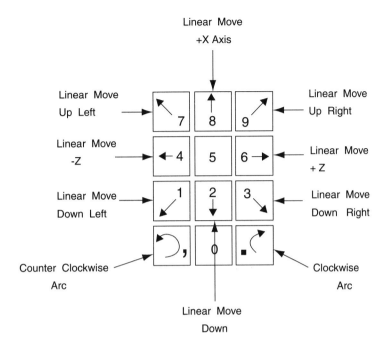

FIGURE 11–47 *Direction keys.*

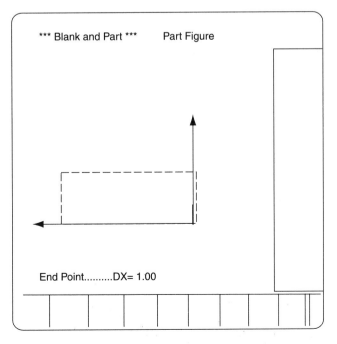

FIGURE 11–48 *DX entry.*

press **INPUT**. The second question deals with length. What is the length of the thread? If we look back at Figure 11–35, the length of the thread is 1.125 inches. Type in **1.125** and press **INPUT**. The third question deals with the lead of the thread. To calculate the lead of the thread, divide 1 by the number of threads per inch. Type in the lead of the thread now and press **INPUT**. The fourth question asks you if there is multiple entrance on the thread, i.e., is it a single lead thread? The control defaults to **1** for a single lead thread. This is correct, press **INPUT**. The last question, thread height, the control figures out on its own. You should see the thread height figures appearing on the screen. Press **INPUT** to accept these values.

The next element will be a linear move to the left ←. This is the diameter element where the thread will reside. Select the **4** key now and press **INPUT**. The control will now ask for surface roughness. The surface roughness is defaulting on the screen. This is allowable so press **INPUT**. The control is now asking for the END POINT Z. The end point of this element is **1.125** inches. Which is 1.25 minus the .125 groove (see Figure 11–35). Type **1.125** and press **INPUT**. Notice that the linear element and the thread element now appearing on the screen.

The next element we need to put in is the Groove. Press the **Grooving** button located on the keypad (see Figure 11–49) and press **INPUT**.

The next question the control asks is ON WHICH ELEMENT? We have not described this element yet. Remember we stopped short of the groove by only going to 1.125 on the last element. The groove will be on the Next Element.

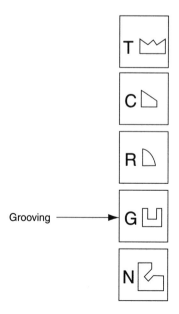

FIGURE 11–49 *Location of grooving button.*

Answer the question by typing a **1** and press **INPUT**. The grooving direction is down. Press the **2** ↓ arrow. Press **INPUT**. The width of the groove is .125. Type in **.125** and press **INPUT**. The grooving depth is .07. Type in **.07** and press **INPUT**. When we talk about depth it is the depth of the groove on one side only. The last four questions deal with a groove shape other than a square groove. We don't need to be concerned with these. Press the **NEXT PAGE** button. The next element will be a linear element to the left ←. This is the element where the groove will reside. Select the **4** key now and press **INPUT**. The control will now ask for surface roughness. The surface roughness defaults on the screen. This is allowable so press **INPUT**. The control is now asking for the END POINT Z. The end point of this element is 1.25. Type **1.25** and press **INPUT**. The groove should now appear on the screen. The next key we need to press is the **Up Arrow** ↑ **key 8.** Press **INPUT**. The surface roughness is defaulting on the screen. This is allowable so press **INPUT**. The control will now ask us for the X End Point or DX . Type in **1.25** for the 1.25-inch diameter, press **INPUT**. The screen should now show our new element (see Figure 11–50).

The next element will be a linear element to the left ←. Select this key now. Now press **INPUT**. Remember, if at any time you press the wrong key, you can erase that by pressing the **CAN** or **Cancel** key which is located on the keypad. Accept the surface roughness default by pressing **INPUT**. The control is now asking for the END POINT Z. The end point of this element is 2.00. Type **2.00** and press **INPUT**. The screen should now show our new element (see Figure 11–51).

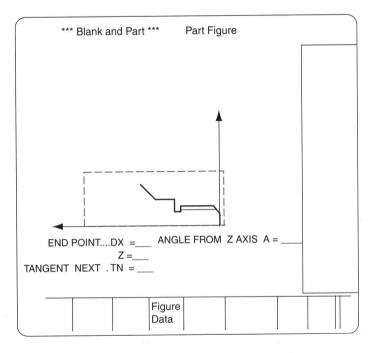

FIGURE 11–50 *The screen shows the new element.*

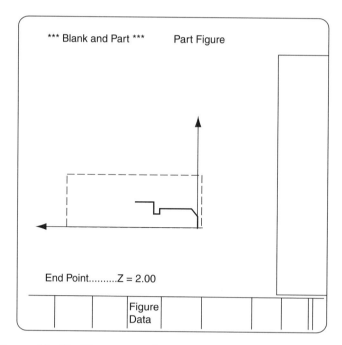

FIGURE **11–51** *The screen after entering the end point Z value.*

The next element we need to press is the **Up Arrow** ↖ **key 7.** Press **INPUT** to accept this element. Accept the surface roughness default by pressing **INPUT**. The control will now ask us for the X End Point or DX. Type in **2.00** for the X axis end point of the angled line, press INPUT. Type in **3.00** for the Z axis end point of the angled line, press INPUT. The rest of the questions would be used only if we didn't know the end points of the line. Skip these questions by pressing the **NEXT PAGE** key. The screen should now show our new element (see Figure 11–52).

To create the next element press the **Up Arrow** ↑ **key 8.** Now press **INPUT**. The surface roughness is defaulting on the screen. This is allowable so press **INPUT**. The control will now ask us for the X End Point or DX. Type in **3.00** for the final line up to the X axis stock boundary. Now press **INPUT**. The screen should now show our new element. Press the NEXT PAGE button to complete the Part Figure sequence. We have now completed the Part Figure, step 2 in Blank & Part (Part Figure). We will now move on to step 3. The shape that was already input will appear again on the screen and the following menu will be displayed on the lower portion of the screen.

FIGURE

1. NEW

2. CORRECT

3. CORRECT (ERASE GRAPHIC)

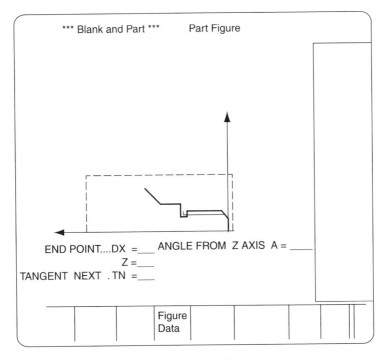

FIGURE 11–52 *Screen asking for X axis end point and Z axis end point.*

Because there are no corrections needed and you want to proceed, press the **NEXT PAGE** button (see Figure 11–53).

HOME POSITION & INDEX POSITION

The cursor should now be blinking number 3 for Home Position & Index Position. Press **INPUT** to accept this default.

The system is now asking you for the home position of the Machine (see Figure 11–54).

HOME POSITION

DXH = Diameter value from the spindle center

ZH = Distance from the Z origin (workpiece face)

Type in **4.00** for the DXH and press **INPUT**. Type in **4.00** for the ZH and press **INPUT**.

The index position is the actual position where the tool turret will index during tool changes (see Figure 11–55).

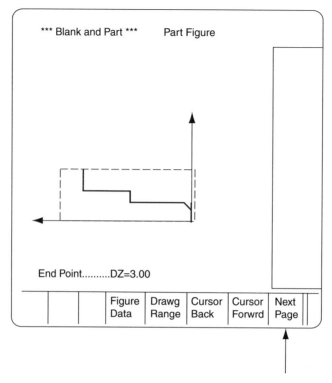

Figure 11–53 *Location of the Next Page button on the screen.*

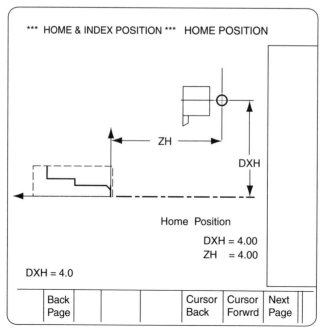

Figure 11–54 *Screen acking for the home position of the machine.*

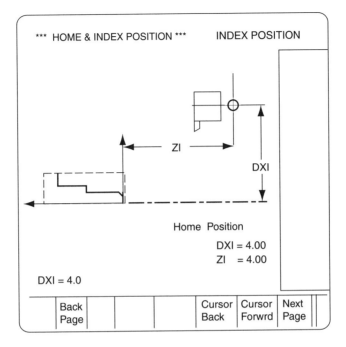

FIGURE 11–55 *Index position screen.*

Type in **4.00** for the DXI and press **INPUT**. Type in **4.00** for the ZI and press **INPUT**.

After you have input these values, press the **NEXT PAGE** button.

DEFINITION OF MACHINING

The screen will change to the menu which you will use to define how you want the part to be machined (see Figure 11–56).

The system is asking you what machining operation you want to perform first. We want to face off the part and rough turn the outer figure. We need to answer the question PROC. 01 = ___, with a number 3 for ROUGHING OF OUTER FIGURE. Type a **3** and press **INPUT**.

The system now needs to know some TOOL DATA information. The screen should now be prompting you for TOOL DATA (see Figure 11–57).

The tool I.D. number we will use for roughing the outer figure is TOOL I.D. NO. 11. Type an **11** and press **INPUT**. The TOOL SELECT is the tool turret position that the roughing tool is in. Type in the tool turret number of the roughing tool and press **INPUT**.

The TOOL OFFSET NO. will be the same as the TOOL SELECT number. Example: Tool 11, offset 11. Type in the offset number and press **INPUT**. The

*** MACHINING DEFINITION ***   KINDS OF MACHINING

PROC. 01   Roughing of Outer Figure   T0101 X4.000 Z4.00

KINDS OF MACHINING............PROC.01= 3

1. CENTER DRILLING   6. SEMI FINISHING OF INNER FIG.
2. DRILLING   7. FINISHING OF OUTER FIGURE
3. ROUGHING OF OUTER FIGURE   8. FINISHING OF INNER FIGURE
4. ROUGHING OF INNER FIGURE   9. GROOVING OR NECKING
5. SEMI - FINISHING OF OUTER F.   10. THREADING

PROC. 01 = 3

| | | | | | | | CURSOR BACK | CURSOR FORWRD | NEXT PAGE |

FIGURE 11–56 *Machining definition screen.*

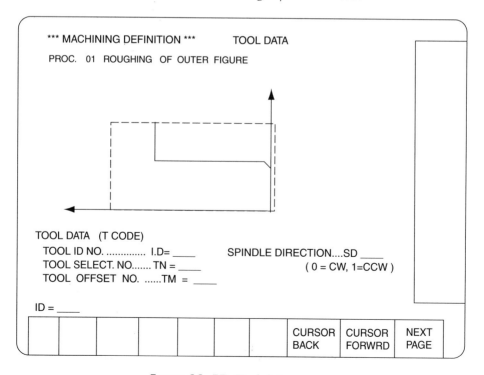

*** MACHINING DEFINITION ***   TOOL DATA
PROC. 01 ROUGHING OF OUTER FIGURE

TOOL DATA (T CODE)
   TOOL ID NO. .... I.D= ____   SPINDLE DIRECTION....SD ____
   TOOL SELECT. NO....... TN = ____   (0 = CW, 1=CCW)
   TOOL OFFSET NO. ......TM = ____

ID = ____

| | | | | | | | CURSOR BACK | CURSOR FORWRD | NEXT PAGE |

FIGURE 11–57 *Tool data screen.*

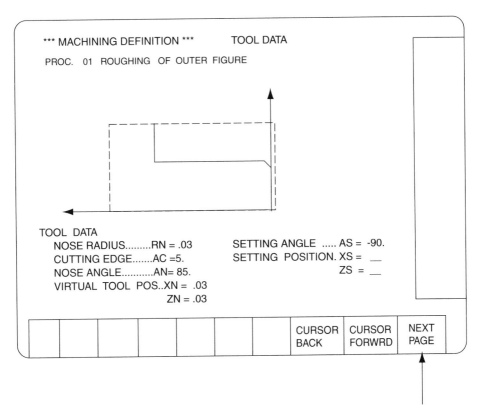

FIGURE 11–58 *To accept the tool data information press the Next Page key.*

spindle direction is clockwise. Type in a **0** for the SD or spindle direction and press **INPUT.** When all of the tool numbers and spindle direction have been input the screen will display the tool data information that was preset in the tool file (see Figure 11–58).

This information can be changed by moving the cursor to the area you wish to change. All of this information can be accepted by pressing the **NEXT PAGE** key located at the bottom right of the screen. Press this key now.

MACHINING START POSITION

The machining start position is the position or point that the tool will go through on its way to and from the home position. Type a **4.00** for the X-axis coordinate value of the Machine Start position and press **INPUT**. Type **4.00** for the Z-axis coordinate value of the Machine Start position (see Figure 11–59) and press **INPUT**.

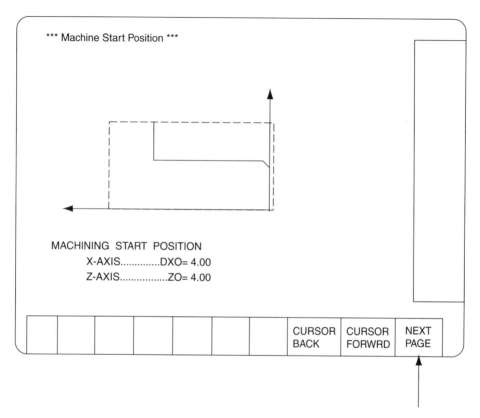

FIGURE 11–59 *Machine start position screen.*

CUTTING CONDITIONS

The system asks you for the cutting conditions required for the machining process. The cutting conditions consist of information that the control will need to properly machine our part. We will change the depth of cut. Press the **CURSOR FORWARD** key until the depth of cut is highlighted. Key in **.05** and press **INPUT** for the depth of cut. We will now need to input the MAX. RPM. Cursor Forward to MAX. RPM. Key in **3000** and press **INPUT**.

CUTTING DIRECTION

The control is asking you for the cutting direction. We are rough facing the part. Press the **2** ↓ or down area key (see Figure 11–60). Press **INPUT**.

CUTTING AREA DEFINITION

The cutting area definition defines the area that you want to machine in this particular operation. The area is defined by the cursors (■) as displayed along

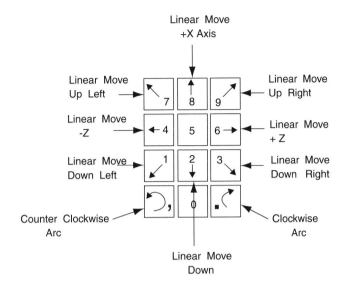

FIGURE 11–60 *Press the 2 or down key.*

the part figure. The cursor is moved using the **CURSOR FORWARD/ CURSOR BACK** keys located at the bottom of the screen. The cutting area START POINT cursor needs to be located at the point where the face of the part meets the chamfer start (see Figure 11–61).

Press the **CURSOR FORWARD** button until the blinking cursor is in the start point position as seen in the Figure 11–61. The control now wants to know in which direction the rough material is located from the blinking cursor. You will note that with the cursor at the start point, the control DIVIDING DI-RECTION is defaulting ↑. This is correct so press **INPUT** to accept this. The control now prompt you for the END POINT.

If the blinking cursor is not already in the end point position shown in Figure 11–61, move it there using the **CURSOR FORWARD** key. The control

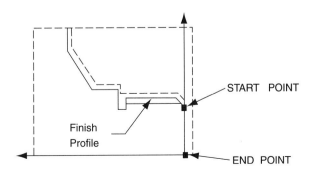

FIGURE 11–61 *Cutting area definition showing start and end points.*

DIVIDING DIRECTION is defaulting →. This is correct, so press **INPUT** to accept this. Press **NEXT PAGE**. After the cutting area has been designated, the designated cutting area is faced off.

The control now wants to know if you are going to use the same tool to cut another area. Remember we only ROUGH FACED the part we need to rough the rest of the figure. Key in a **1** for yes, and press **INPUT**.

CUTTING CONDITIONS

The system asks you again for the Cutting Conditions required for the machining process. The cutting conditions are the same as we had for the rough facing operation. You can accept all of this information by simply pressing the **NEXT PAGE** key. Press this key now.

CUTTING DIRECTION

The control is asking you for the cutting direction for rough turning. We are turning in the Z negative direction. Press the **4 ←** from the keyboard for right-hand turning. Press **INPUT**.

CUTTING AREA DEFINITION

The cutting area definition defines the area that you want to machine in the rough turning operation. The area is defined by the cursors (■) as displayed along the part figure. The cutting area START POINT cursor needs to be located at the point where the face of the part meets the chamfer (see Figure 11–62). Move the cursor there now.

The control now wants to know which direction the rough material is located from the blinking cursor. You will note that with the cursor at the start point,

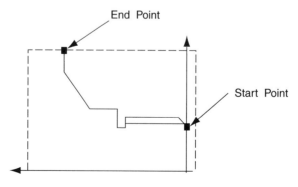

FIGURE **11–62** *The start point needs to be where the face of the part meets the chamfer.*

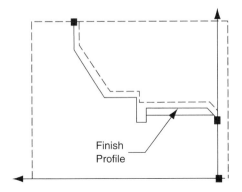

FIGURE 11–63 *The blue dotted line closely follows the part profile.*

the control DIVIDING DIRECTION is defaulting →. This is correct so press **INPUT** to accept this. The control now prompt you for the END POINT. If the blinking cursor is not already in the end point position shown in Figure 11–62, move it there using the **CURSOR FORWARD** key. The control DIVIDING DIRECTION is defaulting ↑. This is correct, so press **INPUT** to accept this. Press **NEXT PAGE**. After the cutting area has been designated, the designated cutting area is rough turned. Notice that the blue dotted line closely follows the profile of the part, leaving just a small gap for finishing (Figure 11–63). The control now wants to know if you are going to use the same tool to cut another area. Lets finish turn the part with a finishing tool. The control should be defaulting to "0." If not key in a **0** for no, and press **INPUT**.

DEFINITION OF MACHINING

The screen will change to the menu that you will use to define your next operation. The next operation will be FINISHING OF OUTER FIGURE, number 7. Type a **7** and press **INPUT** (see Figure 11–64).

The screen should now be prompting you for TOOL DATA.

TOOL DATA

The tool I.D. number we will use for finishing the outer figure is TOOL I.D. NO. 114.

Type a **114** and press **INPUT**. Tool 114 is a 35-degree diamond finishing tool. The TOOL SELECT is the tool turret position that the finishing tool is in. Type in the tool turret number of the finishing tool and press **INPUT**.

The TOOL OFFSET NO. will be the same as the TOOL SELECT number. Type in the offset number and press **INPUT**. Type in a **0** for the SD or spindle

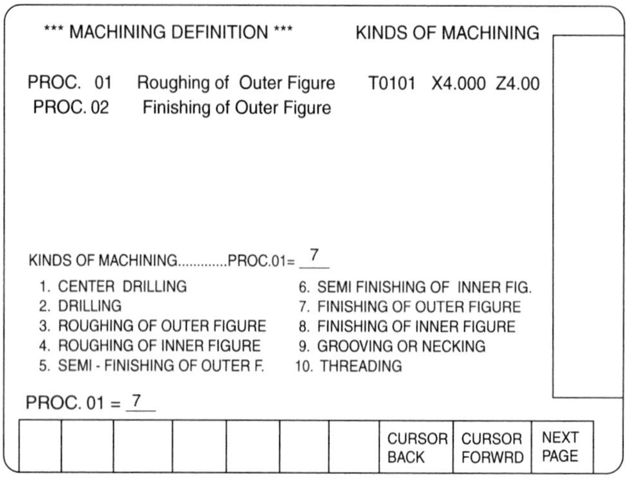

FIGURE 11–64 *Kinds of machining menu.*

direction and press **INPUT**. All of the tool data information can be accepted by pressing the **NEXT PAGE** key located at the bottom right of the screen. Press this key now.

MACHINING START POSITION

Type in a **4.00** for the X axis coordinate value of the Machine Start position, press **INPUT**. Type in **4.00** for the Z axis coordinate value of the Machine Start position, press **INPUT**.

CUTTING CONDITIONS

The system asks you for the cutting conditions required for the finish machining process. The information is based on the material we selected. The controller already has the parameters loaded so the cutting conditions are defaulting on the screen. Use the **CURSOR FORWARD** key to MAX. RPM. Key in **3000** and press **INPUT**.

CUTTING DIRECTION

The control is asking you for the cutting direction. We are finish facing the part. Press the **2** ↓ or down area key located on the keyboard. Press **INPUT**.

CUTTING AREA DEFINITION

The cutting area START POINT cursor needs to be located at the point where the face of the part meets the chamfer (see Figure 11–65).

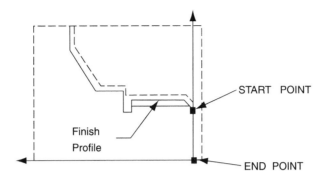

FIGURE 11–65 *The start point and end point must be defined.*

Press the **CURSOR FORWARD** button until the blinking cursor is in the start point position as seen in Figure 11–65. The control now wants to know in which direction the rough material is located from the blinking cursor. You will note that with the cursor at the start point, the control DIVIDING DIRECTION is defaulting ↑. This is correct so press **INPUT** to accept this. If the blinking cursor is not already in the end point position shown in Figure 11–65, move it there using the **CURSOR FORWARD** key. The control DIVIDING DIRECTION is defaulting →. This is correct, so press **INPUT** to accept this. Press **NEXT PAGE**. The control now wants to know if you are going to use the same tool to cut another area. Remember we only FINISH FACED the part we need to finish the rest of the figure. Key in a **1** for yes, and press **INPUT**.

CUTTING CONDITIONS

The system asks you again for the cutting conditions required for the machining process. The cutting conditions are the same as we had for the finish facing operation. You can accept all of this information by simply pressing the **NEXT PAGE** key. Press this key now.

CUTTING DIRECTION

The control is asking you for the cutting direction for turning. Now we will Press the **4** ← from the keyboard for right-hand turning. Press **INPUT**.

CUTTING DIRECTION

The control is asking you for the cutting direction for turning. We will finish turn in the Z negative direction. Press the **4** ← from the keyboard for right-hand turning. Press **INPUT**.

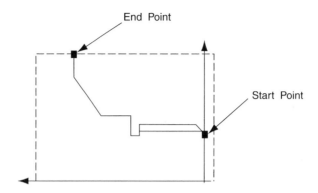

FIGURE **11–66** *Defining the start point ane end point.*

CUTTING AREA DEFINITION

The cutting area definition defines the area that you want to machine in the finish turning operation. The cutting area START POINT cursor needs to be located at the point where the face of the part meets the chamfer (see Figure 11–66).

The control DIVIDING DIRECTION is defaulting →. This is correct so press **INPUT** to accept this. The control now prompt you for the END POINT. If the blinking cursor is not already in the end point position shown in Figure 11–66, move it there using the **CURSOR FORWARD** key. The control DIVIDING DIRECTION is defaulting ↑. This is correct, so press **INPUT**. Press the **NEXT PAGE** key. After the cutting area has been designated, the designated cutting area is finish turned. The control should be defaulting to 0. If not, key in a **0** for no, and press **INPUT**.

GROOVING

The screen will change to the menu which you will use to define your next operation. The next operation will be Grooving, number 9. Type a **9** and press **INPUT**.

The system now needs to know some TOOL DATA information. The screen should now be prompting you for TOOL DATA

TOOL DATA

The tool I.D. number we will use for grooving is TOOL I.D. NO. 690. Type **690** and press **INPUT**. The Tool is a .125 wide carbide grooving tool. Type in the tool turret number of the grooving tool and press **INPUT**. Type in a **0** for the SD or spindle direction and press **INPUT**. Press the **NEXT PAGE** key now.

MACHINING START POSITION

Type in a **4.00** for the X axis coordinate value of the Machine Start position, press **INPUT**. Type in **4.00** for the Z axis coordinate value of the Machine Start position, press **INPUT**.

CUTTING CONDITIONS

The controller already has the parameters loaded so the cutting conditions will be defaulting on the screen. Use the **CURSOR FORWARD** key to MAX. RPM. Key in **3000** and press **INPUT**.

CUTTING AREA DEFINITION

The cutting area START POINT cursor needs to be located at the leading edge of the groove (see Figure 11–67).

Press the **CURSOR FORWARD** button until the blinking cursor is in the start point position as seen in Figure 11–67. The control now wants to know in which direction the rough material is located from the blinking cursor. You will note that with the cursor at the start point, the control DIVIDING DIRECTION is defaulting ↑. This is correct so press **INPUT** to accept this. The control now prompt you for the END POINT.

If the blinking cursor is not already in the end point position shown in Figure 11–67, move it there using the **CURSOR FORWARD** key. The control DIVIDING DIRECTION is defaulting ↑. This is correct, so press **INPUT** to accept this. Now press the **NEXT PAGE** key. The control now wants to know if you are going to use the same tool to cut another area. We have only one groove. Press **INPUT** to accept the default **0**.

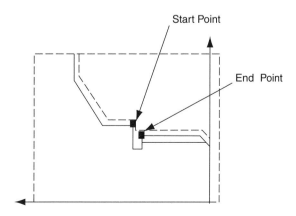

FIGURE 11–67 *The start point needs to be at the leading edge of the groove.*

CUTTING CONDITIONS

You can accept all of this information by simply pressing the **NEXT PAGE** key. Press this key now.

THREADING

The screen will change to the menu which you will use to define your next operation. The next operation will be Threading, number 10. Type a **10** and press **INPUT**.

The system now needs to know some TOOL DATA information. The screen should now be prompting you for TOOL DATA

TOOL DATA

The tool I.D. number we will use for threading is TOOL I.D. NO. 770.

Type **770** and press **INPUT**. The Tool is a .125 wide, 60-degree carbide threading tool. The TOOL SELECT is the tool turret position that the finishing tool is in. Type in the tool turret number of the threading tool and press **INPUT**. Type in a **0** for the SD or spindle direction and press **INPUT**. All of the information can be accepted by pressing the **NEXT PAGE** key.

MACHINE START POSITION

Type in a **4.00** for the X axis coordinate value of the Machine Start position, press **INPUT**. Type in **4.00** for the Z axis coordinate value of the Machine Start position, press **INPUT**.

CUTTING CONDITIONS

The system asks you for the cutting conditions required for the threading process. The cutting conditions consist of information that the control will need to properly thread our part. The information is based on the material we selected. The controller already has the parameters loaded so the cutting conditions will be defaulting on the screen. Use the **CURSOR FORWARD** key to MAX. RPM. Key in **3000** and press **INPUT**.

NC DATA PREPARATION

We will now be executing the final step in FAPT programming. If you are not already in NC Data preparation type in a **5** and press **INPUT**. Type in a program number and press **INPUT**. Press the **START** key located at the bottom of the screen. If the tool path looks correct, the operator must press the register key **before** pressing the START key. This will generate the code and save

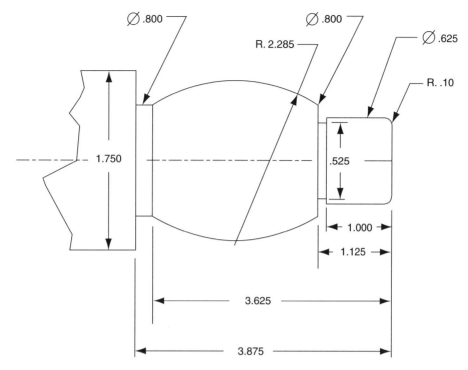

FIGURE 11–68 *Part print for exercise 3.*

the program. After pressing the **START** key the control will now re-start the NC/Data preparation on the screen.

Our third attempt at programming will be Figure 11–68. As you follow along with this example, please refer back to the print. This will help you to understand what we are attempting to do.

Select **FAPT EXEC** to start programming the sample part in Figure 11–68.

BLANK & PART

Type a **1** and press the **Exec** key.

MATERIAL TYPE

Type a **1** for 1018 steel and press the **INPUT** key.

SURFACE ROUGHNESS DEFAULTS

We will use a finish feed rate default of 0.008 inch per revolution. This is acceptable for our part finish requirements so press **INPUT** to accept this surface roughness default.

DRAWING FORMAT

The machine is defaulting to the drawing format number 2. This is the one we want, so press **INPUT**.

BLANK FIGURE & BASE LINE

The stock we are using for this simple part is a solid cylinder. Type a **1** in for the BF __ and press **INPUT**.

Estimate the length of the rough stock for our part. Make sure to add on what is going to be held in the chuck. Type in **7.0** and press **INPUT**. Next is the diameter of the rough stock. Type in a **1.75** and press **INPUT**.

Let's face off 0.030 inches. Type in **0.030** and press **INPUT**.

PART FIGURE

Starting at the origin, the element symbol we need to press is the **Up Arrow** ↑ key number **8**. Select this key now and press **INPUT**. The surface roughness is defaulting to 3 on the screen. This is allowable so press **INPUT**. The starting point in X is 0.0. Type in a **0.0** for the SDX. Press **INPUT**. SZ is the Z value of the start point. Type in **0.0** and press **INPUT**. The part exists to the left of the Z0.0. The control should be defaulting to 1. Press **INPUT**. Our first turned diameter is .625 inch. Type in **.625** for the 1-inch diameter and press **INPUT**. Remember, we don't have to consider the corner radius yet. We will let the control figure that out on the next figure element. The right end of the part should now appear on the screen in red. The control now prompts us for the next element. Our next element will be the corner radius. Press the R **corner radius** key and press **INPUT**. The control now wants to know how big the corner radius is. Type in **.100** and press **INPUT**. The next element will be a linear move to the left ←. Select the **4** key. Now press **INPUT**. The surface roughness is defaulting to 3 on the screen. This is allowable so press INPUT. The control is now asking for the END POINT Z. The end point of this element is 1.00. Which is 1.125 minus the .125 groove. Type **1.00** and press **INPUT**.

The next element we need to put in is the groove. Press the Grooving button located on the keypad. The control now prompts us for surface roughness. Accept the default by pressing **INPUT**. The next question the control asks is: ON WHICH ELEMENT? We haven't described the element yet. Remember, we stopped short of the groove by only going to 1.000 on the last element. The groove will be on the Next Element. Answer the Question by typing a **1** and press **INPUT**. The grooving direction is down. Press the ↓ arrow. Press **INPUT**. The width of the groove is .125. Type in **.125** and press **INPUT**. The grooving depth is .05. Type in **.05** and press **INPUT**. Press **NEXT PAGE** button located at the bottom of the screen. The next element will be a linear element to the left ←. This is the element where the groove will reside. Select

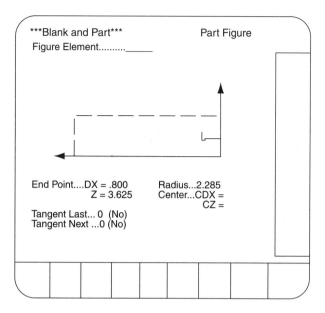

Blank and Part   Part Figure
  Figure Element..........____

End Point....DX = .800   Radius...2.285
              Z = 3.625   Center...CDX =
                                             CZ =
Tangent Last... 0 (No)
Tangent Next ...0 (No)

FIGURE 11–69 *The screen after entering the radius value and selecting the CCW arc key.*

the key **4** now and press **INPUT**. The control will now ask for surface roughness. The surface roughness is defaulting on the screen. This is allowable so press **INPUT**. The control is now asking for the END POINT Z. The end point of this element is 1.125. Type in **1.125** and press **INPUT**. The groove should now appear on the screen.

Next element we need to press is the **8 Up Arrow** ↑ key. Now press **INPUT**. The surface roughness is defaulting on the screen. This is allowable so press **INPUT**. The control will now ask us for the X End Point or DX . Type in **.800** for the .800 inch diameter, press **INPUT**. The screen should now show our new element.

The next element will be the 2.285 radius. Select the [⤴] **CCW** arc key, press **INPUT**. Accept the surface roughness default. The next screen should resemble Figure 11-69.

The first piece of information the control needs are the arc end points. DX is X axis end point. Type in **.800** and press **INPUT**. The Z arc end point is 3.625. Type in **3.625** and press **INPUT**. Tangent Last is the next question. This arc is not tangent to the last element or the next element. Type in a **0** for both of the tangency questions. What is the radius? Type in **2.285** for the radius and press **INPUT**. The control now wants to know where the arc center is. In our case we have input three pieces of information about the radius. The end point on the X axis, the end point on the Z axis, and the radius. This is all the information the control needs to have to generate this arc. Press the **NEXT PAGE** key.

The next element will be a linear element to the left ←. Select this key now. Now press **INPUT**. The surface roughness is defaulting to 3 on the screen. This is allowable so press **INPUT**. The control is now asking for the END POINT Z. The end point of this element is 3.875. Type **3.875** and press **INPUT**. The screen should now show our new element.

The next element we need to press is the **8 Up Arrow** ↑ key. Now press **INPUT**. The surface roughness is defaulting to 3 on the screen. This is allowable so press INPUT. The control will now ask us for the X End Point or DX . Type in **1.75** for the 1.75-inch diameter, press **INPUT**. The screen should now show our new element.

We have now completed the part figure, step 2 in Blank & Part (Part Figure). We will now move on to step 3 by pressing the **NEXT PAGE** key. The shape already input on the screen will appear again and the following menu will be displayed on the lower portion of the screen.

> **FIGURE**
> **1. NEW**
> **2. CORRECT**
> **3. CORRECT (ERASE GRAPHIC)**

Because there are no corrections needed press the **NEXT PAGE** button at the bottom of the screen.

HOME POSITION & INDEX POSITION

Type in **4.00** for the DXH and press **INPUT**. Type in **4.00** for the ZH and press **INPUT**.

Type in **4.00** for the DXI and press **INPUT**. Type in **4.00** for the ZI and press **INPUT**.

DEFINITION OF MACHINING

The screen will change to the menu that you will use to define how you want the part to be machined. Type a **3** and press **INPUT**. The tool I.D. number we will use for roughing the outer figure is TOOL I.D. NO. 11. Type an **11** and press **INPUT**. Type in the tool turret number of the roughing tool and press **INPUT**. Type in the offset number and press INPUT. Type in a **0** for the SD or spindle direction and press **INPUT**. When all of the tool numbers and spindle direction has been input the screen will display the tool data information that was preset in the tool file. All of this information can be accepted by pressing the **NEXT PAGE** key located at the bottom right of the screen. Press this key now.

MACHINING START POSITION

When all of the tool data is input and complete, the system asks you for the machining start position. The machining start position is the position or point that the tool will go through on its way to and from the home position. It is important that this point lies well above and to the right of the part or rough stock. Type a **4.00** for the X-axis coordinate value of the machine start position and press **INPUT**. Type **4.00** for the Z-axis coordinate value of the machine start position and press **INPUT**.

CUTTING CONDITIONS

The system asks you for the cutting conditions required for the machining process. Carefully check this information. Change only the desired items. We will change the depth of cut. Press the **CURSOR FORWARD** key, located at the bottom of the screen, until the depth of cut is highlighted. Key in **.05** and press **INPUT** for the depth of cut. We will now need to input the MAX. RPM. **CURSOR FORWARD** to MAX. RPM. Key in **3000** and press **INPUT**.

CUTTING DIRECTION

The control is asking you for the cutting direction. We are rough facing the part. We rough face the part in an X negative or ↓ down direction. Press the **2** ↓ or down area key located on the keyboard. Press **INPUT**.

CUTTING AREA DEFINITION

The cutting area definition defines the area that you want to machine in this particular operation. The cutting area START POINT cursor needs to be located at the point where the face of the part meets the start of the corner radius.

Press the **CURSOR FORWARD** button until the blinking cursor is in the start point position as seen in the Figure 11–70.

The control now wants to know in which direction the rough material is located from the blinking cursor. The control calls this the DS or DIVIDING DIRECTION. You will note that with the cursor at the start point, the control DIVIDING DIRECTION is defaulting ↑. This is correct so press **INPUT** to accept this. The control now prompt you for the END POINT. If the blinking cursor is not already in the end point position shown in Figure 15–69, move it there using the **CURSOR FORWARD** key. The control DIVIDING DIREC-TION is defaulting →. This is correct, so press **INPUT** to accept this. Press **NEXT PAGE**. After the cutting area has been designated, the designated cutting area is faced off. This can be confirmed by looking at the graphic display. The blue dotted line represents the material that still needs to be cut.

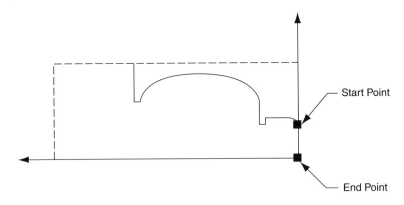

FIGURE 11–70 *Press the* **CURSOR FORWARD** *button until the cursor is in the start point position.*

The control now wants to know if you are going to use the same tool to cut another area. Remember we only ROUGH FACED the part and we need to rough the rest of the figure. Key in a **1** for yes, and press **INPUT**.

CUTTING CONDITIONS

The system asks you again for the cutting conditions required for the machining process. The cutting conditions are the same as we had for the rough facing operation. You can accept all of this information by simply pressing the **NEXT PAGE** key. Press this key now.

CUTTING DIRECTION

The control is asking you for the cutting direction for rough turning. We are turning in the Z negative direction. We faced the part with ↓ down direction now we will Press the **4** ← from the keyboard for right-hand turning. Press **INPUT**.

CUTTING AREA DEFINITION

The cutting area definition defines the area that you want to machine in the rough turning operation. The cutting area START POINT cursor needs to be located at the point where the face of the part meets the corner radius. Move the cursor there now.

The control now wants to know which direction the rough material is located from the blinking cursor. You will note that with the cursor at the start point, the control DIVIDING DIRECTION is defaulting →. This is correct so press **INPUT** to accept this. The control now prompt you for the END POINT. If the

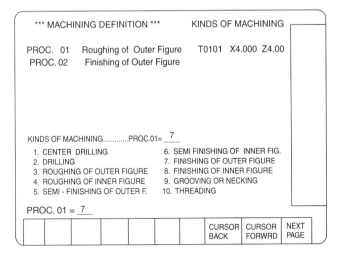

FIGURE 11–71 *Kinds of machining menu.*

blinking cursor is not already in the end point position, move it there using the **CURSOR FORWARD** key. The control DIVIDING DIRECTION is defaulting ↑. This is correct, so press **INPUT** to accept this. Press **NEXT PAGE**. After the cutting area has been designated, the designated cutting area is rough turned. Now we have to finish turn the part with a finishing tool. The control should be defaulting to 0. If not key in a **0** for no, and press **INPUT**.

DEFINITION OF MACHINING

The screen will change to the menu which you will use to define your next operation. The next operation will be FINISHING OF OUTER FIGURE, number 7. Type a **7** and press **INPUT** (see Figure 11–71).

The system now needs to know some TOOL DATA information. The screen should now be prompting you for TOOL DATA

TOOL DATA

The tool I.D. number we will use for finishing the outer figure is TOOL I.D. NO. 114. Type a **114** and press **INPUT**. Tool 114 is a 35-degree diamond finishing tool. Type in the tool turret number of the finishing tool and press **INPUT**.

The TOOL OFFSET NO. will be the same as the TOOL SELECT number. Type in the offset number and press **INPUT**. Type in a **0** for the SD or spindle direction and press **INPUT**. Pressing the **NEXT PAGE** key to move to the next operation.

MACHINING START POSITION

The machining start position is the position or point that the tool will go through on its way to and from the home position. We will use the same position numbers for the machine start position as you did for the start position of the roughing tool. Type in a **4.00** for the X axis coordinate value of the machine start position, press **INPUT**. Type in **4.00** for the Z axis coordinate value of the machine start position, press **INPUT**.

CUTTING CONDITIONS

The controller already has the parameters loaded so the cutting conditions will be defaulting on the screen. Use the **CURSOR FORWARD** key to MAX. RPM. Key in **3000** and press **INPUT**.

CUTTING DIRECTION

The control is asking you for the cutting direction. We are finish facing the part. Press the **2** ↓ or down arrow key located on the keyboard. Press **INPUT**.

CUTTING AREA DEFINITION

The cutting area START POINT cursor needs to be located at the point where the face of the part meets the chamfer (see Figure 11–72).

Press the **CURSOR FORWARD** button until the blinking cursor is in the start point position as seen in Figure 11–72. The control now wants to know in which direction the rough material is located from the blinking cursor. You will note that with the cursor at the start point, the control DIVIDING DIRECTION is defaulting ↑. This is correct so press **INPUT** to accept this. The

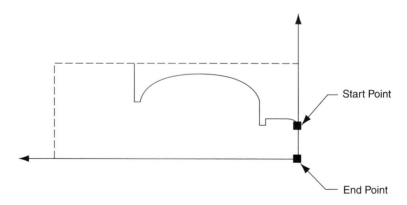

FIGURE 11–72 *Cutting area start point and end point.*

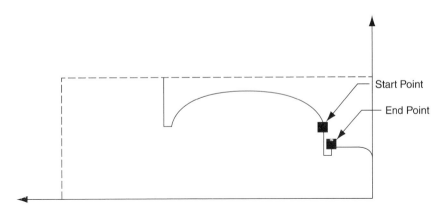

FIGURE 11–73 *The cutting area start point cursor
needs to be at the leading edge of the groove.*

control now prompt you for the END POINT. If the blinking cursor is not already in the end point position shown in Figure 11–73, move it there. The control DIVIDING DIRECTION is defaulting →. This is correct, so press **INPUT** to accept this. Press **NEXT PAGE**. The control now wants to know if you are going to use the same tool to cut another area. Remember we only FINISH FACED the part we need to finish the rest of the figure. Key in a **1** for yes, and press **INPUT**.

CUTTING CONDITIONS

The system asks you again for the cutting conditions required for the machining process. The cutting conditions are the same as we had for the finish facing operation. You can accept all of this information by simply pressing the **NEXT PAGE** key. Press this key now.

CUTTING DIRECTION

The control is asking you for the cutting direction for turning. We are finish turning in the Z negative direction. We faced the part with down direction now we will press the **4** ← from the keyboard for right-hand turning. Press **INPUT**.

CUTTING DIRECTION

The control is asking you for the cutting direction for turning. We will finish turn in the Z negative direction. We rough turned in 4 ← direction and we will finish turn in the same direction. Press the 4 ← from the keyboard for right-hand turning. Press **INPUT**.

CUTTING AREA DEFINITION

The cutting area START POINT cursor needs to be located at the point where the face of the part meets the chamfer. You will remember this is the exact same position we used in rough turning. The control now wants to know which direction the finish material is located from the blinking cursor. This the DIVIDING DIRECTION. The control DIVIDING DIRECTION is defaulting →. This is correct so press **INPUT** to accept this. The control now prompt you for the END POINT. If the blinking cursor is not already in the end point position move it there using the **CURSOR FORWARD** key. The control DIVIDING DIRECTION is defaulting ↑. This is correct, so press **INPUT**. Press the **NEXT PAGE** key. After the cutting area has been designated, the designated cutting area is finish turned. This can be confirmed by looking at the blue dotted line. Note the blue dotted line is the same as the profile of the part. The control now wants to know if you are going to use the same tool to cut another area. The control should be defaulting to 0. If not, key in a **0** for no and press **INPUT**.

GROOVING

The screen will change to the menu which you will use to define your next operation. The next operation will be grooving, number 9. Type a **9** and press **INPUT**.

The system now needs to know some TOOL DATA information. The screen should now be prompting you for TOOL DATA

TOOL DATA

The tool I.D. number we will use for grooving is TOOL I.D. NO. 690.

Type **690** and press **INPUT**. The Tool is a .125 wide carbide grooving tool. Type in the tool turret number of the grooving tool and press **INPUT**. Type in a **0** for the SD or spindle direction and press **INPUT**. When all of the tool numbers and spindle direction has been input the screen will display the tool data information that was preset in the tool file. All of this information can be accepted by pressing the **NEXT PAGE** key. Press this key now.

MACHINING START POSITION

When all of the tool data are input and complete, the system asks you for the machining start position. The machining start position is the position or point that the tool will go through on its way to and from the home position. We will use the same position numbers for the machine start position as you did for the start position of the roughing tool. Type in a **4.00** for the X axis coordinate value of the machine start position, press **INPUT**. Type in **4.00** for the Z axis coordinate value of the machine start position, press **INPUT**.

CUTTING CONDITIONS

The controller already has the parameters loaded so the cutting conditions will be defaulting on the screen. Use the **CURSOR FORWARD** key to MAX. RPM. Key in **3000** and press **INPUT**.

CUTTING AREA DEFINITION

The cutting area START POINT cursor needs to be located at the leading edge of the groove (see Figure 11–73).

Press the **CURSOR FORWARD** button until the blinking cursor is in the start point position as seen in the figure above. The control now wants to know in which direction the rough material is located from the blinking cursor. You will note that with the cursor at the start point, the control DIVIDING DIRECTION is defaulting ↑. This is correct so press **INPUT** to accept this. The control now prompt you for the END POINT.

If the blinking cursor is not already in the end point position shown in Figure 11–73, move it there using the **CURSOR FORWARD** key. The control DIVIDING DIRECTION is defaulting ↑. This is correct, so press **INPUT** to accept this. Press **NEXT PAGE**. The control now wants to know if you are going to use the same tool to cut another area. We only have 1 groove. INPUT the default **0.**

CUTTING CONDITIONS

The system asks you again for the cutting conditions required for the machining process. The cutting conditions are acceptable for the operation. You can accept all of this information by simply pressing the **NEXT PAGE** key. Press this key now.

MACHINE START POSITION

Type in a **4.00** for the X axis coordinate value of the machine start position, press **INPUT**. Type in **4.00** for the Z axis coordinate value of the machine start position, press **INPUT**.

NC DATA PREPARATION

We will now be executing the final step in FAPT programming. Type in a **5** and press **INPUT**. Type in a program number and press **INPUT**. Press the **START** key located at the bottom of the screen. If the tool path looks correct, the operator must press the register key **before** pressing the **START** key. This will generate the code and save the program. After pressing the **START** key the control will now re-start the NC/Data preparation on the screen.

DAINICHI FAPT PROGRAMMING
PROJECT 1

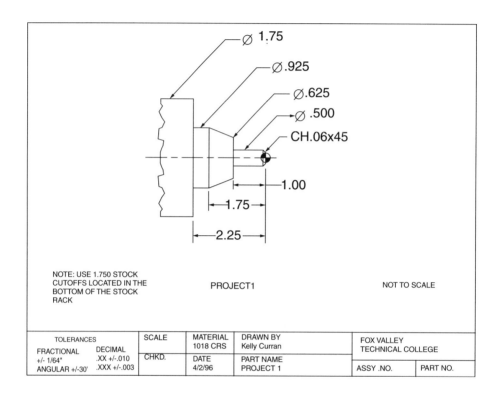

Ø 1.75

Ø.925

Ø.625

Ø .500

CH.06x45

1.00

1.75

2.25

NOTE: USE 1.750 STOCK
CUTOFFS LOCATED IN THE
BOTTOM OF THE STOCK
RACK

PROJECT1

NOT TO SCALE

TOLERANCES		SCALE	MATERIAL 1018 CRS	DRAWN BY Kelly Curran	FOX VALLEY TECHNICAL COLLEGE	
FRACTIONAL +/- 1/64" ANGULAR +/-30'	DECIMAL .XX +/-.010 .XXX +/-.003	CHKD.	DATE 4/2/96	PART NAME PROJECT 1		
					ASSY .NO.	PART NO.

DAINICHI FAPT
PROJECT 2

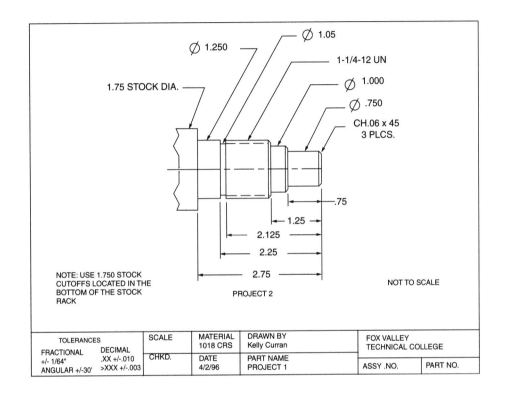

Ø 1.05

1-1/4-12 UN

Ø 1.250

Ø 1.000

1.75 STOCK DIA.

Ø .750

CH.06 x 45
3 PLCS.

.75

1.25

2.125

2.25

2.75

NOTE: USE 1.750 STOCK
CUTOFFS LOCATED IN THE
BOTTOM OF THE STOCK
RACK

PROJECT 2

NOT TO SCALE

TOLERANCES		SCALE	MATERIAL 1018 CRS	DRAWN BY Kelly Curran	FOX VALLEY TECHNICAL COLLEGE	
FRACTIONAL +/- 1/64" ANGULAR +/-30'	DECIMAL .XX +/-.010 >XXX +/-.003	CHKD.	DATE 4/2/96	PART NAME PROJECT 1	ASSY .NO.	PART NO.

DAINICHI FAPT
PROJECT 3

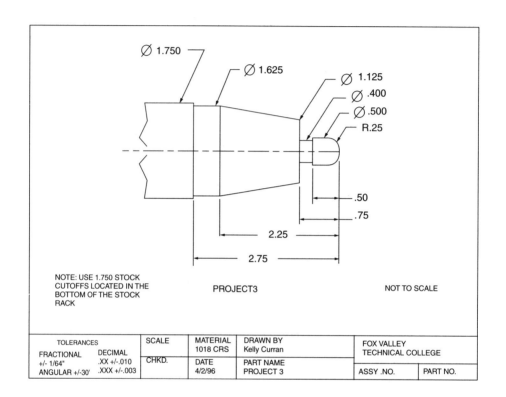

Ø 1.750

Ø 1.625

Ø 1.125

Ø .400

Ø .500

R.25

.50

.75

2.25

2.75

NOTE: USE 1.750 STOCK
CUTOFFS LOCATED IN THE
BOTTOM OF THE STOCK
RACK

PROJECT3

NOT TO SCALE

TOLERANCES		SCALE	MATERIAL 1018 CRS	DRAWN BY Kelly Curran	FOX VALLEY TECHNICAL COLLEGE	
FRACTIONAL +/- 1/64" ANGULAR +/-30'	DECIMAL .XX +/-.010 .XXX +/-.003	CHKD.	DATE 4/2/96	PART NAME PROJECT 3	ASSY .NO.	PART NO.

DAINICHI FAPT PROGRAMMING
PROJECT 4

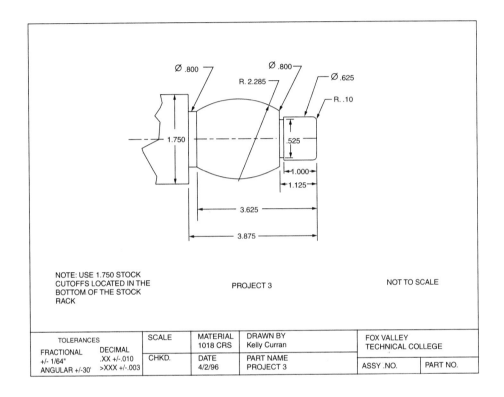

Ø .800

Ø .800

R. 2.285

Ø .625

R. .10

1.750

.525

1.000

1.125

3.625

3.875

NOTE: USE 1.750 STOCK
CUTOFFS LOCATED IN THE
BOTTOM OF THE STOCK
RACK

PROJECT 3

NOT TO SCALE

TOLERANCES		SCALE	MATERIAL 1018 CRS	DRAWN BY Kelly Curran	FOX VALLEY TECHNICAL COLLEGE	
FRACTIONAL +/- 1/64" ANGULAR +/-30'	DECIMAL .XX +/-.010 >XXX +/-.003	CHKD.	DATE 4/2/96	PART NAME PROJECT 3	ASSY .NO.	PART NO.

Chapter 12

ELECTRICAL DISCHARGE MACHINING (EDM)

INTRODUCTION

Electrical discharge machining (EDM) has become a very important method of machining. EDM can produce very accurate, complex parts that in many cases cannot be economically produced by any other method. This chapter will cover the fundamentals of electrical discharge machining and the programming of wire EDM machines.

OBJECTIVES

Upon completion of this chapter, the reader will be able to:

- *Describe the wire-feed EDM process.*
- *Describe three benefits of the wire-feed EDM process.*
- *Explain the basic components of a wire-feed EDM.*
- *Explain terminology such as "dielectric," "UV axis," and "XY axis."*
- *Explain how to set up the workpiece on the machine and align the wire.*
- *Write simple two- and four-axis wire-EDM programs.*

INTRODUCTION TO EDM

EDM was once considered a nontraditional machining method. Today EDM has become a very accepted and widely used technology. For many complex and intricate shapes that are very difficult and time consuming to machine (see Figure 12–1), EDM can be effectively used. EDM at one time was mainly used in tool and die shops to produce dies, but today it is gaining acceptance in production processes as well.

A wire EDM is just like an extremely accurate band saw. Instead of a blade, the wire EDM uses wire and electricity to cut. Wire diameter is usually between .002 and .013 inches, and wire can be made from brass, brass alloys, and molybdenum.

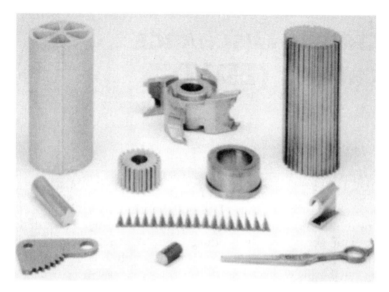

FIGURE 12–1 *Several examples of complex parts that were machined on a wire EDM machine. (Courtesy Fanuc LTD.)*

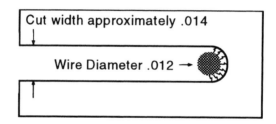

FIGURE 12–2 *Overcut produced by a wire EDM.*

Brass wire is the most commonly used material. The most common size is .008 inches. If we add the normal .001 overcut, the width of cut will be approximately .010 inches. The wire never actually touches the material it is cutting. There is a gap of approximately .001 inch. (See Figure 12–2 for an example.)

Smaller wire diameters can be used to cut smaller radii and very intricate shapes. Here, tungsten or molybdenum wires are used because their high tensile strength and high melting point allow smaller wire diameter.

The wire comes on spools that weigh between 2 and 100 pounds. A spool holds a very long continuous length of wire that permits a machine to run in excess of 500 hours on one spool of wire. The machine has a supply spool and a take-up spool (see Figure 12–3). The new wire is put in the supply position and the wire then travels through guides (see Figure 12–4) and the workpiece to the take-up spool.

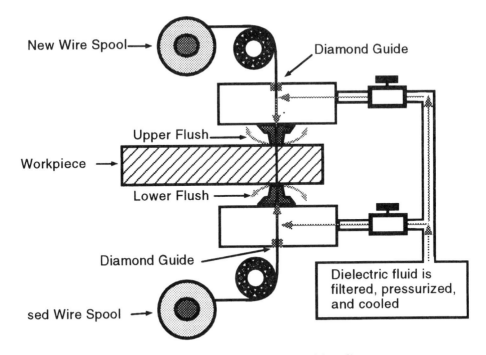

FIGURE 12–3 *A typical wire EDM machine diagram.*
Note the wire guide above and below the workpiece
and the flushing from top and bottom.

Some machines have a box to catch the wire instead of a take-up spool (fewer moving parts). The wire is thus continuously new, which helps assure accurate cutting. Most machines use take-up spools for the used wire. The rate of wire feed can be very slow (1 meter/minute) or quite fast (10 meters/minute).

EDM machines can cut with extreme accuracy; some machines are accurate to within ±.0002. EDM produces very little heat and no machining forces on the material, so it doesn't distort pieces during production. Wire EDMs provide very good surface finishes, which can eliminate the need for further machining. Some EDMs are capable of surface finishes of 15 RMS or less.

EDM is effective on a variety of materials, too. EDM can cut anything that is conductive, including steel, aluminum, super alloys, and even tungsten carbide. When we cut carbide, a non-conductive material, with EDM, we are actually eroding the binder material. The carbide nodules are flushed away. Small finish cuts will improve the surface finish. Skim cuts also improve the surface quality. The workpiece can be hardened before EDM machining. In fact, EDM can cut hardened D2 tool steel 20 to 40 percent faster than it can cut cold rolled steel.

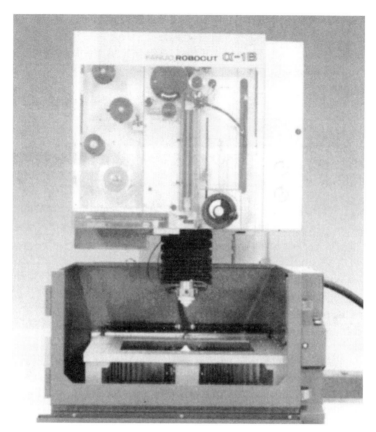

FIGURE **12–4** *Wire EDM machine. Note the upper and lower wire guides and the open-frame table. (Courtesy Fanuc LTD.)*

CUTTING WITH EDM

Wire wear and cutting rate depend on the characteristics of the workpiece material. Speed is measured in square inches cut per hour, and manufacturers rate equipment by cutting speed. Speed of cut is usually rated on 2.25 inch thick, hardened D2 tool steel under ideal cutting conditions. The important characteristics that determine cutting speed include the melting point of the material, the conductivity of the material, and the length and strength of the electrical pulses. Aluminum, for example, has a low melting point and is a good conductor, so it cuts much faster than steel.

EDM cuts by creating sparks between the wire and the workpiece. These rapid, high-energy sparks cause small pieces of the workpiece to melt and vaporize. Fluid is run between the wire and the workpiece. This fluid is dielectric, meaning non-conductive. This fluid serves several purposes, such as shielding the wire from the workpiece. The fluid acts like a resistor until enough voltage

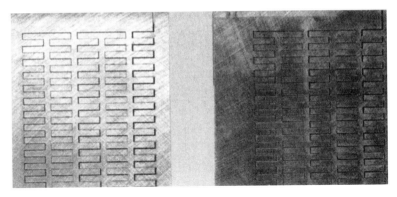

FIGURE 12–5 *AC powered cut (left) and DC (right). Note the corrosion on the part on the right. (Courtesy Fanuc LTD.)*

is applied. With sufficient voltage, the fluid ionizes and a spark melts and vaporizes a small piece of the workpiece material. These pulses occur thousands of times each second along the length of the wire within the workpiece.

The sparks can be DC or AC current. DC current allows a very small current to flow through the dielectric fluid, which can cause electrolytic corrosion (see Figure 12–5). AC power supplies do not have this problem. An AC power supply can help prevent rust formation on iron workpieces. The AC method also helps to substantially reduce minute cracking that can occur during cutting, making dies more durable. Carbide retains its hardness better with the AC method because the binder is prevented from being eluted (washed out or extracted).

After the metal has been melted, the fluid cools the vaporized metal and carries it out of the cut. Good particle removal by the flushing is essential to efficient cutting. The fluid also keeps the workpiece and the wire cool.

A very important characteristic is the flushing of the vaporized material so a new energy column can form. The contaminated fluid runs through a filter to remove the vaporized particles and is then reused. The fluid also runs through a chiller to maintain the temperature, which helps maintain the machine's accuracy.

The dielectric fluid is normally deionized water. Regular water is run through an ion exchanger that removes all the impurities. Regular water is a conductor, deionized water is a good insulator. This is vital because we don't want a short circuit between the wire and the workpiece. The amount of deionization is measured by the specific resistance of the water. The lower the resistance, the faster the cutting. A higher resistance is desirable for materials such as higher density graphites and carbides.

The flow rate is crucial to efficient, accurate cutting. Two valves control the fluid flow. One valve controls the top flow, and the other valve controls the fluid flow from the bottom (see Figure 12–3). Initial adjustment is made by opening the top valve until you get a steady stream of fluid. You then begin opening the bottom valve until the fluid appears to flair out at the top of the workpiece.

The color of sparks also can be used to adjust flow rate. Blue sparks are desirable. Red sparks indicate an insufficient flow rate. Too high a flow rate is also undesirable. It can cause wire deflection, which can cause inaccuracy when cutting tapers.

Some machines now have programmable flushing that can accommodate different conditions for entering, leaving, skim cuts (finish cuts), or even changing workpiece thickness.

The electrical control on the machine maintains a gap of .001 to .002 inches between the wire electrode and the workpiece. The wire never touches the workpiece. If it did, it would short out and the machine would sense it and make a correction.

TYPES OF WIRE EDM MACHINES

There are three main types of wire EDM. The first is a simple two-axis EDM. This machine is like a simple XY table that can make only simple right-angle cuts (see Figure 12–6).

The second type of wire EDM is a simultaneous four-axis EDM (see Figure 12–7). This EDM can cut tapers through the workpiece, but the top and bottom of the cut must be the same shape. For example, if the top of the cut is a square, the bottom shape must also be a square, although it can be a different size.

The third type of EDM is the independent four-axis system. In this type of system the top shape of the cutout can be different than the bottom shape. This is especially useful for extrusion dies that are used to shape material into predefined shapes.

The four-axis wire EDM has a UV axis and an XY axis. The UV axis guides the wire above the workpiece, and the XY axis guides the wire below the workpiece. The UV and the XY axes can be controlled independently to cut tapers on workpieces (see Figures 12–7, 12–8, and 12–9).

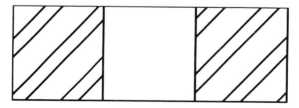

FIGURE 12–6 *A simple XY machine cut. Note that there is no taper in the hole.*

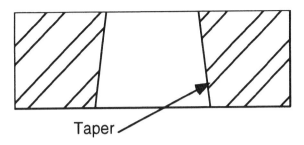

FIGURE 12–7 *A workpiece that required a tapered hole. This would require a four-axis wire EDM.*

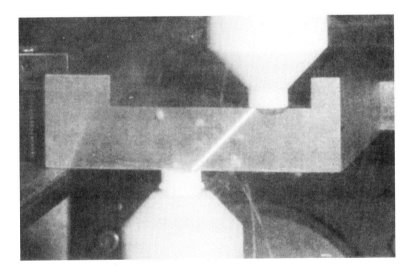

FIGURE 12–8 *The UV wire guide (top) and the XY wire guide (bottom). (Courtesy Fanuc LTD.)*

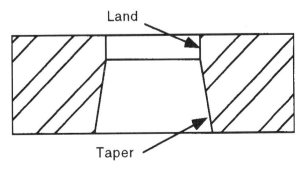

FIGURE 12–9 *A part that requires two cuts. The straight cut through the piece would be made first, and then the taper would be cut.*

PARTS OF THE WIRE-FEED EDM

The main parts of a wire-feed EDM are the bed, saddle, column, UV axis, XY axis, wire-feed system, dielectric fluid system, and machine control.

Wire-feed machines have several servo systems to control various aspects of the machining process. The electrical current level must be accurately controlled, and the feed rate of the axis and the gap between the wire and the actual workpiece must be accurately controlled to approximately .001 to .002 inches. If the wire were to contact the workpiece the wire could be broken or arcing could occur, which would damage the workpiece. The servo control senses the current and adjusts the drive motors to speed up or slow down to retain the proper gap between wire and workpiece. Sometimes called *adaptive control,* this control can also adjust the feed rate to compensate for workpiece thickness and changing cutting conditions (see Figure 12–10).

FUZZY LOGIC CONTROL

A new technology will drastically improve the productivity of EDM machining. Fuzzy logic is already being used on sinking-style EDM machines. By attempting to think more like a human would, fuzzy logic can consider several variables at once and weight their importance depending on the conditions at the time. It then makes adjustments based on that analysis. This is done continuously to assure near-optimal cutting conditions. It would be like having your best operator watching the cutting process, monitoring all the variables, and then, depending on the current conditions, making any adjustments that would improve the cutting speed or finish. In tests, fuzzy logic has outperformed operators consistently.

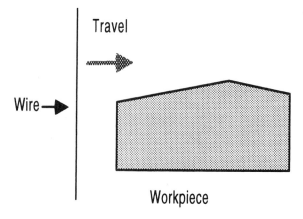

FIGURE 12–10 *How adaptive control can adjust the feed rate for varying workpiece conditions.*

This technology is also employed in many consumer products, such as video cameras. Fuzzy logic stabilizes the picture even if the operator is not steady. Fuzzy logic decides whether the operator intended to move the camera or is just unsteady. To do this, the fuzzy controller considers the distance and speed of any movement. Quick, jerky, short moves are unintended motion, and the fuzzy controller stabilizes the picture. With slow, longer movement, the fuzzy controller assumes the person wants to move. The fuzzy controller constantly looks at the variables and adjusts when necessary. Fuzzy logic is sure to become even more important in industrial control.

IMPORTANT MACHINING CONSIDERATIONS

One of the problems of EDM machining is called *recast.* This very thin region of the metal at the cut is affected by the tremendous heat that the cutting causes. EDM removes metal by generating extreme heat on the metal, which causes it to vaporize and melt away. This extreme heat changes the molecular structure on the surface of the part that is cut. The recast layer is material that has been vaporized and has reattached itself to the remaining material and solidified before it was flushed away. These particles stick to the surface of the cut and are what most would refer to as recast. The thickness is generally .0001 to .0002 inches. These particles can generally be easily removed by bead blasting.

The heat-affected zone below these particles is a thin region that was heated to the melting point and then cooled very rapidly. This region can be very hard and brittle, which makes it very susceptible to surface cracks because it cools more rapidly than the region below. These cracks can, in some cases, cause catastrophic failure in the part. The thin region immediately below this layer cools at a slower rate and thus is annealed.

The effects of heat can be minimized. The wire EDM process works at high current for short on-time cycles, which helps minimize the heat-affected zone. When properly run, a wire EDM can produce a part with a heat-affected zone as thin as .00004 inch. Skim cuts (finish cuts) that remove the heat-affected zone also help reduce these effects. The energy should be reduced on skim cuts, and they should remove the heat-affected zone from the previous pass.

When machining corners, you need to understand how the wire EDM cuts. The wire EDM will leave a sharp edge on the inside of the turn and a radius on the outside of the turn (see Figure 12–11). The smallest radius possible will be equal to the radius of the wire plus the spark gap. In the example shown, the smallest radius possible would be .007 inches: .006 inches (radius of the wire) plus .001 inches (the smallest spark gap possible). The wire also tends to deflect in the middle because of the electromagnetic field generated. Some machines have special codes to help reduce this effect.

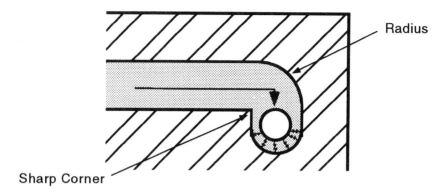

FIGURE 12–11 *The effects of turning a corner. Note the difference between the inside and outside corners.*

MACHINE SETUP

The workpiece is clamped to the table, which must be open in the middle to allow the wire to travel through the table (see Figure 12–10). Many attachments are available to help clamp the workpiece to the table. Remember, the wire must be able to cut through the workpiece. You also have to make sure that the piece that is cut out will not move and short the wire. This could potentially damage the workpiece and/or the machine.

The workpiece must be properly aligned on the table. Accurate alignment is fundamental to producing a quality part. The operator must then choose and thread the appropriate wire for the job. Once the wire has been chosen, the operator mounts the spool of wire on the machine. The wire is then threaded through the machine. The wire path will vary depending on the make and model of the machine. If the wire is not threaded properly, it will cause a short to the machine and the machine will not operate. Many machines feature automatic threading.

The wire tension must then be set using a wire tension gage (tensiometer). The amount of tension used depends on the type and diameter of the wire (see Figure 12–12). Machine sensors monitor the wire for breakage. If the wire should break, the cutting cycle is stopped automatically. The operator then must rethread the wire and then hit the cycle start button.

The wire must then be aligned in a vertical orientation. A setup block, used to align the wire, is normally made of granite with two metal contacts on it. The setup plate is clamped to the table with the metal contacts facing the cutting area. The machine then does the vertical alignment. The machine control monitors the wire and moves the U and V axes until the wire is vertical. The controls that set the cutting conditions must be set very low during this step.

Desired Wire Tension					
Wire Diameter		Copper		Molybdenum, Brass, and Zinc Coated	
inches	mm	ounces	grams	ounces	grams
.002	.05	3.5	100	7	200
.003	.07	5.3	150	10.5	300
.004	.1	7	200	14	400
.005	.12	8.8	250	17.6	500
.006	.15	10.5	300	21.2	600
.007	.17	12.3	350	24.7	700
.008	.20	14	400	28.2	800
.009	.22	15.8	450	31.7	900
.010	.25	17.6	500	35.2	1000
.011	.27	19.4	550	38.8	1100
.012	.30	21.2	600	42.3	1200

FIGURE 12–12 *The correct settings for wire tension.*

Next, the operator must align the wire in relation to the actual workpiece. Sometimes the wire is aligned with the edges of the part, and in some cases the wire must be aligned in relation to a hole.

EDGE DETECTION

Many machines have an automatic edge detection function. The operator locates the wire close to an edge and turns on the wire tension control switch (the wire should be running and the flushing should be on). The operator then starts the edge-finding control. The machine moves in the direction the operator has chosen until the control senses continuity (a short circuit) between the wire and the workpiece. This process is repeated several times, and the machine then averages the results to accurately determine the edge location.

HOLE LOCATION DETECTION

The operator threads the wire through a hole in the workpiece and aligns the table so that the wire is approximately in the center of the hole. The operator chooses the hole detection mode on the control and then turns on the wire tension control switch (the wire should be running and the flushing should be on). The operator chooses a slow feed rate and presses the desired axis button. The machine then begins the hole detection sequence and accurately centers the wire in the hole.

SLOT LOCATION DETECTION

The operator threads the wire through a slot in the workpiece and aligns the table so that the wire is approximately in the center of the slot. The operator chooses the slot-detection mode on the control and turns on the wire tension control switch (the wire should be running and the flushing should be on). The operator chooses a slow feed rate and presses the desired axis button. The machine begins the slot detection sequence and accurately centers the wire in the slot.

TEST SQUARE

The test square program, often run before cutting the actual part, cuts a .100-inch square. This permits the operator to make any offset adjustments that may be required. It also allows the operator to check the surface finish and make any changes to the cutting conditions before the actual part is cut.

SKIM CUTS

Finish cuts produce better accuracy and surface finish. On a wire EDM machine, they are called *skim cut*. After the part has been machined, the operator can enter a small offset value for the wire. This is often less than .001 inch. Very little material is removed in a skim cut, so the cutting speed will be faster. The same program is then run, and a very accurate part with excellent surface finish is produced (see Figure 12–13). Most parts, however, can be produced with one cut.

Here are some of the changes that can be made for skim cuts.

1. *increase the no-load voltage*
2. *decrease the on-time*
3. *increase the wire tension*
4. *reduce the water flow*
5. *change to AC power*

PROGRAMMING

The next program would machine the die shown in Figure 12–14. Figures 12–17 and 12–18 contain G- and M-code tables.

    N005 #0007;

This line is the program name (#0007).

    N010 G90;

This line puts the control in absolute mode.

    N015 M22;

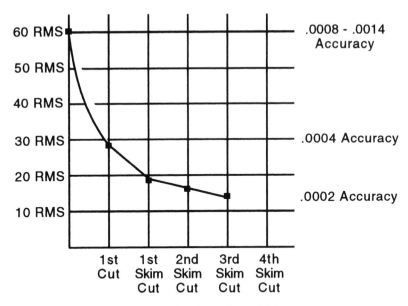

FIGURE 12–13 *Graph of number of skim cuts versus the surface finish and accuracy possible.*

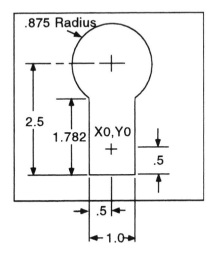

FIGURE 12–14 *Simple example of a die to be machined using a wire EDM machine.*

This line turns PWB on.

N020 M23;

This line turns the auto servo override off.

N025 G92 X0 Y0;

This line establishes home position (initial wire position) at X0 Y0.

> N030 G42 D1 S1;

This line calls for a wire diameter compensation right and uses an offset that is specified in variable D1. The S1 tells the controller to use burn condition file 1.

> N035 G01 X0 Y-.50;

This line moves the wire to X0 Y-.50 (bottom of the die opening).

> N040 X-.50;

This line moves the wire to a position of X-.50 Y -.5.

> N045 Y1.125;

This line moves the Y axis to a position of X-.5 Y1.125.

> N050 G02 X.5 Y1.125 I.5 J.875;

This line makes a clockwise circular move to cut the circular opening in the die. The end point of the move is X.5 Y1.125. The distance from start point of the arc to the center of the arc in the X direction is .5 in the X direction, so the I value is .5 (I.50). The distance from the start point of the arc to the center of the arc in the Y direction is .875, so the J value is .875 (J.875).

> N055 G01 Y-.50;

This line moves the wire to a position of X.5 Y-.5.

> N060 M01;

This line is the optional program stop code.

> N065 X0;

This line moves the wire to a position of X0 Y-.5.

> N070 G40 Y0;

This line cancels moving the wire to X0 Y0 (home position) and cancels the offset.

> N075 M30;

This line is the program end and reset command.

A FOUR-AXIS PROGRAM EXAMPLE

The next example is a more complex part: a four-axis part (see Figures 12–15 and 12–16). There are two programs: a main program and a subprogram. The

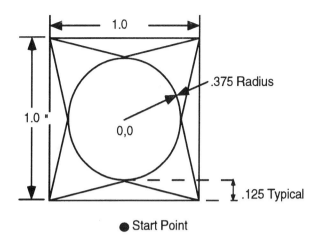

FIGURE 12–15 *Sample four-axis part print.*

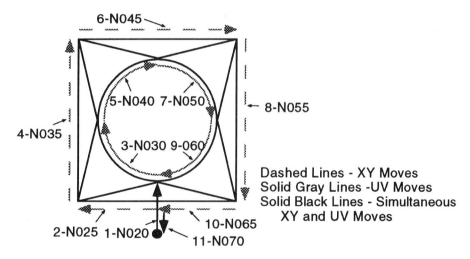

FIGURE 12–16 *Path of the wire for each line of the program.*

main program will call the subprogram four times. There will be one rough cut and three finish cuts. The subprogram contains the part geometry. The main program controls the machining and offsets. These programs could have been written as just one program, but this way is substantially shorter.

N005 #0002;

This line specifies that this is program 2. This is the main program that will call subprogram 0001 three times.

N010 G90;

This line puts the control in absolute mode.

N015 M22;

This line puts PWB on. This code is used for rough cutting. The machine slows down before changing directions to help insure that the wire is truly straight in the vertical direction. The center of the wire lags slightly behind the die guides opposite the direction of cut. The M22 slows the feed rate down before direction change and speeds back up afterward. This code is used mostly during roughing because the wire tension is reduced to prevent breaking.

N020 M23;

This line turns auto override off. Auto override is only used for skim cuts. Finish cutting is much faster than rough cutting. In order to maintain maximum feed rate and true geometry, M23 is used. When M23 is active the machine will read ahead in the program four lines. When the machine is within 12 microns of the direction change, the servos are slowed down to maintain the part shape. After a direction change, when the machine is 12 microns away, it will speed up again.

025 D1 S1;

This line tells the control to use offset number 1. This will be the roughing pass. The S1 means that the machine should use burn condition file 1.

N030 M98 P1;

This line calls and runs program number 1, which contains the moves (geometry) to cut the part.

N035 M21 M24;

This line turns PWB off and turns auto override on.

N040 D2 S2;

This line tells the control to use offset number 2, used for the first skim pass. The S2 tells the controller to use burn condition file 2.

N045 M98 P1;

This line calls and runs program number 1 again. Program number 1 will now be used with the new offset to make the first skim cut.

N050 D3 S3;

This line tells the control to use offset number 3. This offset will be used for the first skim pass. The S3 tells the controller to use burn condition file 3.

N055 M98 P1;

This line calls and runs program number 1 again. Program number one will now be used with the new offset to make the second skim cut.

N060 D4 S4;

This line tells the control to use offset number 4, used for the second skim cut. The S4 tells the controller to use burn condition file 2.

N065 M98 P1;

This line calls and runs program number 1, which contains the moves to cut the part. The new offset value assures that the part will be cut to the right size.

N070 M30;

This line is a program end and reset to the beginning.

N005 #0001;

This line is the number (name) of this program.

N010 G90;

This line puts the control in the absolute mode.

N015 G92 X0 Y-.75;

This line establishes the home position. In this case it tells the control that the wire is at X0 and Y -.75.

N020 G01 G41 Y-.5 V.125;

This is a linear move (move 1). The Y axis (bottom axis) moves to -.5 and the V axis Z (top axis) moves to -.375. Remember that the V axis is an incremental distance from the Y axis position. In this case it is .125, which is actually -.375 on the part. This establishes the taper on the part that runs from the circle top to the square bottom. This line starts cutter (wire) compensation also.

N025 X-.50 Y-.50 U.50 V.125;

This line moves to X -.5. Y is already at -.5. U and V do not change (move 2). Their incremental values in this line (U.5 V .125) are the incremental distances from the new X and Y, so the U and V do not change position. This line kept the wire in the same position on the top (UV) and moved the wire on the bottom to the left bottom corner (X-.5 Y-.5).

N030 G02 U.125 V.50 K.50 L.50;

This line makes a 90-degree clockwise arc on the lower left corner of the part (move 3). This line will keep the wire in the XY plane stationary at the lower left-hand corner. The UV axis (top axis) will move the wire to cut a 90-degree arc. Remember that the U and V values will be incremental from where the X and Y axes end up. The X and Y axes do not move in this line, so their values are X-.5 and Y-.5. The U value is the end point for the arc in the X direction. In this case the value is U.125. This means that the U end point is .125 incremental from the X position, and the actual position would be -.375. The V value is the end point for the arc (circle) in the Y direction. In this case the value is V.5, which means that the V value is .5 incremental from the Y position. The K value is the incremental distance from the X position to the arc center, and here the value of K is .5. This means that the distance from the X position to the center of the arc is K.5. The L value is the incremental distance from the Y position to the arc center. In this case the value of L is .5. This means that the distance from the Y position to the center of the arc is L.5.

    N035 G01 X-.50 Y.50 U.125 V-.50;

Line N035 moves the XY axis but keeps the UV axis stationary (move 4). The X axis stays at X-.5 and the Y axis moves to Y.5. The U and V values then are the incremental distances from the new X and Y position. The U position is .125. This means that the U will not move; it will end up at an incremental distance from X of .125. The V axis will not move either; it will end up at an incremental distance of -.5 from the Y position.

    N040 G02 U.50 V-.125 K.50 L-.50;

Line N040 makes a 90-degree clockwise arc on the upper left corner of the part (move 5). This line will keep the wire in the XY plane stationary at the upper left-hand corner. The UV axis (top axis) will move wire to cut a 90-degree arc. Remember that the U and V values will be incremental from where the X and Y axes end up. The X and Y axes do not move in this line, so their values are X-.5 and Y.5. The U value is the end point for the arc in the X direction. In this case the value is U.5. This means that the U end point is .5 incremental from the X position. In this case the actual position would be -0. The V value is the end point for the arc (circle) in the Y direction. In this case the value is V-.125, which means that the V value is -.125 incremental from the Y position. The K value is the incremental distance from the X position to the arc center. In this case the value of K is .5, which means that the distance from the X position to the center of the arc is K.5. The L value is the incremental distance from the Y position to the arc center. In this case the value of L is -.5, which means that the distance from the Y position to the center of the arc is L-.5.

    N045 G01 X.50 Y.50 U-.50 V-.125;

This line moves the XY axis, but keeps the UV axis stationary (move 6). The Y axis stays at X.5 and the X axis moves to X.5. The U and V values then are the incremental distances from the new X and Y position. In this case the U position is -.5. This means that the U will not move; it will end up at an incremental distance from X of -.5. The V axis will not move either; it will end up at an incremental distance of -.125 from the Y position.

N050 G02 U-.125 V-.50 K-.50 L-.50;

This line makes a 90-degree clockwise arc on the upper right corner of the part (move 7). This line will keep the wire in the XY plane stationary at the upper right-hand corner. The UV axis (top axis) will move wire to cut a 90-degree arc. Remember that the U and V values will be incremental from where the X and Y axes end up. The X and Y axes do not move in this line, so their values are X.5 and Y.5. The U value is the end point for the arc in the X direction. In this case the value is U-.125. This means that the U end point is -.125 incremental from the X position. The V value is the end point for the arc (circle) in the Y direction. In this case the value is V-.5, which means that the V value is -.5 incremental from the Y position. The K value is the incremental distance from the X position to the arc center. In this case the value of K is -.5, which means that the distance from the X position to the center of the arc is K-.5. The L value is the incremental distance from the Y position to the arc center. In this case the value of L is -.5. This means that distance from the Y position to the center of the arc is L-.5.

N055 G01 X.50 Y-.50 U-.125 V.50;

This line moves the XY axis, but keeps the UV axis stationary (move 8). The X axis stays at X.5 and the Y axis moves to Y-.5. The U and V values then are the incremental distances from the new X and Y position. In this case the U position is -.125. This means that the U will not move; it will end up at an incremental distance from X of -.125. The V axis will not move either; it will end up at an incremental distance of .5 from the Y position.

N060 G02 U-.50 V.125 K-.50 L.50;

This line makes a 90-degree clockwise arc on the lower right corner of the part (move 9). This line will keep the wire in the XY plane stationary at the lower right-hand corner. The UV axis (top axis) will move wire to cut a 90-degree arc. Remember that the U and V values will be incremental from where the X and Y axes end up. The X and Y axes do not move in this line, so their values are X.5 and Y-.5. The U value is the end point for the arc in the X direction. In this case the value is U-.5. This means that the U end point is -.5 incremental from the X position. The V value is the end point for the arc (circle) in the Y direction. In this case, the value is V.125, which means that the V value is .125 incremental from the Y position. The K value is the incremental distance from the X position to the arc center. In this case the value of K is -.5, which means that the distance from the X position to the center of the arc is K-.5. The L value is the incremental distance from the Y position to the arc center. In this case the value of L is .5, which means that the distance from the Y position to the center of the arc is L.5.

N065 G01 X0 Y-.50 U0 V.125;

Line N065 moves the XY axis, but keeps the UV axis stationary (move 10). The X axis moves to X0. The Y axis stays at Y -.5. The U and V values then are the incremental distances from the new X and Y position. In this case the U position is 0. This means that the U will not move; it will end up at an incremental distance from X of 0. The V axis will not move either; it will end up at an incremental distance of .125 from the Y position.

    N070 G01 G40 Y-.75 V0;

This line cancels the offset and moves the wire back to home position (move 11).

    N075 M30;

End of program.

G Code	Function
G00	Rapid Traverse
G01	Linear Interpolation
G02	Circular Interpolation Clockwise
G03	Circular Interpolation Counterclockwise
G04	Dwell
G20	Inch Programming
G21	Metric Programming
G40	Wire Diameter Compensation Cancel
G41	Wire Diameter Compensation Left
G42	Wire Diameter Compensation Right
G48	Automatic Corner Rounding On
G49	Automatic Corner Rounding Off
G50	Wire Inclination Angle Cancel
G51	Wire Inclination Angle Left
G52	Wire Inclination Angle Right
G60	Same Radii Top and Bottom When Tapering
G61	Conical Corner R
G70	Edge Finding
G71	Circle Center Finding
G72	Finding the Center of a Groove
G75	Rapid Positioning in Relative Coordinate System
G76	Rapid Positioning to the Face Positioning Point
G77	Rapid Positioning to a Specified Point
G78	Positioning to the Work Corner Point
G79	Calculation and Setting the Work Inclination Angle
G90	Absolute Programming
G91	Incremental Programming
G92	Coordinate System Setting
G94	Constant Feed by Program
G95	Servo Feed

FIGURE 12–17 *Typical G-code table.*

M Code	Function
M00	Program Stop
M01	Optional Program Stop
M02	Program End
M13	Manual Feed Override Off
M15	Set Taper Cutting Mode
M21	PWB Off
M22	PWB On
M23	Auto Over RIde Off
M24	Auto Over Ride On
M30	Program End and Reset
M31	Cut Time Display Reset
M40	Discharge Off
M41	Discharge On
M42	Wire Feed Off
M43	Water Flow Off
M44	Wire Tension Off
M50	Cut Wire (AWF)
M60	Connect Wire (AWF)
M70	Retrace Start
M80	Discharge On
M81	EDM Power On
M82	Wire Feed On
M83	Water On
M84	Wire Tension On
M96	Reverse Cut in Mirror Image End
M97	Reverse Cut in Mirror Image Start
M98	Subprogram Call
M99	Subprogram End

Figure **12–18** *Typical M-code table.*

CHAPTER QUESTIONS

1. Describe the wire EDM process.
2. List at least five advantages of wire EDM machining.
3. Describe the characteristics of wire that are important for wire EDM machining.
4. Why might a wire other than brass be used?
5. If a .010-inch wire is being used, what is the approximate width of cut?
6. How do manufacturers rate the speed of their machines?
7. Describe the purpose and importance of flushing in the wire EDM process.
8. What is a two-axis wire EDM capable of?
9. What is the difference between a four-axis and independent four-axis system?
10. Why is it so important that the gap is controlled in EDM machining?
11. Describe how a wire EDM is typically set up to run a job.
12. What is a skim cut and why are they used?
13. Study the part in Figure 12–19 and write a two-axis program to make the part.

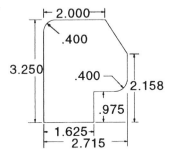

FIGURE 12–19 *Use with question 13.*

14. Study the part in Figure 12–20 and write a four-axis program to machine
    the part.

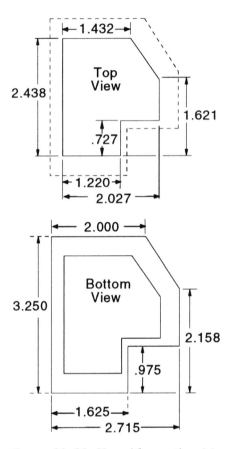

FIGURE 12–20 *Use with question 14.*

15. Study the part in Figure 12–21 and write a four-axis program to machine the part.

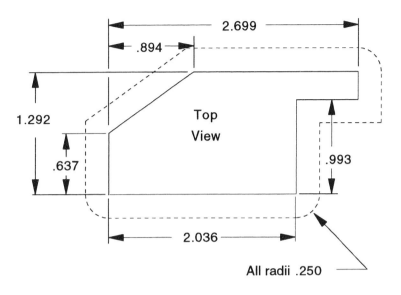

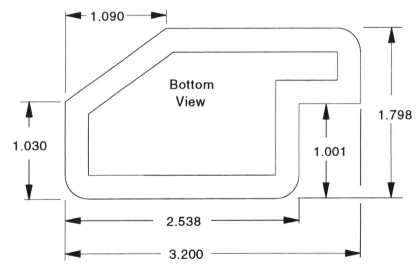

FIGURE **12–21** *Use with question 15.*

Chapter 13

....

FUNDAMENTALS OF COMMUNICATIONS

INTRODUCTION

If enterprises are to become more productive, they need to improve processes. This requires accurate data, and communications are vital. Programs need to be downloaded/uploaded to CNC machines. Production devices hold very valuable data about their processes. This chapter examines how this data can be acquired.

OBJECTIVES

Upon completion of this chapter, the reader will be able to:

- *Describe the four levels of plant communications and characteristics at each level.*
- *Define such terms as "serial," "synchronous," "RS-232," "RS-422," "device," "cell," "area," "host," and "SCADA."*
- *Describe how computers can communicate with CNC machines.*
- *Describe what software and hardware are required to communicate from a computer to a CNC machine.*

INTRODUCTION TO COMMUNICATIONS

The computer has revolutionized manufacturing by making automation flexible and affordable. Computers control processes across the plant floor. In addition to producing product, computers also produce data, which can be more profitable than the product. This may not seem obvious; however, most processes are very inefficient. If we can use the data to improve processes, we can drastically improve profitability. The inefficiencies are not normally addressed because people are busy with other, more pressing problems. (A good friend in a small manufacturing facility said it best: "It's hard to think about fire prevention when you're in the middle of a forest fire.") With very few changes, the data can be gathered and used to improve quality, productivity, and uptime. Huge gains are possible if this data is used.

To be used, the data must first be acquired. Many managers today will say that they are already collecting much of the data. Why should they invest in electronic communications when they are already gathering data from the plant floor manually? The reasons are many. Often, data gathered manually is very inaccurate. The data is not real-time either. The information must be written down by an operator, gathered by a foreman, taken "upstairs," entered by a data processing person, printed into a report, and distributed.

This all takes time. It can often make data many days or weeks late. If mistakes in entry were made on the floor, it is often too late to correct them. The reports that are produced are often mazes of meaningless information: too much extraneous information to be useful to anyone. The lateness and inaccuracy of the data gathering makes it almost counterproductive.

This communication is quite easily achieved through electronic communication. Many of the data required already exists in the smart devices on the factory floor. Much of the data that people write on forms in daily production already exists in plant floor production devices.

The other communication that is required is real-time information to the operator: current and accurate CNC part programs, accurate orders, accurate instructions, current specifications, and so on, which is often lacking in industrial and service enterprises today. CNC programs are usually revised if they are used for production runs. We must always make sure that the current revisions are being used. Often improvements are made on the floor after the program is initially released. These should be uploaded to the computer and saved.

The improvements in computer hardware and software have made communication much easier. There are many software communications packages that make it easy to communicate with CNC machines and other production equipment. Communications will increase in importance as American manufacturing tries to improve its competitive position.

LEVELS OF PLANT COMMUNICATION

Plant floor communications can be broken down into levels. Whether you break them into four levels or into five, the basic concepts are the same. In this book we use a four-level model. These four levels are device, cell, area, and host (see Figure 13–1). Each level is a vital link. The device level is the production level. As we move up the pyramid, management of production becomes the task.

The easiest way to understand this model is to relate it to people. Figure 13–2 shows the human model of factory communications. The workers are at the device level. The foreman is at the cell level. The factory supervisors are at the area level, and the plant manager is at the host level. If we think of the typical duties that each person would have at each level, it is easy to understand the function of each level in the electronic model. This chapter covers each in more detail.

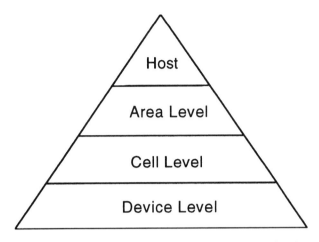

FIGURE 13–1 *Four-level model of plant communication.*

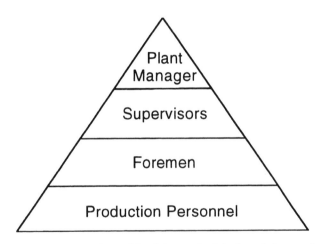

FIGURE 13–2 *Four-level model of the typical industrial organization of people. It is necessary to communicate information down to the workers and also to send information back up to the top of the organization. Compare this model with the model in Figure 13–1.*

DEVICE LEVEL

The device level is the lowest level of control on the plant floor. Think of devices as pieces of equipment that produce or handle product (see Figure 13–3). Some examples of devices are robots, conveyers, computer-controlled machine tools (CNC machines), hard automation, and automated storage/automated retrieval systems (AS/RS). These devices have several things in common. They

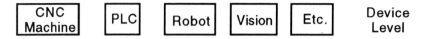

FIGURE 13–3 *The device level. Devices are production-oriented
equipment that add value to a product.*

are all directly involved with the product. Some move the product, while others
add value during production. Each device produces valuable data during production. Machines know how many parts have been produced, how long it
takes to produce the product (cycle time), uptime and downtime, and so on.

In the human model the device level relates to production personnel. Production personnel add value to the product. They handle the product and add
value in some manner. They also fill out paperwork so that management can
monitor quality, productivity, and so on.

In the electronic model the device might be a CNC machine. The CNC produces machined parts very efficiently. In addition to producing parts, the
CNC is also producing data. It can track cycle times, piece counts, downtime,
etc. If it were used, this data could drastically improve productivity.

Unfortunately, very few manufacturers use the data. In the rest of the cases
the data invisibly and continually spill out the back of the device onto the
floor or into the infamous and fictitious bit bucket. The fact is that data generated in devices is not used. This is a huge lost opportunity.

One of the other things that devices have in common is a need for data. Each
device needs a program to tell it what, when, and how many to produce. The
programs will be different for each device, but they serve the same function.
In the human model, people need the same information: What should I work
on? How many should I make? What should I do next? Is there enough raw
material? This information is crucial to efficient production. Unfortunately,
inaccurate, untimely information is more the rule than the exception in enterprises. In most factories, people are used to coordinate the devices. People
start and stop the machines, count pieces, monitor quality, monitor performance, and watch for problems. The importance of accurate, timely information is just as vital when people are involved.

When you think of the device level, think of task-specific equipment that is
adding value to the product. Think of production-type tasks. The device level
is where value is added and thus where the enterprise makes its money. This
level creates the wealth to support the rest of the organization. Some people
may be uncomfortable with that thought; however, the Japanese view the
production level as the most important and they do anything they can to
make that level more efficient. They use the concept of *Kaizen,* or continuous
improvement. Accurate, real-time data can be crucial to real improvement.

CELL LEVEL

The cell is a logical grouping of devices used to add value to one or more products. A cell will typically work on a family of similar parts. A cell consists of various dissimilar devices (see Figure 13–4). Each device typically has its own unique type of program and communication protocol. Devices are typically unable to communicate with each other. A CNC machine, for example, typically cannot communicate with a robot.

Compare this to the human model. The foreman is the cell controller. The foreman's job is to show up early for work and find out what needs to be produced that day. The foreman must then choose the appropriate production person to perform each task that needs to be accomplished. Of course, all people have different personalities, but a good foreman can communicate well with each individual. The foreman monitors each employee's performance and, by coordinating efforts, produces that day's product. The foreman makes sure that people cooperate to get the overall job accomplished.

The purpose of the cell controller is to integrate various devices into a cooperative work cell. The cell controller must then be able to communicate with each device in the cell. Even if the devices are not able to communicate with each other, they must be able to communicate with the cell controller. The cell controller must be able to upload/download programs, exchange variable information, start/stop the device, and monitor the performance of each device. There can be many foremen in a plant, and there can be many cell controllers in a plant. Each can control a group of devices. Each cell controller can talk to other cell controllers and also up to the next level.

Many devices have the ability to communicate. For example, we may need to upload/download programs or update variables. Most devices offer serial communications capability, using the asynchronous communications mode, and have an RS-232 serial port available. Although you would think that any device with an RS-232 port would easily communicate with any other device with an RS-232 port, this is not always true. Devices have their own protocol.

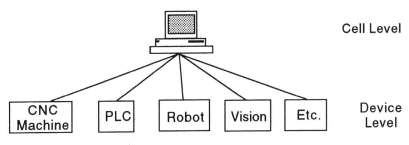

FIGURE 13–4 *Typical cell control scheme. One cell controller is controlling several devices.*

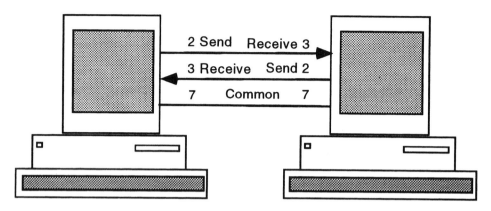

FIGURE 13–5 *This simple RS-232 wiring scheme shows the simplest of RS-232 connections.*

CNC machines will normally communicate using asynchronous RS-232 communications.

The RS-232 standard specifies a function for each of 25 pins. It does not say that any of the pins must be used, however. Some manufacturers use only three, as in Figure 13–5. Some device manufacturers use more than three pins, so some electrical handshaking can take place.

HANDSHAKING

Handshaking is the passing of a signal between one device and another to be sure they are both ready to talk. One sends a signal that says it has some information for it, and the other device responds by saying it is ready to receive it.

HARDWARE HANDSHAKING

Hardware handshaking works by sending electrical signals between computers. Handshaking implies a cooperative operation. In hardware handshaking one device will look for the presence of an electrical signal on one of the communications lines to see if it is ready to send the next character. It does this by setting pin 4 (RTS-the request to send pin) high.

The receiving device sees the request to send pin high, and if it is ready to receive, it sets the clear to send pin 5 (CTS) high. The first computer then knows that the cable is connected, the computer is on, and it is ready to receive. Some devices can be set up to hardware handshake; others cannot. Fortunately cables can be wired so that two devices with different capabilities can still communicate. This configuration, called a *null modem cable,* will be covered later in this chapter.

SOFTWARE HANDSHAKING

Software handshaking is done by sending control codes (characters) in the data to be sent. The most common scheme is known as XON/XOFF. The receiving device sends either the XON or the XOFF depending on whether it is ready to receive data. The XOFF control character (usually control-S) tells the sending device to pause in transmitting data. When the receiver is ready to receive data again it sends the XON character (usually control-Q).

In Figure 13–5, no hardware handshaking is taking place. The first computer sends a message regardless of whether there is another device there. The computer cable could be unplugged or the computer turned off, and the sending computer would not know.

Fortunately when a device is purchased it is generally capable of communicating with an IBM® personal computer. The user usually just has to open the device manual to the section on communication to find the wiring configuration for the cable. It is still difficult and expensive to communicate when a wide variety of devices are involved.

SIMPLEX COMMUNICATIONS

In simplex communications data can move only in one direction. This is very uncommon.

HALF-DUPLEX COMMUNICATIONS

Half-duplex transmission allows communications in both directions, but in only one direction at a time.

FULL-DUPLEX COMMUNICATIONS

Full-duplex communications allows communications in both directions at the same time. Full duplex communication is much more efficient than half-duplex communication.

When a message is sent using asynchronous communication, the message is broken into individual characters and transmitted one bit at a time. The ASCII system is used normally. In ASCII, every letter, number, and some special characters have a binary-coded equivalent. There are two types of ASCII. In 7-bit ASCII, there are 128 possible different letters, numbers, and special characters. In 8-bit ASCII, 256 are possible.

Each character is sent as its ASCII equivalent. For example, the letter "A" would be 1000001 in 7-bit ASCII (see Figure 13–6). More than 7 bits are needed to send a character in the asynchronous model, however. Other bits

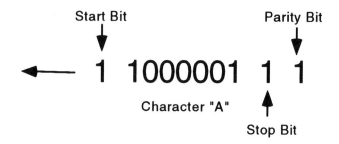

FIGURE 13–6 *Transmitting the letter A in the asynchronous serial mode of communications, assuming odd parity. There is an even number of ones in the character A, so the parity bit is a one to make the total odd. If the character to be sent has an odd number of ones, the parity bit would be a zero. The receiving device counts the number of ones in the character and checks the parity bit. If they agree, the receiver assumes that the message was received accurately. This is rather crude error checking because two or more bits could change state and the parity bit could still be correct, but the message could be wrong.*

Start Bit	Data Bits	Parity Bits	Stop Bits
1 Bit	7 or 8	1 Bit - Odd, Even, Mark, Space, or None	1, 1.5, or 2 Bits

FIGURE 13–7 *Transmitting a typical ASCII character.*

are used to make sure the receiving device knows a message is coming, that the message was not corrupted during transmission, and that the character has been sent. The first bit sent is the start bit (see Figure 13–7). This tells the receiver that a message is coming. The next 7 bits (8 if 8-bit is used) are the ASCII equivalent of the character. Then there is a bit reserved for parity. Parity is used for error checking. The parity of most devices can be set up for odd or even, mark or space, or none.

New standards will help integrate devices more easily. RS-422 and RS-423 were developed to overcome some of the weaknesses of RS-232. The distance and speed of communications are drastically higher in RS-422 and RS-423, which were developed in 1977.

RS-422 is called *balanced serial*. RS-232 has only one common. The transmit and receive lines use the same common, which can lead to noise problems. RS-422 solves this problem by having separate commons for the transmit and receive lines. This makes each line very noise immune. The balanced mode of communications exhibits lower crosstalk between signals and is less susceptible to external interference. Crosstalk is the bleeding of one signal over onto another, which reduces the potential speed and distance of communications. This is one reason that the distance and speed for RS-422 are much higher. RS-422 can be used at speeds of 10 megabits for distances of over 4000 feet, compared to 9600 baud and 50 feet for RS-232.

RS-423 is similar to RS-422 except that it is unbalanced. RS-423 has only one common, which the transmit and receive lines must share. RS-423 allows cable lengths exceeding 4000 feet. It is capable of speeds up to about 100,000 bps.

RS-449 is the standard that was developed to specify the mechanical and electrical characteristics of the RS-422 and RS-423 specifications. The standard addressed some of the weaknesses of the RS-232 specification. The RS-449 specification specifies a 37-pin connector for the main and a 9-pin connector for the secondary; however, the RS-232 specification does not specify what type of connectors or how many pins must be used.

These standards are intended to replace RS-232 eventually. There are so many RS-232 devices that it will take a long time, but it is already occurring rapidly in industrial devices. Adapters are cheap and readily available to convert RS-232 to RS-422 or vice versa (see Figure 13–8). This can be used to advantage if a long cable length is needed for a RS-232 device (see Figure 13–9).

RS-485 is a derivation of the RS-422 standard. It is unbalanced, however. The main difference is that it is a multidrop protocol, which means that many devices can be on the same line. This requires the devices to have some intelligence, however, because the devices must each have a name so that each knows when it is being talked to.

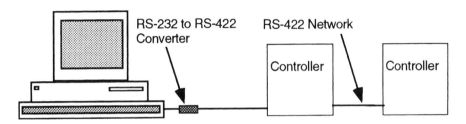

FIGURE 13–8 *Use of a converter to change RS-232 communications to RS-422 communications. The computer is then able to communicate with other devices on its network. The devices on the right are on an RS-422 network.*

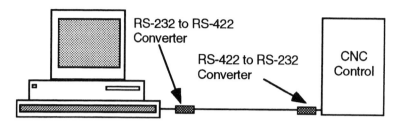

FIGURE 13–9 *Use of two converters to extend the cable length. Although RS-232 is only reliable to about 50 feet, the use of two converters allows 4000 feet to be covered by the RS-422 and then converted to RS-232 on each end. However, the speed will be limited by the RS-232.*

TYPES OF CELL CONTROLLERS

Programmable logic controllers (PLCs) and computers can both be used for cell control applications. A programmable logic controller is a special purpose computer that is programmed using ladder logic. Ladder logic is very easy for electricians to work with. PLCs are also very easy to wire with sensors, switches, and output devices.

PLCs as Cell Controllers

The PLC offers some unique advantages as a cell controller. It is easily understood by plant electricians and technicians. If the devices in the cell need to communicate in primitive mode, it is very easy to do with a PLC. If there are other PLCs of the same brand in the cell, it is easy for the PLC to communicate by using the data highway of that brand of PLC.

The PLC is not very applicable as a cell controller when there is more than one brand of device in the cell. Typical PLCs do not offer as much flexibility in operator information as computers do, although this is changing rapidly. Graphic terminals and displays are becoming very common for PLCs.

Computers as Cell Controllers

The computer offers more flexibility and capability than the PLC. The computer has much more communications capability than any PLC. In fact, all device manufacturers want their devices to communicate with a microcomputer. It is difficult to sell a device that does not communicate with an IBM microcomputer. Most device manufacturers do not care to communicate with other brands of PLCs, robots, and so on, but they all want to talk to an IBM compatible. It is much easier for the computer to act as cell controller when it can communicate with each device in the cell.

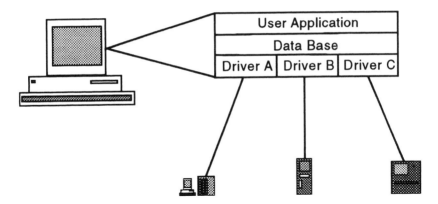

FIGURE 13–10 *How a typical SCADA software package works. The user application defines which variables from the devices must be communicated. These are collected through the drivers and stored in a database that is available to the application. Once the computer has the desired data, it is relatively easy to make the data available to other devices.*

This communication is usually accomplished by the use of SCADA (supervisory control and data acquisition) software. The concept is that software is run in a common microcomputer to enable communications to a wide variety of devices. The software is typically like a generic building block (see Figure 13–10).

The programmer writes the control application from menus or in some cases graphic icons. The programmer then loads drivers for the specific devices in the application. A *driver* is a specific software package written to handle the communications with a specific brand and type of device. They are available for most common devices and are relatively inexpensive.

The main task of the software is to communicate easily with many brands of devices. Most enterprises do not have the expertise required to write software drivers to communicate with devices. SCADA packages simplify the task.

In addition to handling the communications, SCADA software makes it possible for applications people to write the control programs. Thus, the people who best know the application write it without learning complex programming languages. Drivers are available for all major brands of PLCs and other common manufacturing devices.

In general, an applications person would write the specific application using menu-driven software. The software is easy to use. Some applications are like spreadsheets and some use icons for programming. Instead of the specific I/O numbers that the PLC uses, the programmer uses tagnames.

For example, the application might involve temperature control. The actual temperature might be stored in register S20 in the PLC. The applications programmer would use a tagname, such as "temp1," instead of the actual number.

Device	Actual #	Tagname
PLC	Reg20	TEMP1
PLC	Reg12	CYCLETIME1
PLC	S19	TEMP2
PLC	N7:0	QUANTITY1
Robot	R100	QUANTITY2

FIGURE **13–11** *Sample tagname table.*

This makes the programming transparent. *Transparent* means that the application programmer does not have to worry very much about what brand of devices are in the application. A table is set up that assigns specific PLC addresses to the tagnames (see Figure 13–11). In theory, if a different brand of PLC were installed in the application, the only change required would be a change to the tagname table and the driver.

Fortunately, there is more software available every day to make the task of communications easier. The software is more friendly, faster, more flexible, and more graphics oriented. The data gathered by SCADA packages can be used for statistical analysis, historical data collection, adjustment of the process, or graphical interface for the operator.

AREA CONTROL

Area controllers are the supervisors (see Figure 13–12) that look at the larger picture. They receive orders from the host and then assign work to cells to accomplish the tasks. They also communicate with other area controllers to synchronize production. Area controllers use synchronous communications methods and are attached via local area networks (LANs).

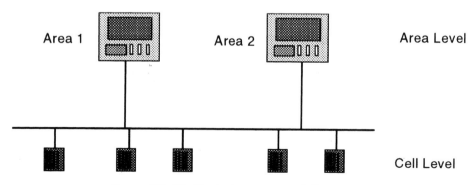

FIGURE **13–12** *Typical area control diagram.*

LOCAL AREA NETWORKS (LANS)

Local area networks are the backbone of communications networks. The topic of LANs can be broken down into various methods of classification. We examine three: topology, cable type, and access method.

TOPOLOGY

Topology refers to the physical layout of LANs. There are three main types of topology: star, bus, and ring.

Star Topology

The star style uses a hub to control all communications. All nodes are connected directly to the hub node (see Figure 13–13). All transmissions must be sent to the hub, which then sends them on to the correct node. One problem with the star topology is that if the hub goes down, the entire LAN is down.

Bus Topology

The bus topology is a length of wire into which nodes can be tapped. At one end of the wire is the head end (see Figure 13–14), an electronic box that performs several functions. The head end receives all communications, then remodulates the signal and sends it out to all nodes on another frequency. *Remodulate* means that the head end changes the received signal frequency to another frequency and sends it out for all nodes to hear. Only the nodes that are addressed pay attention to the message. The other end of the wire (bus) dissipates the signal.

Ring Topology

The ring topology looks like it sounds. It has the appearance of a circle (see Figure 13–15). The output line (transmit line) from one computer goes to the input line (receive line) of the next computer, and so on. In this straightforward topology, if a node wants to send a message, it just sends it out on the transmit line. The message travels to the next node. If the message is addressed to it, the node writes it down; if not, it passes it on until the correct node receives it.

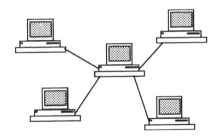

FIGURE **13–13** *Star topology.*

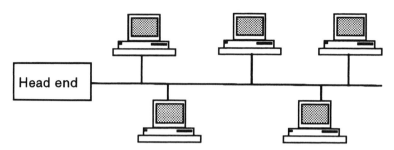

FIGURE 13–14 *Typical bus topology. Each node (communication
device) can speak on the bus. The message travels to the head
end and is converted to a different frequency. It is then sent
back out, and every device receives the message. Only
the device for which the message was intended
pays attention to the message*

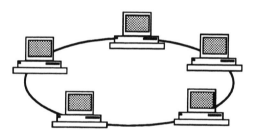

FIGURE 13–15 *Ring topology.*

The more likely configuration of a ring is shown in Figure 13–16. This style is
still a ring; it just does not appear to be. This is the convenient way to wire a
ring topology. The main ring (backbone) is run around the facility and interface boxes are placed in line at convenient places around the building. These
boxes are often placed in "wiring closets" close to where a group of computers
will be attached. These interface boxes, called multiple station access units
(MSAUs), are just like electrical outlets. If we need to attach a computer, we
just plug it into an outlet on the MSAU.

The big advantage of the MSAU is that devices can be attached/detached
without disrupting the ring or communications.

CABLE TYPES

There are four main types of transmission media: twisted pair, coaxial, fiber
optic, and radio frequency. Each has distinct advantages and disadvantages.
The capabilities of the cable types are expanding continuously.

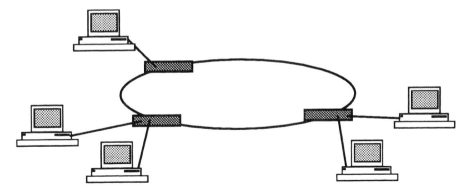

FIGURE **13–16** *Typical ring topology, although it looks more like a star topology. The multiple station access units (MSAUs) allow multiple nodes to be connected to the ring. The MSAU looks like a set of electrical outlets. The computers are just plugged into the MSAU and are then attached to the ring.*

Twisted Pair

Twisted-pair wiring is, as its name implies, pairs of conductors (wires) twisted around each other along their entire length. The twisting of the wires helps make them more noise immune. The telephone wires that enterprises have throughout their buildings are twisted pair.

There are two types of twisted-pair wiring: shielded and unshielded. The shielded type has a shield around the outside of the twisted pair, which helps to make the wiring noise immune. The wiring that is used for telephone wiring is typically unshielded. The newer types of unshielded cable are more noise immune than in the past. Higher speeds are being accomplished continuously.

Unshielded cable is very cheap and easy to install. In some cases companies may be able to use spare telephone twisted pairs to run the LAN wiring instead of running new cable. Remember that much of the twisted-pair wire in buildings today is the older, less noise-immune type.

The shielded twisted pair is now used commonly for speeds up to 16 megabits. Some companies supply LAN cards for unshielded twisted pairs that can also run at 16 megabits. Committees are studying the feasibility of up to 100 megabits for shielded twisted-pair wiring, and there should soon be a standard for 100 megabits on shielded twisted-pair wiring.

Coaxial Cable

Coaxial cable (coax) is a very common communication medium used by cable television. It is broadband, which means that many channels can be transmitted

FIGURE **13–17** *Coaxial cable. Note the shielding around the conductor.*

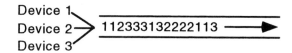

FIGURE **13–18** *How time-division multiplexing works. Each device must share time on the line. Device 1 sends part of its message and then gives up the line so that another device can send, and so on.*

simultaneously. Coax has excellent noise immunity because it is shielded (see Figure 13–17).

Broadband technology is more complex than *baseband* (single channel). With broadband technology there are two ends to the wire. The head end receives all signals from devices that use the line, then remodulates (changes to a different frequency) the signal and sends it back out on the line. All devices hear the transmission, but pay attention only if it is intended for their address.

Frequency-division multiplexing is used in broadband technology. The transmission medium is divided into channels, each with its own unique frequency. Buffer frequencies between each channel help with noise immunity. Some channels are for transmission and some are for reception.

Time-division multiplexing, also called *time slicing,* is used in the baseband transmission method. There are several devices that may wish to talk on the line (see Figure 13–18). Because we cannot wait for one device to finish its transmission completely before another begins, they must share the line. One device takes a slice of time, then the next does, and so on. There are several methods of dividing the time on the line.

Fiber-Optic Cable

Fiber-optic technology is also changing very rapidly. Figure 13–19 shows a drawing of the cable. The major arguments against fiber are its complexity of installation and high cost; however, the installation has become much easier, and the cost has fallen dramatically to the point that when total cost is considered, fiber is not much different for some installations than shielded twisted pair.

The advantages of fiber are its perfect noise immunity, high security, low attenuation, and high data transmission rates. Fiber transmits with light, so that it is unaffected by electrical noise. The security is good because fiber does not create electrical fields that can be tapped as can twisted pair or coaxial

FIGURE 13–19 *Fiber optic cable with multiple fibers through one cable.*

cable. Because the fiber must be physically cut to steal the signal, it is a much more secure system.

All transmission media attenuate signals, which means that the signal gets progressively weaker the farther it travels. Fiber exhibits far less attenuation than other media. Fiber can also handle far higher data transmission speeds than can other media.

The FDDI (fiber distributed data interchange) standard developed for fiber cable calls for speeds of 100 megabits. This seemed very fast for a short period of time, but many think that it may be possible to get 100 megabits with twisted pair, so the speed standard for fiber may be raised.

Plastic fiber cable is also gaining ground. It is cheap and easy to install. The speed of the plastic is much less, but is constantly being improved.

Radio Frequency

Radio-frequency (RF) transmission has recently become very popular. The use of RF has exploded in the factory environment. The major makers of PLCs have RF modules available for their products that use radio waves to transmit the data. The systems are very noise immune and perform well in industrial environments. RF is especially attractive because no wiring needs to be run. There are even RF LANs available for office environments. This is a major advantage for office areas where changes are made frequently.

ACCESS METHODS
Token Passing

In the token-passing method, only one device can talk at a time. The device must have the token to be able to use the line. The token circulates among the devices until one of them wants to use the line (see Figure 13–20). The device then grabs the token and uses the line.

Token passing offers very reliable performance and predictable access times, which can be very important in manufacturing. Here's how token passing works:

> *The device that would like to talk waits for a free token.*
>
> *The sending station sets the token busy bit, adds an information field, adds the message it would like to send, and adds a trailer packet. The header packet contains the address of the station for which the message was intended. The entire message is then sent out on the line.*

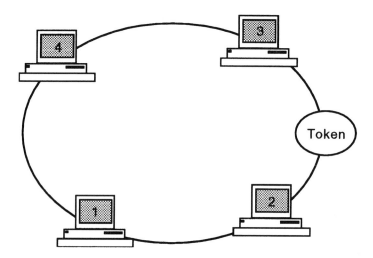

FIGURE **13–20** *Token-passing method.*

Every station examines the header and checks the address to see if it is being talked to. If not, it ignores the message.

The message arrives at the intended station and is copied. The receiving station sets bits in the trailer field to indicate that the message was received. It then regenerates the message and sends it back out on the line. The original station receives the message back and sees that the message was received. It then frees the token and sends it out for other stations to use.

Collision Detection

In collision detection, any device that wishes to speak must listen to the carrier signal on the transmission line. If the line is not busy, the device may use the line to communicate. If two or more devices try to use the line at the same time, there is a collision. The collision is detected and the devices back off for a while and try again.

This principle, called CSMA/CD (carrier sense multiple access/collision detection), is somewhat analogous to access to a highway. The more cars there are on the highway, the more difficult it is to get on the highway. There are also more collisions as the traffic increases. With data transmission, it is far less serious. The device just retransmits after a short wait. Typical industrial communications involve low line-use percentages, and CSMA/CD performs well at these levels.

The future of data transmission will involve much more transmission of graphics. Graphics involve very large files, which will increase line-use levels substantially.

HOST LEVEL

The host-level controller is generally a mainframe. This computer is responsible for the business software, engineering software, office communications software,

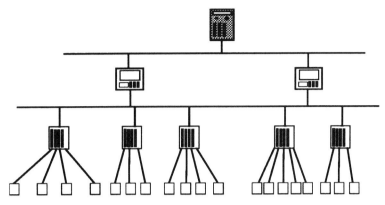

FIGURE 13–21 *Typical host-level control,
also called the enterprise level.*

and so on. The line between mainframe, minicomputer, and personal computer is rapidly blurring. The trend is definitely toward distributed processing.

The business software is generally an MRP package. MRP stands for *manufacturing resource planning.* This software is used to enter orders and bills of materials, check customer credit and inventory, and so on. The software can then be used to generate work orders for manufacturing, orders for raw materials and component parts, schedules, and even customer billing. MRPs are being used more and more for planning and forecasting. (They are typically called MRPII now because of the increased emphasis on planning and forecasting.)

The host level is going to be used more and more to optimize operation of the enterprise. Data from the factory floor will be gathered automatically from devices or from operator interface terminals. The host level's task will be to analyze the data to help improve the productivity of the overall system. The host level must also schedule and monitor the daily operations.

The real key to the future will be this data collection. Better business decisions can be made if accurate, real-time data is available in a format that people can understand and use.

Figure 13–21 shows the levels of enterprise communications. The task of integrating these devices is rapidly becoming easier. Software and hardware are making remarkable advances in ease of use. The future is bright.

CNC COMMUNICATIONS

The PC-based CNC control has drastically changed CNC communications. Users can install a network card into the PC-based CNC and attach the CNC to their company network. This is typically an Ethernet network, although it could be any type of network that a PC can support. The CNC machine is then just another node on the company network and can send and receive files or data from any other computer or CNC machine on the network. The CNC operator can even "surf" the Internet.

Older CNC machines must be connected serially to one computer. A serial cable must be run from the computer's serial port to the CNC machine's serial port. The cable configuration is the most crucial link. Only three lines (wires) are required for simple RS-232 communications. A few other pins on the connectors often need to be jumpered together to override handshaking. The best source for the wiring configuration for any CNC machine will be the maintenance manual for the machine. If the configuration is not shown there, ask a factory technician to fax you a copy.

The next consideration is making sure both devices are set to the same communication parameters. You should normally check your machine's manual for its parameters. They are fixed on some machines and cannot be changed. In this case we just set the computer's parameters to what the machine requires. First the line speed must be set; 9600 baud is very common for devices. The number of data bits, which will be either 7 or 8, must then be set. Next the parity is set. The parity will normally be odd, even, or none. The number of stop bits must then be set. This will normally be 1, 1.5, or 2. Study the communication section of the machine's manual to see if there are other parameters that you need to set.

You will then set the required parameters in the computer software. You will probably be using CNC programming software to communicate with the machine. The process will normally require you to create a file with the specific parameters required for your machine. Once you have entered the parameters, you will save them in a file that you can use from then on to automatically set the parameters for the machine.

The next step would be to turn on the CNC machine and go into the communications mode. On many machines this will be called *tape I/O* because of the use of tape on older machines. Attach the communications cable to the CNC machine and the serial port on the computer. You must then tell the machine whether you are going to upload or download a file. If you are uploading to a computer, ready the computer and the computer will wait until the machine starts sending the program file. If you are downloading, ready the machine and start the download from the computer. The machine will wait for the program file to be transmitted.

CABLING CONFIGURATIONS

Cables are easy to make. The computer end of the cable will be either a 9-pin or a 25-pin connector. Check the particular CNC machine to see if a male or a female connector is required. The machine connector will normally be a 25-pin

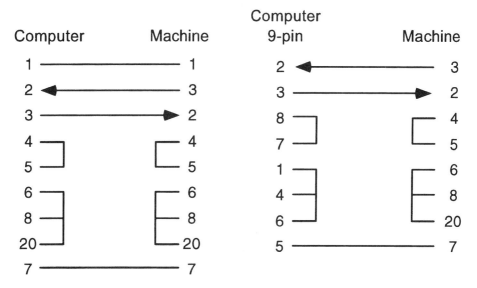

FIGURE 13–22 *Schematic for a cable between an IBM computer and a Bridgeport CNC machine.*

connector. Next find a schematic of the required cable (see Figure 13–22 for a sample communication cable). If your CNC manual does not have it, call the CNC manufacturer and have them fax it to you. Use twisted-pair shielded wire for the cable. The twisted wires are more noise immune and the shielding also helps make the wire more noise immune.

The maximum length recommended for RS-232 communications is 50 feet. Exceeding that can lead to inaccurate communications. Most communications problems are usually cable related, so do a good job of soldering.

If you do not have CNC programming software available for communications, use simple communications software, such as Procomm, for downloading and uploading files. The process is the same. Start the communication software and set the parameters to match your CNC machine. Set up the CNC, attach the cable to the CNC and the computer, tell the software which file you want to upload to or download, and start transmitting.

A SIMPLE COMMUNICATION NETWORK

In many shops there may be more than one computer on which programming is done. There may also be a plotter, a printer, and multiple CNC machines with which you must communicate. In many shops this means constantly switching cables between devices and computers, which leads to confusion and cabling problems, as well as headaches. A simple solution is a switchbox between the computers and devices (see Figure 13–23).

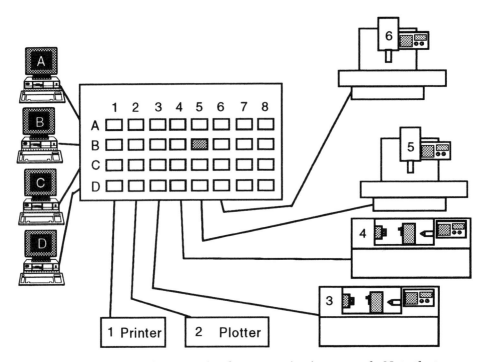

Figure 13–23 *Setting up a simple communication network. Note that four computers were connected to the input ports of the switch box. One printer, one plotter, and four CNC machines were connected to the output ports. All the operator needs to do is choose the row that the computer is on and the column that the machine is on and push the button that is intersected by the two. In the diagram the operator needs computer B to talk to machine 5.*

CHAPTER QUESTIONS

1. In what ways are hardware communication hierarchies like human organization?
2. What are the primary characteristics of the device level?
3. What are the main functions of the cell control level?
4. Describe the term *serial communications*.
5. Describe the term *asynchronous*.
6. Why should you collect so much data from the plant floor?
7. Why is it important to be able to upload and download CNC programs?
8. What is required to communicate between a CNC machine and a computer?
9. Describe how to communicate between a computer and a CNC machine.
10. Find a schematic for one of your CNC machines and draw a wiring diagram. Make sure you indicate what type of connectors will be required on each end.

Chapter 14

··

FUNDAMENTALS OF STATISTICAL
PROCESS CONTROL

INTRODUCTION

Statistics can be a very practical, simple tool when properly applied. The most important fundamental in the effective use of statistics is data. Data must be accurate and relevant. Types and evaluation of data will be examined in this chapter.

OBJECTIVES

Upon completion of this chapter, the reader will be able to:

- *Explain the importance of accurate, relevant data.*
- *Explain terms such as "attribute," "variable," and "histogram."*
- *Explain the advantages and disadvantages of attribute and variable data.*
- *Organize data to make it easier to understand.*
- *Code data and construct histograms and frequency distributions.*
- *Explain terms such as "assignable," "chance," "special," and "common."*
- *Explain the rules of variation.*
- *Define and calculate averages.*
- *Define and calculate measures of variation, such as range and standard deviation.*
- *Explain terms such as "normal distribution" and "bell curve."*
- *Predict scrap rates etc. by using statistical means.*

INTRODUCTION TO STATISTICAL PROCESS CONTROL

Today's manufacturing environment is very competitive. Enterprises have to be competitive in quality and price, which means that production processes have to be very efficient and repeatable to make consistent parts in a timely manner. Statistical process control is one way to improve and keep machining processes under control so that they produce accurate, cost-effective parts.

How much money is wasted in industry because of the cost of making things wrong? Most experts agree that the average industry spends between 15 and 20 percent on the costs associated with doing things wrong. A very successful company might make about 5 percent profit, but most companies are throwing away three to four times that much! Pretend we are a small company. We make a widget that costs us 95 cents. We sell the widget for $1 and make 5 cents on every one. What types of expenses go into the 95 cents that it costs to make our widget? Certainly the material cost and the labor cost to make it, and also some of the cost of the machine, tooling, and maintenance. We have to pay the foreman, supervisor, inspectors, sales people, secretaries, president, and so on. We have to pay for the building, utilities, and many other things.

What does it cost if we make a defective widget? Does it cost us our 5 cent profit? Of course! Does it cost us any more than 5 cents? Yes, we lost the 95 cents we had invested in that part, and probably more. We now have to fill out a scrap ticket, our foreman will have to take time to deal with it, and we might have to order more raw materials. How many parts do we have to make to make up for this loss? We have to make at least 20 parts: We make 5 cents on each part, so 20 * 5 will make up for our loss. We don't make any money on those 20 parts, however; we just try to recover our loss. Also while we are making the 20 parts we can't be making anything else, so we lose potential profit there too. As you can see, the costs of scrap are very high! Statistics help reduce these costs.

Data is crucial in making manufacturing successful because statistical process control cannot work without accurate data. If we have inaccurate data because of inaccurate measuring equipment (out of calibration) or because people are using it improperly, we cannot make quality parts.

There are several important considerations for data collection.

> *Make sure that the data is accurate. Consider the gauges, methods, and personnel.*
>
> *Clarify the purpose of collecting the data. Everyone involved should realize that the purpose is quality improvement. We are not collecting data to make people work harder or get them in trouble.*
>
> *Take action based on the data. When we have learned statistical methods, we will make changes or adjustments to processes based only on the data.*

The use of data and statistical methods also helps depersonalize problems. Too often, problems become personalized and politicized. No one wants to be blamed for a problem. We would typically prefer to blame it on engineering, or purchasing, or some other department.

When we start discussing problems based on data it depersonalizes a problem. Methods that use data identify problems with processes. The problem should then be discussed in terms of data, not in terms of whose fault it is.

TYPES OF DATA

There are two types of data: attribute and variable.

ATTRIBUTE DATA

Attribute is the simplest type of data. The product either has the characteristic (attribute) or it doesn't. If blue is the desired attribute and the product is chairs, we would have two piles of products after we inspected them. One pile would have the desired attribute (blue), and the other pile would have chairs of any other color.

Attribute data can also be go/no-go type data. Go/no-go gauges are often used to check hole sizes. If the go end of the gauge fits in the hole and the no-go end doesn't, we know the hole is within tolerance.

This type of inspection gives two piles of parts: parts that are within the tolerance and parts that are not (good parts and bad parts). Attribute data does not tell us how good or how bad the parts are. We only know they are good or bad.

Attribute data can provide useful information for decision making.

Attribute data is easy to gather and analyze. Measurements are relatively straightforward for some attributes. A go/no-go gauge is almost infallible; however, some attributes are harder to measure. If the attribute is blue color, it would be easy unless we needed a certain shade. Then it becomes more difficult and subject to error. We can lessen the possibility of error by having a sample to compare against. For example, we might have a sample of the desired shade, one that is too light, and one that is too dark. The inspector can compare the part to the sample if there is any doubt.

Attribute data can also be called *discrete data* because it is either good or bad.

VARIABLE DATA

Variable data can be much more valuable. It not only tells us *whether* parts are good or bad, it tells us *how* good or how bad they are.

Variable data is generated using measuring instruments such as micrometers, verniers, and indicators. If we are measuring with these types of instruments, we generate a range of sizes, not just two (good or bad) as with attribute data. We could end up with many piles of part sizes.

This is beneficial because we know how good or bad the parts are. Were they very good or not so good? We can also use this data later to make predictions and decisions about processes.

Errors are possible when we ask people to measure with variable gauges. Instruments must be checked against standards on a regular basis to assure they are accurate. This is called *calibration*. We must also make sure all operators are using the same inspection method and that they thoroughly understand the tool they are using for inspection.

Variable data could also be called *analog data* because it can assume a range of sizes, not just good or bad.

Gauges must be appropriate for the job. A good rule of thumb is that the gauge should be able to measure at least 1/10 of the tolerance. For example, if our blueprint specification was 1 inch ±.005, our total tolerance would be .010. Our gauge should at least be able to measure to within .001 inch. Variable data can yield more information about a process.

CODING DATA

Numbers can become unmanageable when they have many digits. For example, look at the numbers below. Can you add them in your head and find the average?

1.7431 1.7426 1.7437 1.7438 1.7439

Even though there are only five numbers, it is cumbersome. It would be easier if we could work with one-digit numbers. Coding is the way we can do this (see Figure 14–1).

Consider the following blueprint specification:

1.000 ± .005

Let the desired size of 1.000 be equal to zero.

1.000 = 0

If we get a part whose size is 1.000, think of it as zero.

1.005=5	1.006=6	.999=-1	.994=-6
1.009=9	1.002=2	1.001=1	1.000=0
.999=-1	.997=-3	1.002=2	1.0001=1

FIGURE **14–1** *Coded sample part sizes.*

We need to measure to the third place to the right of the decimal point (.001). Let that place equal ones. If a part is 1.002, we will call it a 2 because it is two larger than our desired size of 1.000 (zero).

   1.002 = 2

If the next part were .999, we would call it –1.

   .999 = –1

This is much easier to do than writing down long decimal numbers.

How do you decide which place to assign the value of 1? You simply look at the specification and the number of decimal places. If the specification has three decimal places (5.000, for example), you would make 1 equal one thousandth of an inch. If there were four decimal places a 1 would equal .0001.

Which numbers would you rather work with, the actual size or a coded value? If you needed to find the average of the numbers, you could just figure the average for the coded values. Because coded values are much easier to work with, most companies use coded values when they use statistical methods. It also helps if we are discussing a process. If I say the last part was 2.505, you really have no idea about how good the part was if you don't know the blueprint specification; however, if I say the last piece was 5, you know that it was five above the desired size. Coded values are much easier to work with.

GRAPHIC REPRESENTATION OF DATA

Look at the data in Figure 14–2.

What conclusions can you draw? It can be confusing to look at large amounts of data.

It is easier to deal with if we have a picture of it (graphical representation). It would also be easier if it were coded. One method of presenting data graphically is called a *histogram.*

Look at the data below. These numbers represent coded values for part sizes that were produced.

   1, 4, 1, 4, 1, 3, 2 ,3, 3, 2, 2, 2, 3, 1, 2, 1, 1, 0, 0, 1, 2, 0, 1, 0, 0, 1, 0, 0, 3, 0, 1, 1, 0

1.023	1.021	1.025	1.021	1.021	1.019	1.023	1.023	1.019	1.022	1.017
1.021	1.020	1.024	1.023	1.012	1.021	1.014	1.022	1.020	1.023	1.023
1.019	1.018	1.022	1.022	1.019	1.021	1.021	1.018	1.019	1.021	1.018

FIGURE **14–2** *Some sample part sizes.*

Now look at Figure 14-1. This histogram makes it easy for us to see patterns in data. You can instantly see that the most common size was zero. We could see that the average part size was also about zero. This histogram is much easier to analyze than the list of data.

The only thing we lose in the histogram is the time value. We do not know if the zeroes were the first parts or the last, or if they were distributed throughout the production.

BASICS OF VARIATION

Is it true that all things vary and there are no two things exactly alike? Yes. Fingerprints, snowflakes, and so on—no two are exactly alike. The same is true of a machine that makes steel washers. If we look at the washers, they all look the same. If we measure them with a steel rule, they all measure the same. But if we use a micrometer, we would find differences in every one.

The appropriate measuring instrument will find differences between any two things.

The first rule of variation is straightforward: *No two things are exactly alike.*

Rule number two, *variation can be measured,* is also straightforward. No matter what our product or process, no two will be alike (rule 1), and we will be able to measure and find differences in every part (rule 2).

This assumes that we use an appropriate tool to measure the parts. It also assumes that we have been trained in the correct use of the measuring tool.

The third rule of variation is *individual outcomes are not predictable.* What would happen if we flipped a coin 10,000 times? We would predict that we would get approximately 5000 heads and 5000 tails. We could predict that approximately 50 percent of flips would be heads and 50 percent tails. This would be a very accurate prediction.

What if we flipped the coin 10 times? Our prediction might not be as accurate. We might get seven heads and three tails (not a very good prediction). But if we flipped a larger amount of times, we could predict quite accurately that heads and tails would each occur about 50 percent of the time. Assuming we didn't cheat, 50 percent would be heads and 50 percent tails.

Can we predict that the next flip will be a tail? No! We can never predict an individual outcome.

For example, if we were running a lathe and the last piece was 1.001, can we predict that the next piece will be 1.001? No, because we cannot predict individual outcomes.

Rule number four states *groups form patterns with definite characteristics.*

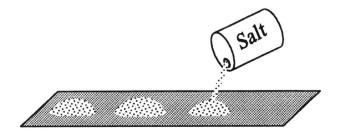

FIGURE 14–3 *A simple process.*

Think of a simple process involving a large salt container (see Figure 14–3). The process is to dump three ounces of salt from a point 6 inches above the table. The product of the process is the salt piles that result. If we were very careful, the piles of salt would appear to be the same. The diameter and height would look identical. We could predict the size of the pile if we were to run the process again. In fact, we could make very accurate predictions. We could not predict where one grain salt would fall, however (rule 3). But rule 4 tells us we can make predictions about groups.

Would all of the piles be the exact same size? No. Rule 1 tells us no two things are exactly alike.

Why wouldn't each salt pile be exactly the same? There are many reasons.

The height of drop varies. The amount of salt dropped varies. How fast the 3 ounces was poured varies. Air currents in the room change. Humidity changes. The shapes of salt grains vary. The surface of the table is not exactly the same all over. Some of these would cause large variations, some small. Can you think of any other reasons?

CHANCE AND ASSIGNABLE VARIATION

There are two types of causes: chance and assignable.

Assignable causes of variation are those causes that we can identify and fix. For example, the height from which the salt was dropped varied. How could anyone maintain exactly 6 inches of height every time? Now that we have

---

No two things are exactly alike

Variation can be measured

Individual outcomes are not predictable

Groups form patterns with definite characteristics

identified it as an assignable cause, we could make a simple stand to maintain the correct height for the process and eliminate height as a cause of variation.

The amount of salt dropped also varied. Could we expect a person to drop exactly 3 ounces of salt every time? The amount of salt is also an assignable cause. How could we remove this cause of variation?

Statistics will be used to identify assignable causes of variation.

Chance causes of variation always exist. Chance causes of variation are those minor reasons for which processes vary. We cannot quantify or even identify all of the chance causes. Room humidity changes may affect the process. Normal temperature fluctuation might affect the process. Normal fluctuations in air currents will affect the process. We really can't separate the effects of these chance causes. Some may even cancel out the effects of others.

You can never eliminate all chance causes of variation. If you are able to identify a cause and its effect, it becomes an assignable cause.

Statistical methods will help us identify assignable causes. Chance causes cannot be separated and evaluated. Chance causes, often called *common causes,* are reasons for variation that cannot be corrected unless the process itself is changed. For example, if we moved our salt process to a temperature- and humidity-controlled room, we would minimize the effects of humidity and temperature. It takes process change to correct common causes.

Assignable causes are often called *special causes.* These are things that go wrong with a process. For example, the operator starts to pour the 3 ounces, but runs out of salt after 1 ounce. The operator refills the container and pours 2 ounces more. This is a special cause of variation. It is something out of the ordinary that occurred in the process. The operator could solve this problem without a process change by just making sure the container is full before each pour.

This is the main difference between special and common causes of variation. Special causes are things not normal to the process. In other words, something has changed in the process. The operator can often identify these problems and correct them without management action or process change. Common causes of variation are causes that are inherent in the process. The only way to correct the effect of the common causes is to change the process. Only management can change processes; therefore, it is management's responsibility to reduce common causes through process changes.

This is why most experts believe that management is responsible for 85 percent of all of the problems in an organization. Only management can change processes. In other words, when there are problems with quality in an organization, the problem in 85 percent of the cases will be due to the process, not the person.

We now know that variation is normal; all things vary. We also know that we cannot make predictions about individual outcomes (next flip of the coin), but we can make predictions about groups of outcomes. (If we flip a coin 1000 times, we will have approximately 50 percent heads and 50 percent tails). We need some ways to describe the characteristics of a group of parts.

AVERAGE (MEAN)

One thing that is helpful in understanding a group of data is what the average or mean is. *Mean* is just another term for average.

For example, if we measured the heights of all persons in a class of 30 students, we could determine the average height. We would add all of the heights up and divide by 30 (the number of students in the class). Statisticians prefer to use the term *mean*. They also have standard notation for data. One person's height, for example, would be called an "x." X stands for one data value. The notation for the average of the data values is $\overline{X}$ (pronounced *x-bar*). Whenever you see a letter with the bar symbol above the letter, it means average or mean. A letter with two bars above it means average of the averages.

Example: heights of five people in inches

$$x = 60, x = 70, x = 75, x = 80, x = 70 \quad \text{Total} = 355/5 = 71 \text{ inches}$$

This means that the average of the individual x's was 71 inches. Remember—average and mean are the same.

MEASURES OF VARIATION

All things vary. It would be desirable to have terms that we could use to describe how much a process varies.

RANGE

One term used to describe variation is range. Range is usually symbolized with the letter "R." The range is simply the largest value in the sample minus the lowest value in the data.

For example:

A person shopping for a new car looked at five cars. Their prices were $8000; $15,000; $9500; $8500; $12,000. The range of values would equal $15,000 minus $8000. The range of car prices that this person looked at was $7000.

Example 2:

A food manufacturer measured five consecutive cans of a product and found the weights to be 7.1, 6.9, 7.2, 7.0, and 7.2 ounces. The range would be the highest value (7.2) minus the lowest value (6.9), or .3 ounces.

Range is a very useful, simple value and is a very useful look at variation. It gives us a good quick look at how much a process or sample varies. If we say a process range is .005, we have a good idea of how much the process varies.

STANDARD DEVIATION

The second term we can use to describe variation is *standard deviation*. Standard deviation can be thought of as the average variation of a single piece. If we had a sample of pieces and the mean was 10 and the standard deviation was 1, we could say that the average piece varies about 1 from the mean of 10. A statistician would define standard deviation as being the square root of the average square deviation of each variate from the arithmetic mean. (Forget this definition, by the way.) It is more useful to think of standard deviation as the *average variation of a single piece.*

There is an elaborate formula to calculate standard deviation. The formula is not as complex as the definition, but it is tedious and time consuming, and mistakes are easy to make.

Fortunately most calculators will calculate standard deviation. All you have to do is input the values, and the calculator outputs the mean and standard deviation.

Sample data:

5, 7, 9, 6, 4, 2, 1, 7, 5, 4

These data values were entered into a calculator, and the standard deviation for the sample was found to be approximately 2.4. The calculated mean was 5.

You should find a calculator that figures standard deviation and try to calculate the standard deviation and mean for the sample.

Calculators can calculate standard deviation for a population or for a sample. Standard deviation for a sample is represented as s or $N - 1$. Standard deviation for a population is represented by σ or N. You will need to use the standard deviation for a sample.

Standard deviation is a very useful term. It can help us visualize processes.

NORMAL DISTRIBUTION

If we chose 25 adult men at random and measured their height, some would be very tall and some very short, but most would be about average size. If we plotted a histogram of their heights, it would look something like the one shown in Figure 14–4.

This distinctive shape is called a *normal distribution* or *bell curve*. A normal distribution looks like a bell.

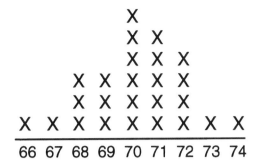

FIGURE 14–4 *A histogram of the heights of 25 men taken at random. The sizes are in inches.*

Most processes produce parts that would be normally distributed. If we measured a dimension on 25 parts from a lathe and then drew a histogram, we would expect a "bell" shape or normal distribution. A few parts would be large, a few small, but most of them would be average.

It should be remembered that we cannot make things exactly alike. There is always variation that makes our processes vary.

The larger the number of parts that we check (or heights of people), the more our histogram would look like a bell shape. More data means more information.

There are some very useful things about normal distributions (bell curves). Consider an example. Assume you measured the height of 50 adult men at random. Their mean height was 5 feet 10 inches (see Figure 14–5). The standard deviation of their heights was also calculated and found to be 2 inches. The bell is normally drawn as being 6 standard deviations wide. In this case the standard deviation was 2 inches, so the bell would be drawn with a width of 6 * 2 or 12. The bell is then broken into six areas. The two areas closest to the middle of the bell each contain 34 percent of all the people's heights. The next two each contain 14 percent of the heights. The last two each contain approximately 2 percent. This is very useful because the same relationship exists for any normal process. The percentages will always be the same.

Bell curves are set up so that if we know the standard deviation, we can make predictions from the data. Thirty-four percent of all adult males' heights should be between the mean and plus one standard deviation. Thirty-four percent of all adult males' heights should fall between the mean and minus 1 standard deviation. Fourteen percent should fall between plus 1 standard deviation and plus 2 deviations. For this example, we know that the standard deviation equals 2 inches, so we could predict that 68 percent of all adult males will be between 5 feet 8 inches and 6 feet tall.

Statisticians have found that 99.7 percent of all part sizes, heights, or whatever we are measuring will be between ±3.

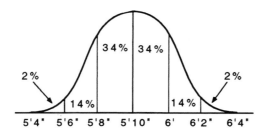

FIGURE 14–5 *Bell curve for the heights of 50 men taken at random. The mean height for this sample was 5 feet 10 inches tall. The standard deviation was calculated and found to be 2 inches. This means that, based on this sample, 34% of all men would be between 5 feet 10 inches and 6 feet tall. We can make all kinds of predictions about men's height. We could say that approximately 14% of all men would be between 5 feet 6 inches and 5 feet 8 inches tall. If our sample was truly random, we could make very good predictions with the data.*

Consider another example. A lathe is producing pins. The outside diameter was measured on 40 pins from the process as they were being run. The standard deviation was measured and found to be .001. The mean (average) was found to be .300.

Consider the bell curve in Figure 14–6. We could now apply these actual part sizes to the bell curve. The mean is .300. This lets us predict that (assuming we don't change the process) 48 percent of all parts should be between .298 and .300.

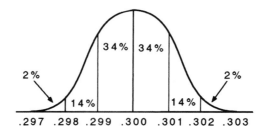

FIGURE 14–6 *A bell curve for a process producing parts of sizes .297 to .303 inches. Note that 34% are between .299 and .300 inches. Another 34% are between .300 and .301. What percentage of parts are between .302 and .303 inches? (2%)*

What percent would be between .298 and .302? (96%); .302 and .306? (approximately 2%); .296 and .298? (approximately 2%).

This is very valuable information. We can make very accurate predictions about how this process will run in the future.

Consider the example of people's heights again. Could we make predictions about the heights of all people in Wisconsin from our data? No, because our data was based on adult males only. We could make pretty good predictions about the adult male population of Wisconsin.

The other point to remember is that the more data we have, the better our prediction can be. If we sampled five people's heights, our prediction would not be very good. The larger the sample size, the better predictions we can make. If we could measure all of the adult males in Appleton, we could make almost perfect predictions for the adult male population of Appleton, but not for Wisconsin or anywhere else.

The term "population" means that we are able to measure all of the items of interest, the entire population. This is a very rare case.

If we produced 1000 parts on a lathe and measured them all, it would still be a sample if we are going to run more parts. So, we will normally be working with samples, not populations.

We can make very accurate predictions from relatively small samples, and it is cheaper to use samples. Time is money, and inspection takes time. Statistical methods will help us make predictions that are almost 100 percent accurate.

CHAPTER QUESTIONS

1. What is attribute data?
2. What is variable data?
3. What are some rules concerning the collection and use of data?
4. What is coding and why is it used?
5. What is a histogram and why is it used?
6. Code the following data:

   Blueprint specification = 1.126. (Hint: 1.126 = 0.)

Blueprint Specification = 1.126				
1.124=	1.128=	1.123=	1.127=	1.121=
1.119=	1.127=	1.125=	1.119=	1.127=
1.129=	1.126=	1.118=	1.121=	1.116=

7. Code the following:

Blueprint Specification = 1.2755			
1.2752=	1.2749=	1.2752=	1.2754=
1.2759=	1.2750=	1.2761=	1.2752=
1.2748=	1.2756=	1.2752=	1.2756=
1.2753=	1.2755=	1.2747=	1.2749=

8. Code the following:

Blueprint Specification = 2.105			
2.109=	2.103=	2.108=	2.113=
2.102=	2.101=	2.096=	2.104=
2.101=	2.100=	2.100=	2.111=
2.098=	2.100=	2.099=	2.109=

9. Code the following:

Blueprint Specification = 5.00				
5.03=	5.06=	5.02=	5.02=	5.07=
5.02=	5.05=	5.01=	5.04=	5.03=
4.98=	4.97=	5.00=	5.01=	4.93=

10. Using the coded data from question 6, construct a histogram.
11. Using the data from question 7, construct a histogram.
12. Define and explain variation.
13. How does an assignable cause of variation differ from a chance cause of variation?
14. Briefly define the following key words.
    a. mean
    b. histogram
    c. normal distribution
    d. bell curve
    e. range
    f. standard deviation
15. Consider the following data. Blueprint specification = 2.250.

#1	#2	#3	#4	#5	#6
2.252	2.249	2.248	2.252	2.246	2.250
2.252	2.249	2.254	2.253	2.248	2.249
2.253	2.243	2.252	2.251	2.247	2.252
2.250	2.248	2.251	2.250	2.246	2.248
2.252	2.249	2.247	2.250	2.249	2.251

    a. Code the data. (Hint: 1 should equal 2.251.)
    b. Find the sample standard deviation.

16. Blueprint specification = 2.0000.

Subgroup 1	Subgroup 2	Subgroup 3
2.0005	2.0005	2.0002
1.9997	2.0004	2.0004
2.0009	2.0001	2.0000
2.0006	2.0004	1.9995
1.9999	2.0002	2.0001

    a. Code the data. (Hint: 2.0005 = 5.)

    b. Find the mean.

    c. Find the sample standard deviation.

17. Draw a bell curve and label with standard deviations and mean. Label the percentages for each deviation. Note: You don't have data concerning the actual mean and standard deviation. Draw a generic bell curve.

18. Consider the following coded data.

    5, 0, 0, 1, 1,4, 2, –1, –1, –1, 2, –1, 0, 0, 0, –1, –2, 1, 2, 0, –2, –1, –1, –3, –2

    Use a calculator.

    a. Calculate the mean.

    b. Calculate the sample standard deviation.

19. Consider the following data.

    9, 0, 1, –1, 2, 8, –1, 0, 0, 3, 7, 4, 2, –3, 8, 2, –4, 3, –5, 5, 3, –2, –2, 0, 3

    a. Calculate the mean.

    b. Calculate the sample standard deviation.

20. Consider the following data.

    –3, 2, –1, –2, –4, –1, –2, 0, –2, 3, 0, –1, –1, 4, 2, 2, 0, 0, 1, 4

    a. Calculate the mean.

    b. Calculate the sample standard deviation.

21. A machining process is studied and the mean and standard deviation were calculated: $x = 5$, $s = 2$ (coded data).

    a. Draw a bell curve.

    b. Draw lines where the 6 standard deviations would be.

    c. Label them with actual sizes from this process. (Hint: 99.7 percent of all parts should lie between –1 and +11.)

    d. Label the percentages.

22. Holes are measured after a drill press operation. The mean was found to be 1 (coded data) and the standard deviation was found to be 3 (coded data).

    a. Draw a bell curve for this data.

    b. If 1000 parts are run, how many will be between –5 and –2?

    c. How many will be between –2 and +4?

    d. If our tolerance was –2 to +7, how many scrap parts would we have?

23. Consider the following coded data.

    3, 3, 3, 3, 5, 4, 2, 4, 2, 1, 5, 4, 3, 2, 4, 6, 1, 4, 3, 2

    a. Construct a histogram.

    b. Does it look like a normal distribution?

24. Consider the following coded data.

    1, 4, 2, 5, 2, 5, 1, 1, 4, 1, 3, 5, 5, 1, 5, 1, 5, 3, 5, 1, 5, 4, 2, 1, 0

    a. Construct a histogram.

    b. Does it look like a normal distribution?

    Note: This one does not look like a normal distribution. This means that something was wrong with our process. Data from a normal process should be normally distributed.

Chapter 15

STATISTICAL PROCESS CONTROL

INTRODUCTION

Control charts are very valuable. If properly used, they can very accurately predict a process's performance. Charts can show when a process has changed, when it should be adjusted, when tools should be changed, and when maintenance should be done, and they can even help find out what has gone wrong with a process.

OBJECTIVES

Upon completion of this chapter, the reader will be able to:

- *Explain at least five benefits of charting.*
- *Explain how charts can improve processes.*
- *Calculate the capability for a process.*
- *Construct and properly use $\bar{X}R$ charts.*

PROCESS CAPABILITY

The real reason to collect data about processes is to use the data to improve our processes. Capability gives us concrete data on how good or bad our processes or machines are.

Consider a simple example. The most common operation on a lathe would be turning an outside diameter to a specified size.

We could perform a study on a lathe to establish the lathe's capability to turn diameters. The term "capability" refers to how close of a tolerance a machine can produce parts. If a lathe can turn diameters to within ±.001, we could say its capability for turning diameters is .002.

To actually conduct the capability study, we would first make sure there is nothing obviously wrong with the machine, tooling, or material. Next, we would run some pieces. Note: We cannot make changes or even adjustments to

Actual diameters and coded values for 25 turned parts				
1st 5	2nd 5	3rd 5	4 th 5	5th 5
1.123 = -2	1.124 = -1	1.123 = -2	1.122 = -3	1.127 = 2
1.128 = 3	1.125 = 0	1.125 = 0	1.124 = -1	1.126 = 1
1.127 = 2	1.128 = 3	1.126 = 1	1.125 = 0	1.127 = 2
1.126 = 1	1.126 = 1	1.127 = 2	1.126 = 1	1.127 = 2
1.127 = 2	1.125 = 0	1.127 = 2	1.127 = 2	1.124 = -1

FIGURE 15–1 *Diameters and the coded values for the 25 diameters that were turned to study this machine.*

our process during the study. We want to see how closely the machine can hold sizes, so we must leave the machine alone, make no changes, and just run the parts. The machine is being studied, not your ability to adjust it. It would ruin the results if it were adjusted or changes were made while it was running.

For this example, assume 25 parts were run and the outside diameter was measured.

The blueprint mean was 1.125.

The actual sizes of the parts are shown in Figure 15–1 (1.125 = 0).

The standard deviation (sample standard deviation) was calculated and found to be 1.67. You should remember (from Chapter 14) that 6 standard deviations is equal to 99.7 percent.

If we multiply 6 * 1.67, we get 10.02. (Remember, this is a coded value.)

This means that if nothing changes in this process, we could run 99.7 percent of all pieces within a range of approximately 10 (coded value). In this case, 1 coded equals .001. We could run 99.7 percent of all pieces on this lathe within approximately .010. (This lathe is obviously not very accurate, but the numbers will be easy to work with for this initial example.)

If the blueprint tolerance is ±.005 (.010 total), we will get almost no scrap (see Figure 15–2).

There is almost no scrap because 6 standard deviations (.010) is equal to our tolerance of .010. (This assumes that we can keep the process exactly on the mean.)

Consider a different job for the same machine. This job has a tolerance of ±.010. This means that the total tolerance is equal to .020 (see Figure 15–3).

The drawing makes it clear that this process should produce no scrap. The blueprint tolerance is twice as wide as the process capability.

For an additional example, assume the same process capability (6 SD = .010). This time the blueprint tolerance is equal to ±.003. Study the diagram shown in Figure 15–4. (One standard deviation has been rounded from .00167 to .0017.)

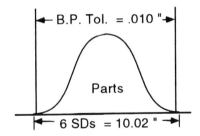

FIGURE 15–2 *Bell curve for a job that has a total tolerance of .010 inches and 6 standard deviations (SDs) that are equal to 10.02 inches.*

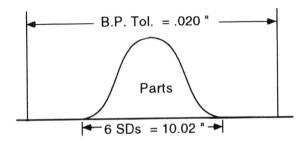

FIGURE 15–3 *Bell curve for a job where the process capability is about 1/2 of the blueprint tolerance. This would be a great job, and it would be very hard to produce scrap.*

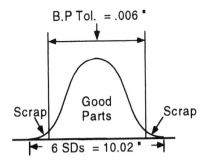

FIGURE 15–4 *Bell curve for a process that would produce approximately 4 percent scrap.*

In this example, the blueprint tolerance is located at about 2 standard deviations. This means that the best this process could do would be about 96 percent good parts (if the process could be kept on the mean). Yelling at the operator will not help reduce the scrap rate. The best that the process can do is 96 percent good parts. Unless the process is changed, it will produce a minimum of 4 percent scrap.

$$CP = \frac{\text{Total Blueprint Tolerance}}{\text{Six Standard Deviations}}$$

FIGURE 15–5 *The formula for simple capability.*

The more parts run, the better our estimate of capability will be. In other words, the more we run, the more confident we could be of our results and the better prediction we could make about running these parts in the future.

Capability is usually not expressed in terms of 6 standard deviations alone.

We would like to be able to compare the process capability to the blueprint specification of the actual job we will be running. This will show how good (or bad) the job is.

One way in which capability is expressed is called *CP* (see Figure 15–5).

You can see that we are comparing the total print tolerance to the process capability (6 standard deviations) and calling the result CP.

Figure 15–6 is used to grade a process for a particular job.

In the first example, the total blueprint tolerance was .010. Six standard deviations was .010. The CP = .010/.010 or 1 (CP = 1). If we look at our capability chart, this job would be classified as a B-C job (not real good, not real bad).

In example two, the blueprint tolerance was .020, and 6 standard deviations was .010.

The CP = .020/.010 or 2 (CP = 2). This job is excellent, and it should be easy to run the job without scrap. In fact, it should be almost impossible to run scrap.

In example three, the blueprint tolerance was .006, 6 standard deviations equaled .010.

The CP = .006/.010 = .6 This job is very, very poor. There is no way we can run this process without producing scrap.

CP is one way to express capability. Another way to express capability is easier to use because it is a percentage. The higher the percentage the worse the job (see Figure 15–7).

Capability
Above 1.33 - A Excellent
1 to 1.33 - BC
Below 1 - D Poor

FIGURE 15–6 *Figure used to grade a job's capability.*

$$CP\% = \frac{\text{Six Standard Deviations}}{\text{Total Blueprint Tolerance}} * 100$$

FIGURE 15–7 *Formula used to find capability in terms of a percentage. The lower the percentage, the better the capability.*

In the first example, 6 standard deviations equaled .010, and total blueprint specification was .010.

.010/.010 * 100 = 100.

The CP% is equal to 100 percent.

This means that our process capability is 100 percent of our tolerance.

In example two, 6 standard deviations was .010, blueprint tolerance was .020.

.010/.020 * 100 = 50%

This means that the capability is equal to 50 percent. The process capability is half of the blueprint tolerance, an excellent process. Remember: the smaller the percentage, the better the job. (The smaller the CP%, the smaller the chance of scrap.)

In example 3, the process capability (6 standard deviations) was .010. The blueprint tolerance is equal to .006.

.010/.006 * 100 = 167%

This is a poor CP%. If our CP% is over 100 percent, we will definitely produce scrap.

Many industries have a CP% goal of 66 percent. They will try to improve processes so that they have a CP% of 66 percent or lower.

There are many advantages to low CP%s:

Better parts (parts are closer to the blueprint mean)

Processes are easier to run

Less scrap

Less frequent inspection is necessary

Higher productivity through quality and process improvement

One example of capability comes from the auto industry. One of the automakers had one model of transmission it wanted to study using statistics.

The automaker made some of the transmissions and purchased some from another automaker. The transmissions were all made to the exact same specifications and tolerances.

The automaker had noticed that the other manufacturer's transmissions seemed to perform more reliably (less warranty work and customer complaints about noise, and so on).

They decided to choose a number of transmissions at random from the other automaker and the same number at random from their production. Inspectors were then assigned to completely disassemble the transmissions and inspect them. They checked everything possible, including torque of nuts and bolts and the sizes of all parts.

They found that all of the parts of the other automaker's transmissions met all of the specifications. It was also found that all parts of all their own transmissions met all of the specifications.

If all of the transmissions met all of the specifications, why did the other automaker's transmissions perform more quietly and reliably?

We would assume that if all parts met specifications, they would perform the same.

This is not true, however. When they analyzed the data, they found that the other automaker had much smaller CPs than their own CPs. (In other words, the other automaker's processes had better capabilities.)

Note that they had exactly the same tolerances. The other automaker just had better processes because they used statistical methods and had improved them.

It should be clear that the blueprint mean (center of the blueprint tolerance) should be the ideal size (the size at which the part performs the best). The other automaker made all parts closer to the mean than they did (see Figure 15–8). Thus, their transmissions were quieter and more reliable.

In fact, the other automaker typically had smaller CPs than many other producers. They accomplished this through process improvement using statistical methods.

Capability, one of the most valuable uses of statistics, tells us much about an enterprise's capability. For example, if we studied all areas of a business and developed capabilities for each machine, we would know what kind of tolerance we could hold on each machine.

It can show us the parts of the business that we need to improve (or even drop). If the lathe department has problems and we cannot hold the tolerances we need to for our customers, this would help us to make the decision to rebuild equipment, buy new equipment, look for work that does not require close tolerances, or decide not to do lathe work anymore.

It also can demonstrate an enterprise's strengths.

This can help us bid on work that is more profitable and show us what not to bid on. It can help direct our maintenance efforts and show where to invest in new equipment. It is very useful for the processes where it is difficult or impossible to shift the mean.

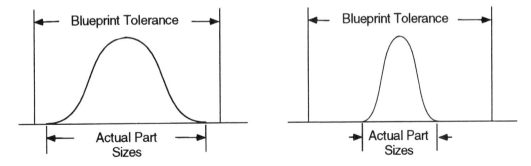

FIGURE **15–8** *Two bell curves for two processes making the same part, and the tolerance is the same. The bell curve on the left shows the process bell curve for one company making the part. The bell curve on the right shows the process bell for another company making the same part. Which company makes better parts? Which company has less variation in their parts? Which company's process runs better? Which company's parts would you want in your transmission?*

Charts can be a very beneficial tool in industry. The charts we will examine next are designed to be accurate 99.7 percent of the time.

BENEFITS OF CHARTING

ADJUSTMENT REDUCTION

Charts will show when a machine needs adjustment. People tend to adjust machines too much. In fact, the more conscientious a worker is, the more he/she will adjust. This is because the worker notices the small variations in a process and tries to adjust the machine; however, you cannot eliminate all variation, and by making unnecessary adjustments you actually make the process variability much worse.

Dr. W. Edwards Deming was one of the most famous quality gurus in history. He is given much of the credit for Japan's amazing manufacturing success. One of Deming's studies involved a paper coating process (see Figure 15–9).

The coating thickness on the paper was very important. Deming was asked to study the process because it was impossible to keep the thickness consistent.

The worker's job was to measure the thickness and make adjustments when necessary. If the coating was too thick, he closed the valve more; if the coating was too thin, he opened it more. The worker was very diligent.

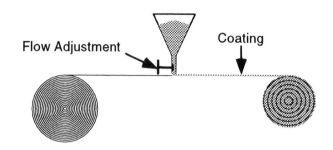

FIGURE 15-9 *Simplified paper coating process. The coating thickness is very important: The more consistent the thickness, the better the product.*

Deming insisted that the process be run with no adjustments to see how good or bad the process was. Everyone believed that the process would run terribly without adjustment. But Deming persisted, and it was run without adjustment.

The process ran very well. There was much less variation in thickness. A chart was put on the machine that showed when adjustment was necessary, and the job became a favorite instead of a problem.

No one knew that variation was normal. The thickness has to vary somewhat. If a machine is adjusted every time a part varies, the process produces much more variation. Charts can make the adjustment decision correctly 99.7 percent of the time.

PROCESS MONITORING

Once a process has been studied and a chart has been constructed, we know how the process should run. The chart will immediately show us when something changes in the process. The chart is 99.7 percent accurate in these tasks.

When we chart, we also are developing historical data on the processes or machines. If we looked at charts over a machine's life, the chart would show the deterioration of the machine or process. If we project this information forward, we could predict when the machine or process will need to be rebuilt or replaced.

CAPABILITY

Charts show a business how good or bad their processes and machines are. This seems strange, but many companies do not know how capable their machines are. Charts help decide which work to run on which machines, and they will also identify which machines need rework or replacement.

Charts will also show a company its strengths and weaknesses. For example, a company might find out it is very good at lathe work, but poor on mill work. It can then either improve its mill work by changing processes, maintaining machines, or replacing machines, or it can devote its capital (money) to producing lathe work and letting some other company do the mill work. Whatever they choose, they will be better off than they presently are.

When a company knows how capable its machines are, it can more accurately bid on jobs. The company will also know, before they get the job, how well it will run.

PROCESS IMPROVEMENT

Charts can become the basis for process improvement. Process improvement means less part variation, and less part variation means higher quality at lower cost. Continual process improvement will yield continually higher quality at continually lower cost.

Imagine the following process—coin flips (heads or tails).

A coin is repeatedly flipped. If it comes up heads, an x is plotted above the centerline; tails, an x is plotted below the centerline.

You would expect that approximately half of the flips would be heads and half tails.

The odds of flipping a heads or tails would be 1/2. What are the odds of flipping two heads in a row? The odds would be 1/2 * 1/2, or 1/4. In other words, there is only one chance in four that two consecutive flips would be heads (or tails). The odds for three in a row would be 1/2 * 1/2 * 1/2, or 1/8.

This should help you understand that we would expect half of our product to be above the centerline and half below. We would also expect that we would not get too many in a row on one side of the centerline. The odds of getting seven in a row above (or below) the centerline are 1/128.

This means that if we are plotting sizes of parts on a chart, we would expect that the sizes would occur randomly above and below the centerline.

There is a very small chance that seven in a row would be on one side of the centerline. A process can be adjusted if seven in a row fall on one side of a centerline. Assume seven part sizes in a row fall below center. The odds are so low that this could happen that we could assume that something has changed in the process. This is the second rule—if seven fall on one side of the centerline, the process has changed and an adjustment or change is necessary. If the average of the seven sizes was calculated, it would give the exact adjustment needed. The chart not only tells when to adjust, but also how much to adjust.

In other words, if seven in a row are above or below center, it means that the mean (average) has shifted up or down.

The only other rule is that a process has changed if seven in a row increase or decrease. This is called a *trend*. If each of the seven in a row gets larger (or smaller), this trend means that the process has changed.

These rules are all based on making the correct decision 99.7 percent of the time.

Rule 1: Do not adjust unless a point falls outside of the limits.

Rule 2: Do not adjust unless seven in a row are on one side of the centerline. (If seven are on one side, the mean has shifted.)

Rule 3: Do not adjust the process unless seven in a row trend up or down. Each one must be larger (or smaller) than the last for the trend. This kind of trend will usually indicate something more than an adjustment is needed. Something has changed in the process.

If you follow these rules, you will make the right decision the vast majority of the time.

CHARTING PROCESSES

The $\overline{X}R$ chart is one of the most widely used charts in industry. The chart is designed to be accurate 99.7 percent of the time.

$\overline{X}R$ means average and range chart. Samples are taken of consecutive parts (usually five) and the average of the sample is plotted on the chart. The range (largest size minus the smallest size) of the sample is also plotted on the chart.

In effect, a chart is really two charts: an average chart with control limits and a range chart with limits.

$\overline{X}R$ CHART CONSTRUCTION

The first step in construction is to gather data from a process. There are often several sizes checked on each piece. One of these must be chosen for a chart. One should try to choose the dimension (or characteristic) that seems most critical. If our process was a lathe part, we might be turning four diameters. Any of the diameters would probably be appropriate to chart. If one diameter were more important, it would be chosen. But if the operations are all very similar, one will be a good indicator of the others.

Once a particular part characteristic has been chosen, data is gathered. It is very important that the process is running well. If we know there is something wrong with the process, it should be fixed before the study is done. The process should be running as well as we think it can (good operators, tooling, etc.).

No adjustments should be made during the study. Because we are trying to study the process, it is imperative that we not change the process through adjustments during the study. We need accurate data on how good the process is, not the operator. The more data that is gathered, the better the results will be. Twenty-five parts will give fairly good results. The data must be consecutive parts.

The data must be accurate. Gauges should be appropriate, and the operator should be proficient in using the gauge. The operator should also understand why the data is being gathered and the importance of accurate data. If this is not done, there is a tendency for a person to be wary and fudge the data. Remember, we are studying the process, not the person.

Rules:

1. *process should be running optimally*
2. *choose an appropriate part characteristic*
3. *do not change the process in any way during the study*
4. *make sure the operator understands the purpose*
5. *use appropriate, well-understood gauges*
6. *gather the data. (It is a good idea to code the data because it is easier to use and understand. It is also harder to make a mistake.)*

See Figure 15–10 for data from a process (coded). As you can see, the data was gathered in groups of five.

The charts we are constructing work only for normal processes. To see if our process is normal, the data must be put into a histogram (see Figure 15–11). A histogram is a simple, quick look that will indicate whether our process is, or is not, normal.

The data in Figure 15–11 looks normally distributed (somewhat like a bell curve). With more data, it would probably look more like a bell. We will assume it is normally distributed. Processes should be normally distributed. If a histogram shows that a process is not normally distributed, something is wrong with the process. The problem in the process must be found and corrected, and new data must be gathered.

We have decided that the process is normal, so we can continue with chart construction. The next step is to find the average (mean) and the range for each group.

Subgroup 1	Subgroup 2	Subgroup 3	Subgroup 4	Subgroup 5
0	2	0	2	2
1	0	-1	1	2
1	-3	-1	-2	1
-1	0	-1	1	3
-2	-1	4	0	-2

FIGURE **15–10** *Part sizes of the 25 parts that were run.*
Subgroup 1 contains the first five parts in order,
subgroup 2 the next five in order, and so on.

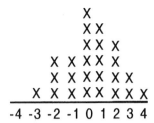

FIGURE **15–11** *Data from Figure 15–10 in a histogram. Each X represents one part. For example, in this histogram there were six parts that had a coded size of 0. The histogram also gives us a rough estimate of what the average would be. It also shows the range of sizes produced. In this case the histogram resembles a bell shape, so the process can be considered normal.*

	Subgroup 1	Subgroup 2	Subgroup 3	Subgroup 4	Subgroup 5
	0	2	0	2	2
	1	0	-1	1	2
	1	-3	-1	-2	1
	-1	0	-1	1	3
	-2	-1	4	0	-2
Total	-1	-2	1	2	6
Average					
Range					

FIGURE **15–12** *The first step in finding the average for each subgroup. Add the five sizes in each subgroup and write the result in the total box.*

First we must find the total for each subgroup. This is done by adding the numbers in each subgroup and writing them in the total column (see Figure 15–12). Study the rest of the subgroup totals.

Next we must find the average of each subgroup. This is done by simply multiplying the total of each column by 2 and writing the result in the average column. Then you must put a decimal point one place from the right (see Figure 15–13). The first column total is 1. Multiply 1 * 2, and the result is 2. Next move the decimal place one place to the left and you get .2. That is the average for the first subgroup; however, this method of averaging works only when there are five values in the subgroup. Study the rest of the subgroup averages.

	Subgroup 1	Subgroup 2	Subgroup 3	Subgroup 4	Subgroup 5
	0	2	0	2	2
	1	0	-1	1	2
	1	-3	-1	-2	1
	-1	0	-1	1	3
	-2	-1	4	0	-2
Total	-1	-2	1	2	6
Average	-.2	-.4	.2	.4	1.2
Range					

FIGURE 15–13 *Averages for each subgroup.*

	Subgroup 1	Subgroup 2	Subgroup 3	Subgroup 4	Subgroup 5
	0	2	0	2	2
	1	0	-1	1	2
	1	-3	-1	-2	1
	-1	0	-1	1	3
	-2	-1	4	0	-2
Total	-1	-2	1	2	6
Average	-.2	-.4	.2	.4	1.2
Range	3	5	5	4	5

FIGURE 15–14 *The ranges for each subgroup were calculated by subtracting the smallest size from the largest size. For example, the largest size in the first subgroup is 1 and the smallest is –2. The range is equal to the largest minus the smallest, or 3.*

The next step is to calculate the range for each subgroup. Remember that the range is simply the difference between the largest and smallest sizes in each subgroup. Look at the first subgroup in Figure 15–14. The largest value is 1 and the smallest value is –2. The difference between these values is 3. Study the other subgroup ranges.

The formulas look complex, but are very simple.

The upper control limit for the averages is $UCLx = \bar{X} + A_2R$.

The lower control limit for the averages is $LCLx = \bar{X} - A_2R$.

You should notice that the formulas are the same, except that the UCL uses a plus sign and the LCL uses a minus sign.

The formulas really just add an amount (A_2R) to the process average or subtract an amount from the process average.

You already know that $\bar{X}$ is the process average (average of the averages). For our example, it is .24. You can find this by adding the five subgroup averages and dividing by 5.

$\bar{R}$ is the average range for our process. We can find it by adding the five subgroup ranges and dividing by 5, or by adding the five ranges and multiplying by 2 and moving the decimal place one place to the left. For this example, the average range is 4.4.

If we substitute these into the formula, we have:

$$UCL_{\bar{X}} = .24 + A_2 * 4.4$$
$$LCL_{\bar{X}} = .24 - A_2 * 4.4$$

Next, we need to know what A_2 is. A_2 is a constant from a table (see Figure 15–15).

The number of parts in our subgroup is 5, so the value for A_2 (n) is .58.

We just substitute .58 into the formula.

$$UCL_{\bar{X}} = .24 + .58(4.4) = 2.792 \quad LCL_{\bar{X}} = .24 - .58(4.4) = -2.312$$

Make sure to perform the multiplication first, then calculate the two limits. They should be $UCL_{\bar{X}} = 2.792$ and $LCL_{\bar{X}} = -2.312$.

This completes the averages portion of the $\bar{X}R$ chart.

The formula for the upper control limit for the range (UCL_R) is $UCL_R = D_4\bar{R}$. The value of $D_4\bar{R}$ is found in the chart in Figure 15–15.

There are five pieces in our subgroups, so we will use a value of 2.11 for D_4.

The formula says multiply D_4 by the average range ().

$$UCL_R = 2.11 * 4.4$$
$$UCL_R = 9.284$$

A process will always have variation. The upper control limit on the range will tell us when the variation is higher than it should be. (If a range is greater than the upper limit, there is a 99.7 percent chance that something changed in the process.)

Notice the upper control limit.

Number in subgroup	Value of A2	Value of D4
3	1.02	2.57
4	.73	2.28
5	.58	2.11
6	.48	2.0
7	.42	1.92
8	.37	1.86
9	.34	1.82
10	.31	1.78

FIGURE 15–15 *This figure shows how the values for A_2 and D_4 are found in the chart. There are five pieces in our subgroups, so the value of .58 is used.*

Figure 15–16 shows a completed chart for this example. The data was transferred to the chart. Notice that the chart really has three areas. The top area contains information about the individual parts: the part name, part number, operation number, and so on, as well as the actual sizes of the parts that were made. The actual part sizes were written down in order in the subgroup areas in groups of five. This area is also used to calculate the averages and ranges for each subgroup.

The middle portion of the chart is used to plot the average for each subgroup. This area is the chart of averages. Notice that the mean and the upper and lower control limits have dotted lines to mark the limits. The actual values for the limits are written on the left side of the chart. If a point that we plot is between the limits, the process is acceptable. The chart will always tell us if something has changed in the process. If a point falls outside the limit, there is a 99.7 percent chance that something has changed in the process.

The subgroup averages are plotted by following the vertical line under each subgroup down to the average portion of the chart. The average for subgroup 1 is –.2. Follow the line under subgroup one down and you will see that a dot was drawn on the line just under 0. The second subgroup's average was –.4. Follow the line down from the second subgroup and you will see that a dot is drawn at –.4. The third subgroup has an average of 1.2. After all of the five averages were plotted, they were connected with lines.

The third area of the chart is the chart of ranges. The range portion of the chart is used to graphically show variation between parts in a subgroup. A small range is desirable and indicates that the variation between parts within the subgroup is small. Note there is only an upper limit on the range chart. We want all of our subgroup ranges to be under the limit. Remember that the

Part Name				Part Number						
Operation Number				Machine						
Blueprint Spec.				Measurements are in						

Individual Readings

Subgroup Number	1	2	3	4	5	6	7	8	9	10	11
Piece Sizes	0	0	0	2	2						
	1	2	-1	1	2						
	1	3	-1	-2	1						
	-1	0	-1	1	3						
	-2	-1	4	0	-2						
Total	-1	-2	1	2	6						
Average	-.2	-.4	.2	.4	1.2						
Range	3	5	5	4	5						
Date											
Time											

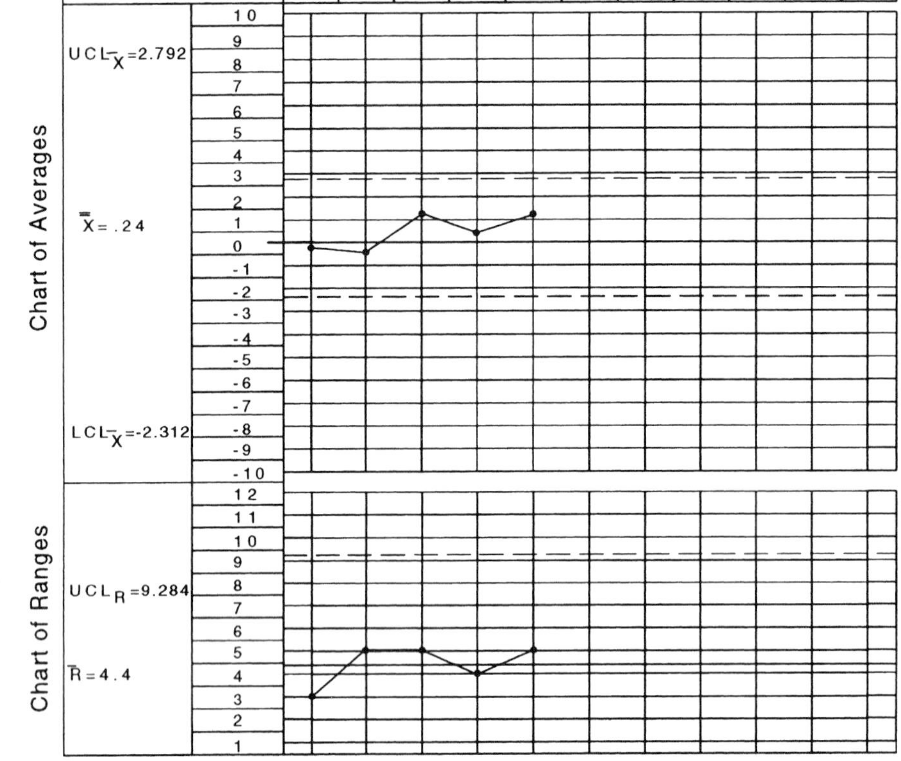

Chart of Averages

$UCL_{\bar{X}} = 2.792$

$\bar{\bar{X}} = .24$

$LCL_{\bar{X}} = -2.312$

Chart of Ranges

$UCL_R = 9.284$

$\bar{R} = 4.4$

FIGURE 15–16 *Completed chart.*

smaller the range the better because it means that there is less variation between the parts in the subgroup.

The subgroup ranges are plotted by following the vertical line under each subgroup down to the range portion of the chart. The range for subgroup 1 is -3.

Follow the line under subgroup 1 down and you will see that a dot was drawn at 3 in the range portion of the chart. The second subgroup's range was 5. Follow the line down from the second subgroup and you will see that a dot is drawn at 5. The third subgroup has a range of 5. After the five ranges were plotted, they were connected with lines.

ANALYZING THE CHART

Are all of the subgroup averages inside of the upper and lower control limits? Are all of the subgroup ranges under the upper control limit?

All of the subgroup averages were inside the limits, as were the subgroup ranges.

Remember, the chart makes correct decisions 99.7 percent of the time. If we use the chart, we will make good decisions 99.7 percent of the time.

However, we have not even looked at blueprint tolerance for this job yet. A chart assures us that a process is running at its best, but even the best may not be good enough for a very tight tolerance.

The operator must still watch the individual piece sizes and scrap the bad parts. For example, how many parts would be scrap for this job if our tolerance was ±.003 inches? You should find one scrap part. Note that the operator must scrap parts that are outside of the blueprint tolerance, but the data must be entered in the chart. Also, notice that the points we plotted did not tell us we had scrap.

Examine the chart in Figure 15–16. Are all of the averages inside the limits? All of the ranges?

The chart would now be ready to be used on this job. Note that we only calculate limits once when we first study the job. From this time on, we would just enter the date and plot points as we run parts.

The chart assures us that the process (or machine) is running at its best. It does not assure us there is no scrap. The operator must compare the individual parts to the blueprint tolerance. If the particular job we are running has a very good CP, we can reduce our inspection. The better the job (wide tolerance compared to six or more), the less we need to inspect. We might be able to sample one subgroup an hour or one a day if we found an excellent job.

Operators should be encouraged to write notes about the job as they run it. Even hunches could prove useful. The more notes on charts the better. If we examine a chart months or even years later, notes will improve our recall.

Now that limits have been set for this particular job on this particular machine, they never have to be recalculated again unless we change the machine or process.

More parts were run on another day. They were entered into the chart, and the averages and ranges were calculated and plotted (see Figure 15–17). Note that we did not have to recalculate limits. If we run the same job on the same machine and with the same tooling, we can use the same limits and the job should run the same. This is very beneficial because we are developing some history on the job and machine. If we were to look at the same job over a long period of time, we might be able to see the variability increase as the machine wears. This might help us plan maintenance and machine replacement.

Note that in subgroup 7 there is an ellipse drawn around the –6 part. This means that there was a tool change done, which explains why the size was so far off. Note that the average for this group was all right, but the range was outside the limit. The range shows that the variability for this subgroup was high because of the tool change, which the operator noted on the chart.

The average for subgroup 8 is well within the limits. The range is fine also.

The average for subgroup 9 is above the upper limit. The operator stopped running parts when the average exceeded the limit. The chart is telling us that something changed. The operator studied the machine, the setup, and the tooling and discovered that the tool was chipped. The operator noted this on the chart and initialed the note. Notes are very important on charts. The range for the subgroup was within the limit.

You should begin to see that a chart will instantly tell us if there is a problem. The chart will make the right decision 99.7 percent of the time. The operator should not make any changes or adjustments to the machine unless the chart indicates a change has occurred in the process.

Rule 1: Do not make any changes to a machine unless the average or range is outside of the limits.

TRENDS

The second rule for charts is that something has changed in the machine or process if there are seven points in a row either above or below the average (see Figure 15–18). This partial chart shows the plots of seven points. Note that all seven are above the average. This indicates something has changed in the process. The operator should stop the machine and see if anything obvious is wrong. If nothing is found, an adjustment should be made.

If you look at the averages and draw an imaginary line through them, you would see that their average is about 1. Because the parts are averaging about 1 over size, the operator should make an adjustment of 1. This is a major advantage of charting. The chart shows us exactly how much to adjust. (This rule also applies to subgroup ranges.)

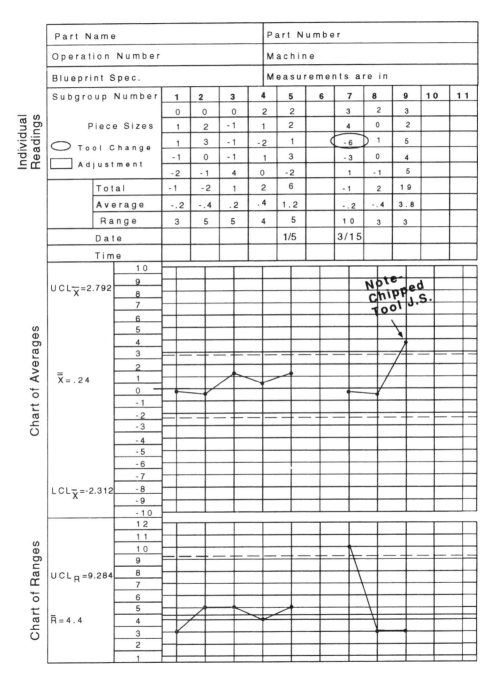

FIGURE 15-17 *Completed chart.*

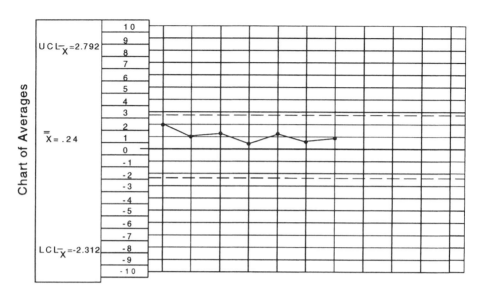

FIGURE 15–18 *Chart with seven points above the process average.*

The second kind of trend is one in which each point increases. If seven points increase in size in a row, something has changed in the process. In Figure 15–19, each of the seven points increased in size. The operator should stop the machine and discover what is wrong before running any more pieces. The rule also applies if seven parts in a row each decrease in size. This rule also applies to the subgroup ranges.

Do not change anything unless:

1. *an average or range is outside of the limits*
2. *seven averages or ranges in a row are above (or below) the average*
3. *seven averages or ranges in a row increase or decrease*

Charts can be invaluable if used correctly. A shop that understands and uses statistical methods will be much more successful than a shop that does not.

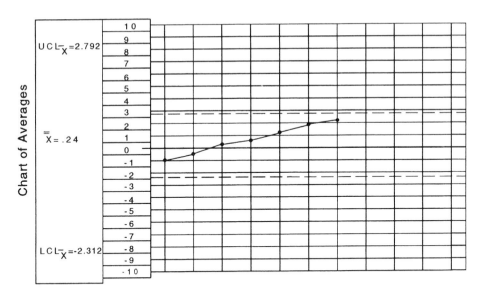

FIGURE 15–19 *Chart with a trend of seven subgroup averages in a row increasing in size.*

CHAPTER QUESTIONS

1. List and explain the three rules for charts that show when a process has changed.

2. Explain at least five benefits of charts. How do they help operators, maintenance, bidding, etc.?

3. These charts make correct decisions _____ percent of the time.

4. Why is it important that no adjustments be made to the machine or process during the initial study?

5. A machine has a standard deviation of .003. We are considering running a job on the machine that has a tolerance of ±.006 (use Figure 15–20).

   a. Calculate the CP.

   b. Will there be scrap? If so, how much? (Hint: Draw a bell curve to help find the answer.)

6. Using the same machine as in question 5, assume the job has a tolerance of ±.010.

   a. Calculate the CP.

   b. Will there be scrap? If so, how much?

7. Standard deviation = .0015; blueprint tolerance = ±.005.

   CP = _____

8. Standard deviation = .001; blueprint tolerance = ±.002.

   CP = _____

9. Based on the data from question 8, will this job run well? If you were asked whether to bid on the job, list at least three alternatives you could give.

10. You are asked to make a recommendation on bidding on a lathe job. The company supplied you with 30 sample pieces to run. The data for the job is shown in Figure 15–21. Code the data and draw a bell curve that compares the capability (6 standard deviations) to the blueprint specifications.

    Blueprint Specification = 2.250 ±.004.

11. Complete an $\overline{X}R$ chart for the process in question 10 (enter the data, calculate control limits, and plot the data).

    a. Standard deviation

    b. Process mean

    c. Average range

    d. Is the data normally distributed? (Check with a histogram.)

    e. Upper control limit averages

    f. Lower control limit averages

    g. Upper control limit range

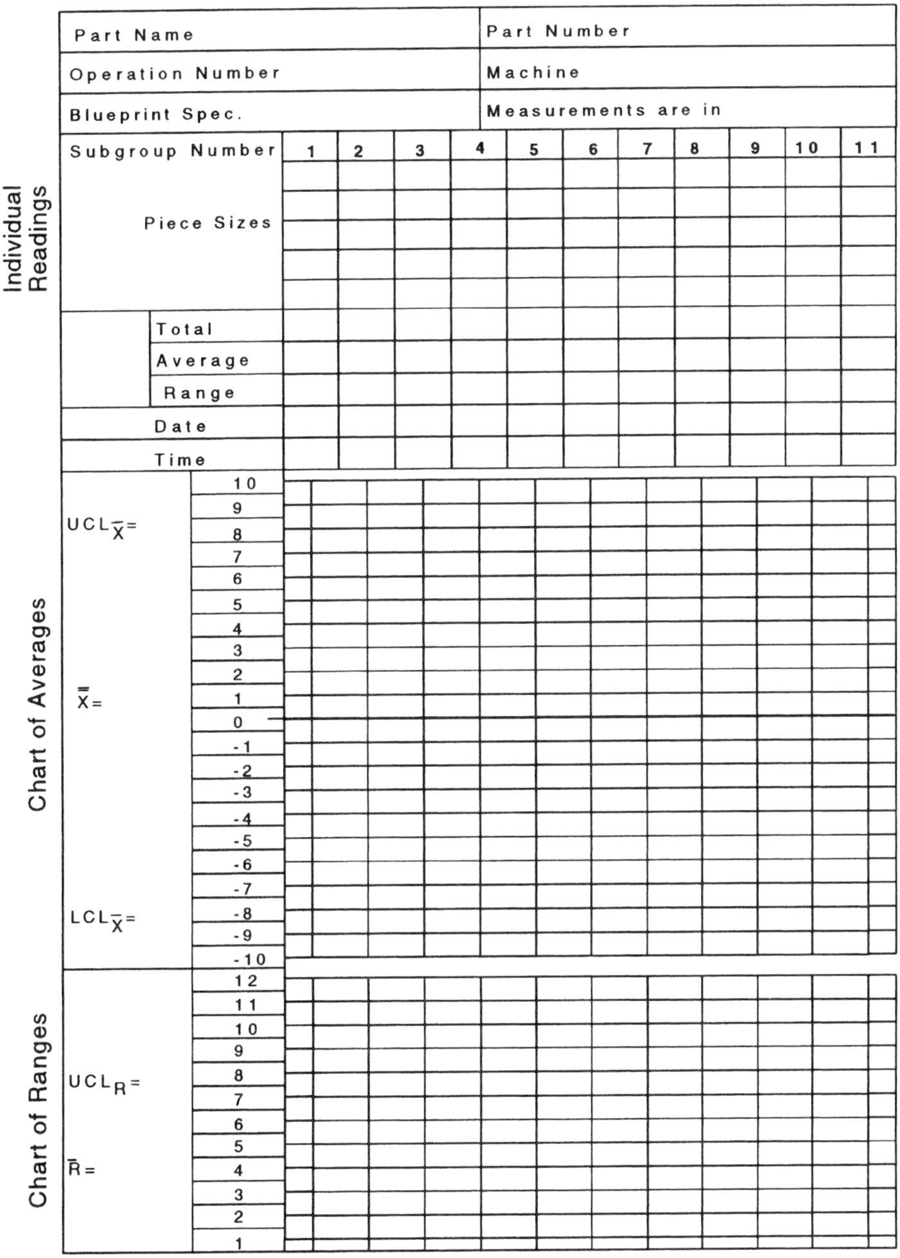
Part Name
Part Number
Operation Number
Machine
Blueprint Spec.
Measurements are in
Subgroup Number
1 2 3 4 5 6 7 8 9 10 11
Individual Readings
Piece Sizes
Total
Average
Range
Date
Time
Chart of Averages
$UCL_{\bar{X}}$=
$\bar{\bar{X}}$=
$LCL_{\bar{X}}$=
10 9 8 7 6 5 4 3 2 1 0 -1 -2 -3 -4 -5 -6 -7 -8 -9 -10
Chart of Ranges
UCL_R=
$\bar{R}$=
12 11 10 9 8 7 6 5 4 3 2 1

FIGURE 15–20 *Blank chart for chapter questions.*

#1	#2	#3	#4	#5	#6
2.252	2.249	2.248	2.252	2.246	2.250
2.252	2.249	2.254	2.253	2.248	2.249
2.253	2.243	2.252	2.251	2.247	2.252
2.250	2.248	2.251	2.250	2.246	2.248
2.252	2.249	2.247	2.250	2.249	2.251

FIGURE 15–21 *Use with question 10.*

#1	#2	#3	#4	#5
1.253	1.254	1.251	1.249	1.247
1.250	1.249	1.245	1.250	1.251
1.247	1.250	1.251	1.256	1.252
1.251	1.253	1.248	1.252	1.251
1.248	1.249	1.252	1.248	1.245

FIGURE 15–22 *Use with question 12.*

12. The data in Figure 15–22 was taken from 25 consecutive pieces. The process was a turning operation on a lathe. The blueprint specification was 1.250 ±.003.

   a. Code the data.

   b. Is the process normal?

   c. Mean

   d. Average range

   e. Complete an $\bar{X}R$ chart (enter the data, calculate control limits, and plot the data).

   f. Compute the capability.

Chapter 16

....

INTRODUCTION TO ISO 9000

INTRODUCTION

ISO 9000 has become a major force in industry. Quality systems developed to meet ISO requirements are very prevalent. Personnel in enterprises must understand the rationale and benefits of ISO systems.

OBJECTIVES

Upon completion of this chapter, the reader will be able to:

- *Describe the ISO 9000 system.*
- *Describe the benefits of implementing a quality system.*
- *Describe some of the requirements of an ISO quality system.*
- *Describe a typical implementation plan.*
- *Given the characteristics of a company, choose which ISO standard is appropriate.*

INTRODUCTION TO ISO 9000 BASICS

ISO (International Standardization Organization) 9000 is a set of standards that were adopted in 1987. They were originally intended to synchronize standards for quality systems in the 12 member countries. They were meant to help specify controls over such activities as design, manufacturing, logistics, and other functions associated with producing quality products and/or services. ISO certification is becoming a requirement for doing business in Europe. Today, all industrialized countries are members of ISO and participate in writing the standards. In fact, many countries have adopted the ISO standard as a national standard. Many countries give the standard a different name or number, but the requirements are the same.

Many companies today are working to improve their systems. Quality is not achieved through luck, by saying the right things, or by slogans or posters, and it never happens by accident.

To achieve quality, an enterprise must provide personnel with everything they need to do their jobs well. These items include well-organized work areas with proper tooling and equipment, good job instructions, clear and accurate specifications, good materials to work with, and the ability to report problems or barriers that prevent them from doing their jobs and have them promptly corrected. A good quality system can achieve these requirements.

A quality system will also achieve other goals for the enterprise. The enterprise will clearly understand the customers' requirements, that the product designed will meet their requirements, that raw materials and processes used will meet their requirements, and that all activities are planned and productive. A quality system will also assure that the enterprise continuously improves.

Most companies started very small. An entrepreneur with a good idea for a product starts a company. The product becomes popular, and the owner begins to add employees. At first there are no systems required because the owner takes care of everything, but as the company grows, the owner can no longer take care of everything. This means that the employees and managers he/she hired make decisions and start to do things their way, because no company procedures exist. As you can see, there can be chaos in a small business that grows.

A sample system is order entry. What happens when an order comes in to a company? In many companies there is no set procedure. Every person who takes an order might do it differently. When an order is taken, it is crucial that the information is correct and complete. When there is no set system, there can be problems. The personnel taking orders might not understand the customer's specifications. They might not be able to make the product or deliver on time. This is only one area where ISO requires a system. This section of ISO is called *contract review*. There are 19 other sections. The requirements for the 20 sections are practical, common-sense things. The 20 sections (elements) will be covered later.

In many companies there are no formal systems. If there are systems, they typically are not followed. Increasing competition and customer demands do not allow a company to be inefficient anymore. Companies need efficient systems that everyone understands and uses to be competitive today. The ISO 9000 series of standards was designed to help enterprises develop and implement systems. Think of ISO 9000 as simply developing and using good business systems.

THE ISO 9000 STANDARDS

The ISO 9000 series actually consists of three systems and other documentation (see Figure 16–1). ISO 9000 is a document that is designed to help a company decide which standard it should choose to implement. 9001 is the broadest standard, designed for enterprises that design and manufacture product. 9002 was designed for enterprises that manufacture, but do not engineer their products, such as a typical small job shop. They use blueprints from a larger company and make parts to the print specifications.

FIGURE 16–1 *The three systems (9001, 9002, and 9003) vary in breadth of coverage. ISO 9003 has the narrowest focus: companies that inspect and test products. ISO 9002 is for companies that inspect/test, produce, and possibly install their products. ISO 9001 is for companies that also do design/development and possibly service their products.*

9003 was designed for enterprises that do not manufacture. They usually buy, inspect, and sell products. An example might be a company that buys nuts and bolts in bulk and then inspects, packages, and sells them. If a company just does inspection and test activities, it would choose 9003. If it does production, but no design, it would choose 9002. If it designs the products it manufactures, it would choose 9001.

LEVELS OF DOCUMENTATION

There are three basic levels of documentation (see Figure 16–2). Think of the system as though it were three small three-ring binders. Each binder contains one of the levels of system documentation.

The first is the policy level. An enterprise must develop a policy manual for this level. These policies are quite generic. In general, they describe how the company wants to do business. This level will include their quality policy and

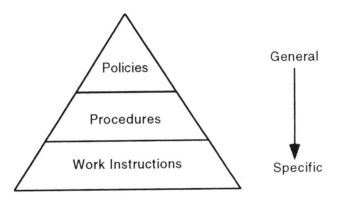

FIGURE 16–2 *Levels of ISO 9000 documentation. As we move down the levels, the information becomes more specific.*

objectives for quality, as well as a statement of management's commitment to the quality policy and objectives.

This level will also include a description of the key positions and their responsibilities within the system. An organizational chart is included to show how various positions interrelate. General descriptions in this level explain how each of the 20 required elements is handled in the company. This manual is quite generic in nature and about 20 to 40 pages in length. Figure 16–3 shows an example of a policy for contract review.

The second level, the procedure level, contains business procedures that describe in detail how the company's systems operate. Procedures detail who, what, where, and how things happen. The procedures are how the policies are actually carried out. Procedures detail how design of new product is to be carried out, when and how all the measuring and test equipment is to be calibrated, and so on. The example cited earlier dealt with how orders are received. One of the procedures at this level would be a contract review procedure. It might state that order entry personnel check every order to make sure that customer specifications are clear and complete, that the schedule is checked to assure on-time delivery, and so on. In general, there might be 30 to 50 procedures in a typical company. Figure 16–4 offers an example of a procedure for the contract review section.

Contract Review Policy   **Revision Level 1**

All orders (contracts) are to be reviewed to ensure that the customer's requirements are adequately defined and understood and that we have the capability to meet the order requirements.

Responsibility

The sales department is responsible for conducting order reviews.

Scope

Order review is the verification that the customer's specifications are adequately defined, documented, and understandable and that we have the capability to meet the requirements of the contract. Contract reviews will follow the Contract Review Procedure (P3).

Records

Sales Representatives will make a record of each review according to the Contract Review Procedure (P3).

FIGURE 16–3 *Sample policy for the contract review section. Policies are very general.*

Contract Review Procedure (P3)   Revision Level 3

3.1 Scope

This procedure covers all contracts for the purchase of products.

3.2 Purpose

The purpose of this procedure is to make sure all contracts for products or services are reviewed to ensure that customer requirements are adequately defined and understood and that products can be shipped in accordance with the requirements of our customer.

3.3 Responsibility

The Sales Representatives are responsible for reviewing all contracts for products and services and for the resolution of conflicts with customers.

3.4 Procedure

3.4.1 When purchase orders are received, the Sales Representative will check to make sure that the following information is correct. The Sales Representative must also check to make sure the product or service can be produced and delivered on time by checking the production schedule.

> *Complete and understandable product specifications*
> *Correct price*
> *Credit terms*
> *Customer's address*
> *Delivery date*

3.4.2 If any discrepancies are found on the order, the Sales Representative will resolve the conflict with the customer before the order is accepted.

3.4.3 If the customer's delivery date cannot be met, the Sales Representative must contact the customer to try to agree on a new delivery date.

3.4.4 The Sales Representative will then sign and date the purchase order when the review is complete and all differences have been resolved.

3.4.5 If any delays occur during production, the Sales Representative will contact the customer to advise them of the situation.

3.5 Records

When the order review and ability to deliver are verified, the Sales Representative must sign, date, and stamp the order ACCEPTED. The stamped copy of the order is the record of the order review. These records will be kept in the sales office for a period of five years unless otherwise specified in the contract.

FIGURE 16-4 *Sample procedure for the contract review section of an ISO quality system.*

Work Instruction WI-27	Revision Level 2

Machine the part using program #23.

Deburr the four mounting holes.

Measure the bore size using the dial bore gauge and record the size on the inspection sheet.

Measure the thickness of the part using a .0001 inch micrometer and record the size on the inspection sheet.

Initial and date the inspection sheet.

If the part is within specification, place the inspection sheet and part in a product bin so that further processing can take place.

If the part does not meet the specification, initial and date a nonconformance sheet and notify your supervisor. Put the nonconformance sheet and the part in an orange bin to prevent further use.

FIGURE 16–5 *Example of a work instruction.*
Note that it is very task specific.

The third level is the work instruction level. At this level work instructions are written to describe how specific tasks are accomplished. For example, a work instruction might describe how to set up a machine, calibrate a micrometer, or fill out a specific form. The difference is that work instructions are very task specific; procedures involve systems. Think of work instructions as being very specific, production-oriented instructions (see Figure 16–5). If a company uses only trained people for jobs, they do not need very many work instructions. Work instructions are required only where their absence would create a quality problem.

BENEFITS OF A QUALITY SYSTEM

There are many benefits to developing and implementing an ISO quality system. When procedures are developed, they should be streamlined so that it becomes easier for people to do their jobs. All personnel must follow the procedures. The procedures assure that product specifications are correct and understood by all and that product is produced and checked to make sure it conforms to product specifications. This helps reduce the amount of scrap and nonconforming product produced. Procedures also ensure that designs are better and more easily manufactured. In short, procedures should help a company do business better.

An ISO system should reduce the number of audits that customers must conduct. For example, if we sell our product to several large companies, they will each want to come in and audit our systems to make sure our systems meet their needs. Each customer has different requirements, so reviews can be painful. With ISO systems customers are more likely to accept your systems and not conduct their own quality system audits. This should reduce the number and the intensity of the audits.

ISO 9000 also helps a company to continuously improve. There are three ways that ISO helps a company achieve continuous improvement. One of these is the internal auditing that is required. A company must conduct internal audits of their system on a regular basis to see if the system is implemented and working. If the audit finds problems, they must be fixed, which improves the company's systems.

Second, ISO leads to continuous improvement when a corrective action process is required. A corrective action procedure assures that once problems are identified they are corrected in a way that prevents them from occurring again. Thus, every time there is a corrective action, the company improves.

Third, management review meetings are required. Generally, these are scheduled approximately once each year. The purpose of these meetings is to review the performance of the quality system over the previous year. The ISO manager presents the audit reports, corrective actions, customer complaints, and other items that may help management get an honest look at past performance and identify areas for improvement. This is one of the only times that management sits down and thinks of the big picture instead of the day-to-day problems.

ISO ELEMENTS

When the experts designed the ISO 9000 system, they identified 20 areas or sections that they considered crucial to a company's success. These 20 areas became the requirements of the standard. Not every company performs all functions in every one of the 20 areas.

The 20 sections are covered next. These are not the actual requirements but a short summary of the important aspects of each section. You should study the actual standard for the actual requirements. It should also be noted that the ISO standard does not tell a company how to meet these requirements. Each company must develop its own system to meet the requirements in a manner that is appropriate to its own situation. If two competitors each develop and implement a quality system, the one who develops the better, more efficient system will be more successful even though both meet the requirements of the standard.

1. MANAGEMENT RESPONSIBILITY

This element is concerned with management's responsibility to plan, document, implement, and support a quality system within the enterprise. ISO recognized the importance of management's commitment to the process. This element requires a quality policy and objectives for quality. It requires that responsibilities, authorities, and interrelations for those involved in quality-related matters be defined and documented. Management commitment is required in several areas. Management must provide adequate resources and training for management, performance of work, and verification activities including internal audits. Management must assure that the quality policy and objectives are clearly understood, implemented, and maintained at all levels of the organization. Management must also appoint an ISO manager who will have overall responsibility for the quality system. Finally, management must schedule management review meetings at an appropriate time interval. These meetings are very useful in improving the overall operation of a company.

2. QUALITY SYSTEM

The supplier must establish, document, and maintain a quality system to make sure that product conforms to specified requirements. This section requires that the supplier develop a quality manual that includes, or makes reference to, documented procedures that describe how things are done in the organization. This section also requires the supplier to do quality planning. This planning is intended to make sure that the supplier has adequate standards, tooling, and inspection equipment necessary to make the product. This section also requires quality records to be kept.

3. CONTRACT REVIEW

This section is concerned with incoming orders (contracts). It requires that contracts are checked to make sure that the customer's specifications are complete, documented, and understood, that the supplier has the capability to meet the requirements, that any differences are resolved, and that records are kept.

4. DESIGN CONTROL

This section is intended to ensure that the product is designed in a manner that assures that the specified requirements are met and that appropriate people get input into the design process. This section requires the preparation of a plan for each design and development activity. It requires design activities to be assigned to qualified personnel who have adequate resources. It requires technical interfaces within the organization so that appropriate input is received from all areas that will be involved with the design and manufacture of the product. It requires that the procedure be documented and that regular reviews are held to verify that the design will meet the specifications. Any changes must be documented and approved before they take place.

5. DOCUMENT AND DATA CONTROL

This section's purpose is to ensure that all documents in the enterprise are current and accurate. The manuals that make up the quality system are covered by this, as well as any other documents. A master list, or control procedure, is required to be sure that only the current revision of documents is being used. This section also requires pertinent documents to be available when they are essential to the functioning of the quality system. Obsolete documents must be removed or protected in some manner so that they are not used. Changes to documents must be reviewed and approved.

6. PURCHASING

The purchasing section is very similar to, but opposite from, the contract review section. Contract review dealt with orders for products; purchasing involves items and materials an enterprise buys. This section requires that purchasing documents clearly describe the product to be purchased, including clear and precise specifications and other relevant data. It also requires that the supplier evaluate and select the companies from which they purchase products, usually in the form of an approved supplier's list (companies from which products can be purchased). The performance of suppliers must be regularly evaluated. This section applies only to purchased products that could affect product quality. It does not apply to things like cleaning supplies and other incidentals that do not affect product quality.

7. CONTROL OF CUSTOMER-SUPPLIED PRODUCT

This section requires that if your customer supplies you with materials or items that are included in the product you make for them, you must establish and document procedures to be sure their items are properly cared for in your plant. An example might be a company that manufactures engines, but sends the engines to another company for painting. This company must make sure that the product is adequately protected and cared for while in their possession. This also applies to raw materials or items that are included in the final product. This section also requires that if any damage is done to the customer's materials it is recorded and reported to the customer.

8. PRODUCT IDENTIFICATION AND TRACEABILITY

This section is intended to make sure that product is clearly identified in the plant during all stages, from beginning to the final product. This requirement will vary considerably depending on the product. Drug manufacturers, for example, are held to a very high standard.

9. PROCESS CONTROL

This section requires that processes that directly affect the product's quality are carried out under controlled conditions. This means that where necessary there will be procedures to specify the manner of production. There must also be monitoring and control of production and criteria for workmanship. The equipment must also be adequately maintained.

10. INSPECTION AND TESTING

This section requires the supplier to establish and maintain documented procedures for inspection and testing activities to ensure that product meets customer specifications. The supplier must ensure that incoming product is not used in production until it has been inspected or otherwise verified as conforming to specifications. This ensures that defective materials are not used in products. This section also requires that inspections and/or testing is conducted according to the quality plan for the product. It requires that product not be released until all inspections have been done and that records are kept to ensure that the inspection and testing were done and were recorded.

11. CONTROL OF INSPECTION, MEASURING AND TEST EQUIPMENT

Calibration of measuring equipment is this section's focus. Quality parts cannot be produced if gauges and inspection equipment are not accurate. This requires equipment to be calibrated according to a regular schedule to assure that the equipment is accurate.

12. INSPECTION AND TEST STATUS

This section addresses the status of product as it moves through production steps. Product inspection status must be clearly identifiable at every stage of production. For example, we should be able to walk over to a box of parts and tell which stage of production they are in, whether the parts were inspected as required, and whether they were good or bad.

13. CONTROL OF NONCONFORMING PRODUCT

This section deals with product that does not meet specification for some reason. It is crucial that it is identified and kept from further processing or shipment to customers. This section requires that a company determine who is responsible for decisions regarding nonconforming product and that they develop a procedure to deal with it.

14. CORRECTIVE AND PREVENTIVE ACTION

One of the most crucial and beneficial requirements of ISO, this section requires that problems in the enterprise are corrected so they never occur again. This is where the big gains can take place. Corrective action can help improve a company's procedures and processes to make it much more productive and profitable. This documented activity ensures that when problems are found they are corrected in a timely manner so that they will be prevented from occurring in the future. Corrective action reports are reviewed by management at regularly scheduled management review meetings.

15. HANDLING, STORAGE, PACKAGING, PRESERVATION, AND DELIVERY

This section requires that product is handled, stored, packaged, preserved, and delivered in a manner to prevent damage or deterioration of the product.

16. CONTROL OF QUALITY RECORDS

This section requires a company to establish and maintain procedures to identify, collect, index, access, file, store, and maintain quality records for an appropriate length of time.

17. INTERNAL QUALITY AUDITS

This is one of the most important requirements of ISO. The company must perform regular audits of the quality system to see if it is implemented and effective. The results are recorded, and if problems are found they go through the corrective action procedure. The results are also an important part of management review meetings.

18. TRAINING

This section requires that management identify training needs in the company. It also requires that needed training is provided and records of the training are kept. This section makes sure that a company has only qualified people performing the work they are asked to do.

19. SERVICING

If a company services its product, it must establish and maintain a documented procedure for performing the service and verifying that the service met the requirements.

20. STATISTICAL TECHNIQUES

This section requires the company to identify the need for statistical techniques in the plant to control and verify process capability and product characteristics. It also requires that the company establish and maintain procedures to implement and control the application of the statistical techniques that are identified as needed.

ISO IMPLEMENTATION

Implementing a quality system is a big job. It can take thousands of hours of labor for a small- to medium-size company. The most successful implementation will involve as many people as possible, especially top management.

The process typically starts when top management or owners hear about ISO and want to learn whether it would be appropriate for their firm. Some education should then take place. Top management must understand the amount of work involved, as well as the potential benefits and pitfalls. Top management must then commit themselves to active participation in the process. Remember: ISO requires commitment from management.

A person is generally chosen at this point to be the ISO manager. This person will have the overall responsibility for system development and implementation. At this point a schedule should also be created for development, implementation, and auditing.

Next, there should be some basic education about ISO, the rationale for deciding to develop and implement ISO, and the need for everyone's participation. This training is intended to take the fear out of ISO and gain participation from all personnel.

The next step is choosing the personnel who will be active in developing and writing the system. The more people the better. Some additional training should be done with these people so that they thoroughly understand what they will be doing. The 20 elements are then divided among the group based on their knowledge and responsibility for the element. Each member then goes out into the company and interviews employees about how these particular systems work. They then write a rough draft of the procedures and, with the ISO manager's help, compare them to the ISO standard to make sure all requirements are met.

As procedures are developed and reviewed, employees should be trained on the use of the procedures and they should be implemented. These procedures should then be audited to see if they are implemented, if personnel are using them, and if they are effective. Once the whole system is in place a more extensive audit should be scheduled.

CERTIFICATION

ISO 9000 systems can be certified by a third-party registrar. A registrar is an organization that has been authorized by a national accreditation board to perform ISO audits and issue certificates that a company is ISO certified. In the United States the accreditation body is called the RAB (Registrar Accreditation Board).

THE CERTIFICATION PROCESS

A registrar should be carefully chosen. The registrar should be familiar with your business. It is also important to choose one who will be respected and recognized by your customers. Once a registrar has been chosen, they will evaluate the quality manual and procedures that you developed. They will determine whether you have met the requirements of the ISO 9000 standard. There is, of course, a charge for this review. They will send the manual back to you with weaknesses or gaps noted. They will not tell you how to fix the problems; that is your job. You must then change your system and resubmit it for evaluation.

When you have successfully developed your system, a major audit is scheduled. This generally will involve a couple of auditors from the registrar and approximately 2 days. The third-party auditors are paid by you to conduct the audit.

The audit process will begin with an opening meeting. With top management and the auditors present, the lead auditor will explain the schedule for the audit. The audit team might then be taken on a quick plant tour.

Next, each auditor will be accompanied by a company representative. Each auditor will go to a different area or department to work on a piece of the audit. The escort will generally be a management person. An auditor will generally spend a couple hours in each area of interest. When a problem is found, the auditor discusses it with the manager.

At the end of each day the auditors will summarize what they have found. There may be things that can be corrected immediately.

At the end of the whole audit, the lead auditor will present a report. The lead auditor will report on findings and minor noncompliances and inform the enterprise if he intends to recommend that they be certified.

If auditors find too many large problems (generally called *findings*) they will not certify you. They will certainly find some problems (called *concerns* or *observations*) that you will need to correct. This certification is generally for 3 years. This does not mean you are done for 3 years, however. Your registrar will return approximately every six months for a smaller audit to keep you on your toes. You, of course, pay for this also. Note that the work doesn't end once a company is certified. People must use the systems that were developed. Internal audits must be continued to see if the procedures are being followed and are effective, and corrective action must occur for problems that are discovered so that they do not occur again. A successful company must use the audits as opportunities to change and improve their systems.

While certification is expensive, there are some major benefits. If you do business in Europe, you might eventually be required to be certified. The largest benefit is probably the fact that the registrar will keep you on track. Everyone knows that the registrar will be back regularly to perform audits, so it keeps everyone on their toes and following the procedures. Certification is an honor. It is the recognition of a third party that your business system meets an international standard.

Many companies may never have to certify. There are many benefits, however, in developing and implementing a quality system. Even if a company does not certify, it would be wise to develop an effective quality system.

CHANGES TO ISO 9000

At the time of printing of this book it appears that the format of the ISO 9000 standard will change. The 20 requirements will be reformatted into four basic sections. At the time of printing it appears that the new system will be ready in late 2000.

The current ISO 9001, ISO 9002 and ISO 9003 standards will be consolidated into a single ISO 9001 standard.

The new process-based structure is more generic than the 20-element structure and adopts a process management approach. Also the new process-based structure is more consistent with ISO 14000 standards on environmental management systems. The 20 elements in the current ISO 9001 will be clearly identifiable in the new format. The major clauses in the revised standards will be as follows:

> *Management responsibility (policy, objectives, planning, quality management system, management review).*
>
> *Resource management (human resources, information, facilities).*
>
> *Process management (customer satisfaction, design, purchasing, production).*
>
> *Measurement, analysis and improvement (audit, process control, continual improvement).*

ISO 9001 does not specify requirements on the layout or structure of an organization's quality management system documentation and neither will the revision. This will allow enterprises to continue to document their quality systems in a way that reflects their own ways of doing business. The revision of the ISO 9000 standards will not require the rewriting of an organization's quality management system documentation. One requirement of the ISO 9000 revision process is that enterprises that currently have ISO 9000 quality systems will find it easy to transition to the revised standards.

QS 9000

The automakers, for the first time ever, adopted a common quality program. It is called QS 9000 and uses ISO 9000 as its base. Many truck manufacturers and steel manufacturers are considering using QS 9000 also. The automaker's QS 9000 involves the ISO 9000 requirements coupled with some industry-specific requirements. The additional specific requirements came from former auto-industry quality programs.

QS 9000 consists of two sections as well as several appendices:

> *The first section consists of the ISO 9000 base requirements with some added requirements that are specific to the automotive industry. There are requirements for continuous improvement, quality and productivity improvement, Advanced Product Quality Planning and Control Plan (APQP), and the Production Part Approval Process (PPAP). Consistency is stressed in this section, including foolproofing (mistake proofing) processes. ISO 9001 is the base requirement.*

> *The second section covers requirements specific to the particular automaker.*

Chapter Questions

1. Describe what the ISO system is and what it is intended to do.
2. Explain the three levels of an ISO system.
3. What are the differences among policies, procedures, and work instructions?
4. Explain at least three benefits of implementing an ISO 9000 system.
5. Consider Acme Manufacturing, a job shop. They receive blueprints from their customers and manufacture, inspect, and ship the products to their customers. Which standard should they think about implementing and why?
6. Explain the ISO certification procedure.
7. Widget Manufacturing designs and manufactures its line of widgets. They sell the product to distributors, who sell it to discount stores. Which ISO system should they consider implementing and why?
8. Explain how a company might develop and implement an ISO system.

Chapter 17

....

FUNDAMENTALS OF CAD/CAM

INTRODUCTION

The rapid advancements in computers and software for computer-aided design (CAD) and computer-aided machining (CAM) have drastically affected CNC programming. The new technology has reduced redundant effort and forced more cooperation between design and manufacturing.

OBJECTIVES

Upon completion of this chapter, the reader will be able to:

- *Describe the advantages of computer-aided part programming (CAPP).*
- *Describe the advantages of using a CAD system.*
- *Describe the importance of levels in a CAD program.*
- *Describe the process of converting CAD data to a program that will run a specific CNC machine.*
- *Describe a typical CAPP system.*
- *Describe the advantages of a CAPP system.*
- *Explain terms such as "post-processor," "off-line," "level," and "job plan."*

INTRODUCTION TO CAD/CAM

This chapter introduces you to technologies that help reduce programming time, complexity, and effort. CAD will be discussed first, CAPP next, and then the link between them is examined.

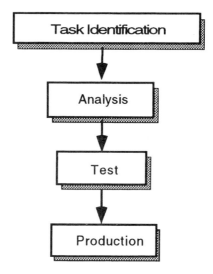

FIGURE **17–1** *The typical steps from initial need to production.*

DESIGN

The design phase of manufacturing involves several steps: defining the task, analysis, the test phase, and production (see Figure 17–1).

TASK IDENTIFICATION

Engineering is undertaken to solve problems, design new parts, design jigs and fixtures for production, etc. Engineering helps make product ideas become reality. In the task-identification phase, the engineering department identifies the problem that needs to be solved.

In the case of a new product, engineering takes the product concept and customer requirements and develops clear specifications for the product. The engineer decides upon the physical and functional product requirements. These requirements must be set so that the product can be produced at a cost that will allow a profit to be made.

The task identification phase is the most crucial phase in the whole process. Design should always involve more than one person. Input must be received from all of the departments that will be involved.

Manufacturing must have input into the design requirements to assure that the part can be easily produced with the machines and processes available. Manufacturing can also help make sure that the machining operations, specifications, and tolerances are realistic.

Sales/marketing should have input to the process to assure that customer needs are really being addressed.

The quality department must have access so that they can develop quality plans that address the needs of the product. Quality must also develop and provide any special gauging that may be required to check the parts. Thus, full cooperation is required to assure that all needs are met in a cost-effective fashion. This will help assure smooth running production of a profitable part.

ANALYSIS

At this point the project engineer has a good idea of the product requirements and potential manufacturing concerns. The project engineer must then analyze all of the input and the requirements and develop an initial product design.

TEST

Prototypes may be built at this point to test the design. These should be tested under the same (or more severe) operating conditions that the customer will use. Building a prototype will also expose potential manufacturing problems and unrealistic specifications. A thorough review should take place at this point and involve all of the relevant personnel. The design must then be adjusted to reflect these improvements.

PRODUCTION

If the previous steps have been done well, production should be fairly smooth. Production is always different than making a few prototypes. It is probable that some operations will need to be modified to optimize product production. Design modifications also may be necessary. The computer simplifies all of these modifications.

COMPUTER-AIDED DESIGN (CAD)

Computer-aided design is a technology that involves a computer in the design process. CAD enables an engineer to develop, change, and interact with a graphic model of a part. Computers are strong in the areas of graphics, calculations, analysis, modeling, and testing.

CAD systems have drastically changed manufacturing. CAD is comprised of two major components: hardware and software.

HARDWARE

Hardware varies a great deal between CAD systems. Mainframes have long been used for CAD systems. The rapid increases in processing speeds have made micro- and minicomputers viable CAD stations. Networks generally are used so that designs can be stored centrally and data can be shared among designers. Many input devices are used. Common input devices include keyboard, mouse, trackball, digitizing pad, joystick, keypad, and light pens. Output devices are typically screen, plotter, printer, and hard drives and disks for storage.

SOFTWARE

There is a wide variety of CAD software available for mainframes, engineering stations, and microcomputers. The packages vary to some degree in their capability. They all have the capability to output generic geometric information in the form of drawing exchange files. These files can be imported into other software to assist in programming the machine tool.

USE OF CAD

CAD, normally used in engineering departments, has drastically changed these departments. Drawings used to be made on paper with pencil or pen and drawing instruments. The drawings were very time intensive to produce. They were then copied, and the copies were sent to the floor for production. The originals were stored in large drawers. Even a small enterprise could have thousands of large blueprints on file. If changes were necessary, the engineer would get the original out of the file drawer, make the changes, copy it, and send the new print to the floor. The computer eliminated the need for all of the physical storage of prints. The computer also allowed for rapid and easy print modifications.

The engineer or designer first draws the part on the screen. This part drawing is the actual part geometry. The sizes and locations are all correct so that the information can be used later to create a program to machine the part.

The designer must work closely with the manufacturing people to establish some standards for design. CAD allows different layers (or levels) to be created. This allows the designer to put different portions of the part geometry on different layers. For example, imagine a simple part that has a 1/2-inch slot milled in it and four holes drilled through it. The designer would put the slot on one layer of the drawing and the holes on another layer. This will allow other software to take the part geometry from each layer and assign different tools to it. Think of different layers as different machining operations that would have different tools assigned to them. Layers are like transparencies (foils). A portion of the total part geometry appears on each transparency. If we put all of the transparencies on top of each other, we see the total part. But we can still take any one transparency and isolate some of the part geometry. The CAD system will print the blueprint so it appears there is only one layer, or individual layers can be printed.

The designer also dimensions the part. This is quite automatic. The designer chooses where he/she wants dimensions placed, and the computer places the actual dimensions. Dimensions would be on a different layer than part geometry.

The computer allows designs to be viewed and tested before the actual part is even manufactured. The part can be tested to make sure it fits with any mating parts. CAD systems can check for interference of parts, which can save a lot of wasted machining and development time. CAD systems can stress-test parts to see if they will meet the strength requirements of the application. The graphics abilities of CAD systems also allow 3-D viewing of parts from any angle.

CAD systems can output files that other software can understand. These are generally called *export files*. The CAD system is told to export the CAD part file in a standard format, often in the form of a DXF file format. DXF stands for *drawing exchange format*. All brands of CAD software can export the part geometry in generic formats such as the DXF format. This allows other software to use the part geometry to create a program that can run the CNC machine.

ADVANTAGES OF CAD

CAD drastically increases the productivity of designers. It helps create better designs because of the ability to see the part and test it before it is manufactured. It reduces redundant effort. CAD allows easy and rapid modifications to be made to prints. It helps assure that prints are current and up to date. As we will see, CAD helps integrate engineering and manufacturing.

COMPUTER-AIDED PART PROGRAMMING (CAPP)

Computer-aided part programming is software used to develop CNC programs. There are many software programs available. CAPP systems can aid in the whole process from part print to CNC machine program; however, there is some confusion over the correct terminology for software used to generate CNC programs. Some call it CAPP. Others use the acronym CAPP to describe computer-aided process planning. Many call the software CAM software (computer-aided machining). This chapter will use CAPP, although CAM is very common.

CAPP enables the programmer to use off-line programming. This means that the machine can be kept running and producing parts while the programmer develops the next program. A CAPP system allows the programmer to develop a model that represents the part and the machining operations. The programmer can then interact with the model graphically to make necessary adjustments and modifications before the CNC code is generated.

The part geometry must be entered first. This can be done by reading in the CAD DXF file for the part or by manually entering the part geometry into the software. The use of the CAD file eliminates redundant effort. If the part data already exists for the part blueprint, it can be used to generate the tool paths for the CNC machine.

The CAPP software reads the DXF file, which contains part geometry and the levels that the geometry exists on. The CAPP software uses a job plan to assign the correct tool to each layer. (This, of course, works best if the designer and CAPP programmer have cooperated or developed some standards.) Figures 17–2 through 17–6 show an example of a CAD file separated into layers. The completed part is shown in Figure 17–2. It appears in this figure to be one drawing showing all part geometry, dimensions, and clamping. The drawing actually consists of four layers. One layer contains the rough part shape and clamping to be used, one contains the dimensions, and the other two each contain geometry for a machining operation.

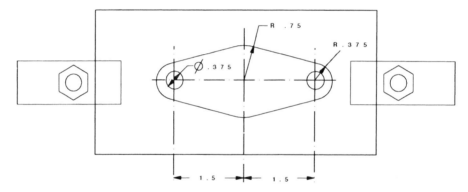

FIGURE **17–2** *Completed CAD drawing of a part and the clamping that should be used for machining. The CAD designer drew the part in four layers.*

FIGURE **17–3** *This layer contains only the raw material shape and the clamping. Note that there are no machining operations associated with this layer.*

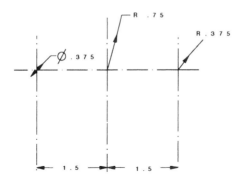

FIGURE 17–4 *This layer contains only the part dimensions. There are no machining operations associated with this layer.*

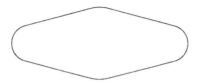

FIGURE 17–5 *This layer contains the geometry for a pocket that needs to be machined in the part. This operation will use an endmill to machine the pocket.*

FIGURE 17–6 *This layer contains two holes and will be used to generate the geometry to centerdrill and drill the holes.*

The machining layers are separate because they will involve different tools and operations. The job plan also knows the material that will be used so it can calculate speeds and feeds for each tool. The programmer can override the speeds and feeds or order of machining operations, as well as machining parameters such as tool selection, depths, and tool offset direction. The use of the CAD file allows even very complex 3-D part programs to be written easily by CAPP software.

CAPP software has excellent graphics capability. The programmer can view the part from different angles and watch the cutter move through its programmed operations to check the program.

Thus far we have been using fairly generic information. The part geometry and tooling information is generic and could apply no matter which machine is used to make the part.

Each machine is different, however. Different brands and models of CNC machines have different capabilities, and some use slightly different coding or program formats. The generic information we have created must be changed so that a particular machine will understand the program. At this point we would choose the machine to machine the part. We then need to post-process the generic part data to create a CNC program to run this particular machine.

POST-PROCESSORS

A post-processor is a translator. This software takes generic information and converts it to code for a specific machine. Imagine that the programmer has already input the part geometry. The programmer has also assigned a job plan that designates specific tools and speeds and feeds to the job. The programmer then must choose the machine to produce the part, which then determines which post-processor will be used. The programmer then runs software that actually writes the CNC code. A post-processor takes the part geometry and job plan and writes code that the specific machine will understand (see Figure 17–7).

CAPP software allows the programmer to easily write and customize post-processors. The programmer usually fills in which codes are used for which purpose and in what format they are used. Once these specifics are entered for

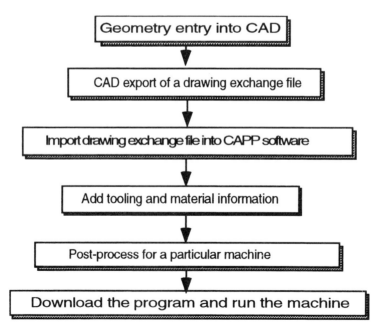

FIGURE 17–7 *The steps from design to machine operation.*

a new machine, the post-processor is tested by producing code, and then any needed modifications can be made to the post-processor.

SIMULATION

CAPP is also beneficial from the standpoint of simulation. A program can be written and tested before it is actually run. Jigs, fixtures, or clamping can be shown during the simulation to determine if there are any potential problems. The simulation can develop a cycle time that can be used to bid on the job. If the simulation shows that the machining time is excessive, the operations can be adjusted to optimize production. This can all be done off-line, which keeps the machine operating. Remember, the only time a machine is making money is when it is actually producing parts. It is a cost at all other times.

DOWNLOADING CNC PROGRAMS

Once the CNC program has been developed, it must be sent to the machine. This is done by manually entering the program into the machine, by inputting a tape or disk, or electronically over a communication cable (see Figure 17–8).

FIGURE 17–8 *Operator entering a program into a CNC. (Courtesy Shoko Ando.)*

THE FUTURE OF DESIGN

Many factors are presently forcing changes upon manufacturing. Customers are demanding ever-increasing quality at ever-decreasing prices. Customers also are demanding just-in-time shipments. As these and many other demands continue, they will force greater cooperation between supplier and customer. This will intensify the need for real-time electronic communication. Instead of mailing blueprints to vendors to bid on, large customers will rely on electronic communication. Vendors will access blueprints with a CAD station in their company. They will then bid electronically on the job, decreasing the leadtime required for new product development.

In the near future real-time, on-line engineering between customer and vendor will be practical. Imagine an engineer at a CAD terminal in the customer's engineering department and an engineer at a terminal in the vendor's plant examining the same blueprint. They can communicate via phone and make changes to the print, which the other engineer can comment on. There will be more and more electronic transmission, bidding, and on-line engineering and closer cooperation with suppliers.

There will also be an ever-increasing need for vendors to do more engineering for their customers. This makes a lot of sense. Imagine a company that builds and sells complex control cabinets for industrial equipment. They are experts at designing control systems that will economically control complex equipment. They do not produce their own cabinets; they purchase them. In the past the company would have designed the cabinet and sent the blueprints out to vendors for bids. Because the company does not have expertise in metal fabrication, wouldn't it make more sense for the vendors to take the cabinet requirements and design a cabinet that would be easy and cost effective to produce? This is a win-win situation. The customer saves engineering time and gets a part that is less expensive and better. The vendor establishes a good working relationship with the customer and gets a part design that is practical and profitable to manufacture.

CHAPTER QUESTIONS

1. What is CAD and how does it differ from CAPP?
2. What are layers and how are they used?
3. Describe how a CAD drawing is changed into a program that can operate a machine tool.
4. List at least four benefits of using CAD software.
5. List at least four benefits of using CAPP software.

Appendix

....

M00	Program stop	Non-modal
M01	Optional stop	Non-modal
M02	End of program	Non-modal
M03	Spindle start clockwise	Modal
M04	Spindle start counter clockwise	Modal
M05	Spindle stop	Modal
M06	Tool change	Non-modal
M07	Mist coolant on	Modal
M08	Flood coolant on	Modal
M09	Coolant off	Modal
M30	End of program & reset to the top	Non-modal
M40	Spindle low range	Modal
M41	Spindle high range	Modal
M98	Subprogram call	Modal
M99	End subprogram & return to main program	Modal

FIGURE A–1 *Common Machining Center M-codes.*

G00	Rapid traverse	Modal
G01	Linear positioning at a feedrate	Modal
G02	Circular interpolation clockwise	Modal
G03	Circular interpolation counter-clockwise	Modal
G28	Zero or home return	Non-modal
G40	Tool diameter compensation cancel	Modal
G41	Tool diameter compensation left	Modal
G42	Tool diameter compensation right	Modal
G43	Tool height offset	Modal
G49	Tool height offset cancel	Modal
G54	Workpiece coordinate preset	
G70	Inch Programming	Modal
G80	Canned cycle cancel	Modal
G81	Canned cycle drill	Modal
G83	Canned peck cycle drill	Modal
G84	Canned tapping cycle	Modal
G85	Canned boring cycle	Modal
G90	Absolute coordinate positioning	Modal
G91	Incremental positioning	Modal
G92	Workpiece coordinate preset	
G98	Canned cycle initial point return	Modal
G99	Canned cycle R point return	Modal

FIGURE A–2 *Common Machining Center G-codes.*

M00	Program stop
M01	Program stop
M03	Spindle start clockwise
M04	Spindle start counterclockwise
M05	Spindle stop
M08	Coolant on
M09	Coolant off
M30	End of program
M41	Low gear range
M42	Intermediate gear range
M43	High gear range

FIGURE A–3 *Common Turning Center M-codes.*

G00	Rapid positioning	Modal
G01	Linear positioning at a feed rate	Modal
G02	Clockwise arc	Modal
G03	Counterclockwise arc	Modal
G28	Zero or home return	Non-modal
G40	Tool nose radius compensation cancel	Modal
G41	Tool nose radius compensation left	Modal
G42	Tool nose radius compensation right	Modal
G50	Workpiece coordinate setting/maximum spindle RPM setting	Modal
G70	Inch programming	Modal
G75	Grooving cycle	
G76	Threading cycle	
G90	Absolute coordinate positioning	Modal
G91	Incremental positioning	Modal
G92	Workpiece coordinate setting	Modal
G96	Constant surface footage	
G97	RPM input	Modal
G98	Feed rate per minute	Modal
G99	Feed rate per revolution	Modal

FIGURE A–4 *Common Turning Center G-codes.*

M Code	Function
M00	Program stop
M01	Optional program stop
M02	Program end
M13	Manual feed override off
M15	Set taper cutting mode
M21	PWB off
M22	PWB on
M23	Auto override off
M24	Auto override on
M30	Program end and reset
M31	Cut time display reset
M40	Discharge off
M41	Discharge on
M42	Wire feed off
M43	Water flow off
M44	Wire tension off
M50	Cut wire (AWF)
M60	Connect wire (AWF)
M70	Retrace start
M80	Discharge on
M81	EDM Power on
M82	Wire Feed on
M83	Water on
M84	Wire tension on
M96	Reverse cut in mirror image end
M97	Reverse cut in mirror image start
M98	Subprogram call
M99	Subprogram end

FIGURE A–5 *Wire EDM M-codes.*

G Code	Function
G00	Rapid traverse
G01	Linear interpolation
G02	Circular interpolation clockwise
G03	Circular interpolation counterclockwise
G04	Dwell
G20	Inch programming
G21	Metric programming
G40	Wire diameter compensation cancel
G41	Wire diameter compensation left
G42	Wire diameter compensation right
G48	Automatic corner rounding on
G49	Automatic corner rounding off
G50	Wire inclination angle cancel
G51	Wire inclination angle left
G52	Wire inclination angle right
G60	Same radii top and bottom when tapering
G61	Conical corner R
G70	Edge finding
G71	Circle center finding
G72	Finding the center of a groove
G75	Rapid positioning in relative coordinate system
G76	Rapid positioning to the face positioning point
G77	Rapid positioning to a specified point
G78	Positioning to the work corner point
G79	Calculation and setting the work inclination angle
G90	Absolute programming
G91	Incremental programming
G92	Coordinate system setting
G94	Constant feed by program
G95	Servo feed

FIGURE A–6 *Wire EDM G-codes.*

Glossary

···

A

A Axis   The A axis is a rotary axis around the X axis.

Absolute   This is a mode in which tool positions are programmed relative to a stationary zero point. The code for absolute positioning mode is normally G90.

Accuracy   Accuracy is the difference between actual machine position and the commanded (or programmed) position.

Adaptive Control   Adaptive control can be used to automatically adjust feeds and speeds based on the actual cutting conditions. Controls with adaptive control can also sense dull or broken tooling and change to new tooling automatically. It is still rare, but is sure to see increased acceptance and usage.

APT (Automatically Programmed Tools)   This is a programming language for CNC machines.

ASCII (American Standard Code for Information Interchange) This is a standard format for the exchange of data between systems. Typically seven bits are used. Each bit can be a 0 or a 1. This allows 128 different combinations. Eight-bit ASCII also exists. Eight bits allow 256 different characters to be represented. Each combination is used to represent one character. For example, 1000001 would represent the letter *a*. When you hit the letter a on the keyboard of your computer, the ASCII equivalent is sent to the microprocessor. Remember that computers can really only work with ones and zeros. We like to work with characters. ASCII is used to help with the conversion.

Auxiliary Functions   These are functions such as coolant on/off, spindle on/off, and pallet change.

Axis   An axis is one of the directions of motion of a machine. In the Cartesian system it is one of the perpendicular lines of the coordinate system. Machines are sometimes classified by the number of axes of motion.

B

B Axis   The B axis is a rotary axis with motion around the Y axis.

Background Editing   This is the ability to edit a program while another program is running. This helps increase productivity because an operator can be producing a part and writing the program for the next part.

Backlash   Backlash is inaccuracy resulting from play, or slop, between a screw thread and the nut. When the table direction is reversed, the table does not move until the slop is taken up in the opposite direction. Every time direction is reversed, backlash becomes a factor. That is why skilled machinists always work in one direction when machining parts that have to be very accurate.

Backlash Compensation   This value is added or subtracted every time a CNC machine reverses direction to compensate for backlash. The value can be changed in the software as the machine wears.

Ballscrew   A ballscrew is a special type of screw that is normally found on CNC axes. Ballscrews are used to convert the rotary motion of a motor to linear motion for an axis. Ballscrews are ground to very close tolerances and use ball bearings in the mating nut. The ball bearings circulate through the nut to reduce wear and friction. Ballscrews drastically reduce friction and increase accuracy.

Bit   Bits are used in the binary number system. Computers can work only in binary. In the binary system each digit is called a bit. A bit can be either a 1 or a 0.

Block   One line of a CNC program. A CNC control reads and executes one block of code at a time. Each block is terminated by a special character called an end-of-block character.

Block Number   This line number in a program is really for the operator's benefit.

Buffer   This is a temporary storage location for CNC program blocks.

Bug   A small problem in a program or system is normally hard to find. One small bug can create big problems and headaches.

Byte   As used in the binary number system, a byte is actually 8 bits. A byte can represent numbers from 0 to 255. Half of a byte is a nibble (4 bits).

C

C Axis   The C axis is a rotary axis with motion around the Z axis.

CAD (Computer-Aided Design)   This is the use of computer software to enter part geometry and specifications. The CAD software is used to help design and produce a blueprint. The CAD data can also be used to help develop a CNC program to produce the parts.

CAM (Computer-Aided Manufacturing)   This very broad term means different things to different people. To machinists it generally means the use of a computer and software to generate programs for a CNC machine. To others it means using computers to assist manufacturing in any fashion.

Canned Cycle   These are machine sequences that are built into CNC controls to make programming easier. One example would be a peck drill cycle. Without the peck drill cycle, a programmer would have to program each individual move. With the peck canned cycle, the programmer inputs only a few values. Typical canned cycles include drill cycles, tap cycles, boring cycles, threading cycles, and roughing cycles.

CAPP (Computer-Aided Part Programming)   This is the use of a computer to input part geometry, which is then converted (post-processed) to create a part program that will run a CNC machine.

Cartesian Coordinates   This system specifies part coordinates by specifying their XYZ coordinates.

Circular Interpolation   This is movement on an arc. On a CNC machine, a circular move is programmed with a G02 or G03 command. A G02 or G03 can be used to program any portion of an arc, from 0 to 360 degrees.

Closed Loop   A closed-loop system is one in which the actual machine position is fed back to the control by the use of a position transducer such as an encoder. This assures that the machine's programmed position is the same as its actual position.

Compact II   A programming language for generating CNC code. English-like commands are used to describe the part geometry, tool path, and machine functions. The compact II code is then post-processed to produce a CNC program that the machine understands.

Contouring This is the movement of two or more axes simultaneously to produce a curve.

Conversational Programming This is an English-like programming language. It is different for every brand of CNC control. Conversational systems use graphics and menus to make writing a program easier. The control prompts the programmer to input information such as the operation, material, and geometry and then generates the actual program automatically.

CPU (Central Processing Unit) This is the brain of the CNC control. It is one or more microprocessors that are used to program and operate the machine.

CRT (Cathode Ray Tube) This display is found on most CNC machines today. It is similar to a television display.

Cutter Compensation Cutter compensation is used to offset a tool. There are offsets for length and offsets for diameter.

Cutting Speed Every material has a desired cutting speed in surface feet per minute depending on the type of cutting tool used. The cutting speed is used to find the desired RPM.

D

Datum A datum is used as a reference for important dimensions. For example, if a hole's location is important in relation to the left and bottom side of the piece, the left and bottom sides would be datums.

Debugging This process finds bugs (small problems) in a program or system. See also Bug.

Diagnostics Diagnostics are provided on many of the newer machines to help the user troubleshoot and find problems relatively easily. Often the user can call the manufacturer's technician on the phone and be led through a troubleshooting sequence that will identify the most likely source of the problem.

Disk A disk is a storage device. The most common disks now are 3.5 inches. One disk can hold many complex programs.

DNC (Direct Numerical Control or Distributed Numerical Control)
Direct numerical control is less common than it once was. When CNC machines first came out, their memories were quite limited. In some cases

computers were used to download the programs a block at a time as the program ran at the machine. The term "DNC" now usually implies distributed numerical control, meaning that several CNCs may be connected to one computer. Programs are then downloaded to machines from the central computer.

Download   This is the process of sending a part program or machine parameters from a computer to a machine.

Dwell   Some canned cycles allow the machine to pause, or dwell, at the bottom of a sequence before it retracts. The dwell is programmable.

E

Edit   A program can be changed, or edited, on the machine control or off-line at a computer.

EIA Code   This standard was established for tape coding. The EIA (Electrical Industries Association) standard is RS-244-B. It is a seven-bit code system based on an eight-channel format. The other standard code system is ASCII (American Standard Code for Information Interchange).

Emergency Stop   These large red buttons on machines let an operator stop all movement immediately. On a CNC machine the emergency stop switch stops all axis motion and cutter rotation immediately. This is different than using the stop switch, which is more like a pause switch. The emergency switch is used for emergency conditions.

Encoder   This position sensor has three rings with slits that light passes through. These slits provide pulses that are used to generate how far an axis moves. The first two rings of slits have the same number of slits, usually between 500 and 2000. These two rings help determine direction of movement and distance of movement. The third ring generates only one pulse, which is used for homing the axis.

End-of-Block Character   This special character indicates the end of a line of CNC code. On a computer it is the "return" or "enter" key.

End-of-Program (EOP)   This is a code that tells the control the end of the program has been reached.

EOB (End-of-Block)   See End-of-Block Character

E-Stop   See Emergency Stop.

F

Feedback   This information is provided to the CNC control from sensors (transducers) on each machine axis. Position feedback is normally provided from an encoder on each axis, and velocity feedback is provided by a tachometer.

Feed Rate Override   This lets the operator change the feed rate that was programmed. While the part is running it is sometimes desirable to increase or decrease the feed rate. This is normally accomplished through a dial or push button.

Fixed Cycle   These cycles are provided to make programming easier for the programmer. They are also called canned cycles. See Canned Cycles.

Fixture Offsets   These offsets can be programmed to account for multiple fixtures or parts on the table. Only one program is needed. The program is offset to machine the other positions.

FPM   Feet per minute.

G

G-Code   Codes that are used to set the mode in which a CNC machine runs. G-codes can be used to specify linear or circular, absolute or incremental, rapid or feed, and so on.

Geometry Offsets   These are offsets used to compensate for different tools on a CNC lathe. The cutting tip of each tool is at a different position and the machine must know where they are. Geometry offsets are used by the machine to calculate the correct positions. One offset per tool is used. Offsets are normally set by touching the tool to a known diameter or with a tool setting arm, which enters the dimensions automatically.

H

Home   When a machine is first turned on, it doesn't know where it is. As the operator goes through a home sequence, the machine moves all axes to a known location. This is the position the machine remembers as its home position.

I

Incremental   This is a method of programming moves. If the incremental system is used, the programmer gives the distance and direction of the move. If absolute mode is used, the programmer gives the actual position to move to.

Interpolation   This the controlled movement of multiple axes along a programmed path. For example, if we needed to go to the video store we wouldn't care about the path taken to get there. We would care only that we arrived there. This would be point-to-point control. If we were running in a marathon, the path would be very important. This is like an interpolated move in a machine. If we were machining the outside shape of a part, the path would be very important. The machine control interpolates a path for the cutter to follow.

J

Jog   The jog function is used to move the axes of a machine. By holding the jog button, an axis may be moved in the positive or negative direction.

L

Linear Interpolation   The machine control calculates the moves and velocities required by each axis to execute a controlled path, in this case a line. See also Interpolation.

M

Machining Center   This is a CNC vertical milling machine with tool-changing capability.

Machine Reference (Zero) Point   This is the origin of the machine's axes coordinate system. They are fixed and cannot be changed.

Macro   This short program performs some task, usually a repetitive task.

Manual Programming This describes a person writing a word-address style program, called G-code programming. This is in contrast to conversational programming, which is more English-like.

M-Code These codes are used to give commands to a CNC machine. They are also called miscellaneous functions. They are used for functions such as M08-coolant on, M03-spindle on clockwise, and M06-tool change.

MDI (Manual Data Input) In this method, the operator can input G-codes into the control and run them.

Microprocessor This is the brain of the CNC controller. A CNC controller will have one or more microprocessors to control the machine. The microprocessors are the same as those found in microcomputers.

Mirror Image Mirror image can be used to invert the axis of the machine. For example, imagine that we have a program to make your right hand, but we need to make a left hand. We just invert the X axis moves (mirror image about the X axis). It is possible to mirror the Y axis also, or both axes at the same time. This is accomplished automatically through codes in the program.

Miscellaneous Function This code controls operations such as spindle direction, spindle on/off, coolant, pallet changes, and tool changes.

Modal A command (code) will stay in effect until another code cancels or changes the mode. For example, a G01 (linear feed) will stay in effect until a G00 (rapid), G02 (clockwise circular), or G03 (counterclockwise circular) is issued to the control.

N

NC (Numerical Control) This is the use of numbers to control the movements and functions of a machine tool. In numerical control codes and numbers are used to tell the machine what to do.

Numerical Control See NC.

O

Off-line Programming This programming is done away from the machine, normally on a microcomputer. One advantage is that the machine can continue to operate while the next part program is being written. Also, it allows a manufacturer to keep skilled operators running machines while a programmer works on a computer generating part programs. Another advantage is that off-line systems allow programmers to use part information that may already exist in the computer. For example, if an engineer designed the part on a CAD system, the part geometry already exists in the computer. The programmer can use that part geometry to generate a part program for the machine.

Offset   Offsets are used to change the path of a tool to compensate for wear or tool geometry.

Open Loop   This system does not have feedback. The term is normally used when talking about position control. Some of the small, educational-type machines are open loop. It basically means that if the machine control tells the machine to move 10 inches in the +X direction, it assumes the machine actually made the move. Because an open-loop machine has no feedback, it doesn't know if its commands were actually performed.

Optional Stop   This code can be used in a program. When the program reaches this command, the machine will stop until the operator pushes the start button again.

P

Parity   This bit is attached to the information that is sent using RS-232 asynchronous communications. It is used for error checking. It can be set in to be odd, even, none, mark (1), or space (0).

Polar Coordinates   Polar coordinates are another method of specifying geometry of parts, normally by specifying an angle and a length.

Preset Tools   Preset tooling is a way to increase production. In some shops the tooling for each job will be set to length ahead of time. The operator just needs to put new tools in and make minor offsets to run the job.

Program Reference (Zero) Points   These points are chosen for convenience by the programmer. In many cases the programmer will choose either the lower left corner as X0, Y0, Z0, the middle of the part, or some feature of the part for a reference.

Program Stop   This is a stop button provided on CNC controls to allow the operator to "pause" the machine operation for non-emergency stops and starts. An E-stop is used to stop the machine immediately in case of an emergency condition.

Prove Out   This is done to test a new program or a program that has been modified. The program is run the first time to check the machine moves and speeds and feeds to be sure the program is safe, efficient, and accurate.

R

Rapid Traverse   This is the rapid movement of an axis.

Repeatability   This is a measure of how closely a machine can repeatedly come back to the same position.

Reset   This key is normally used to reset the control after an error has occurred.

Resolver   This analog device is used on some machines as a position feedback device.

Retrofit   This conversion of a manual machine to computer control normally involves replacing the lead screws with ballscrews and adding motors, drives, and a machine control.

RS-232-C   In this standard for serial communications, each character is sent as its ASCII equivalent one bit at a time. RS-232 has a cable length limitation of about 50 feet. If a cable is longer than 50 feet, there can be occasional transmission problems.

RS-244   This standard has been established for odd-parity punched tape. The standard specifies the width and thickness of the tape and also the size, spacing, and location of the punched holes.

RS-274-D   This standard set of codes was developed for controlling CNC machines.

RS-358   This standard for even-parity punched tape specifies tape characteristics such as thickness, width, location, spacing, and size of the punched holes.

S

Sequence Number   This number is used to identify line numbers in a program, primarily for the programmer's benefit. Sequence numbers begin with an "N" followed by a number. Many programmers increment each line number by 5 or 10 in case lines need to be inserted later.

Skim Cut   This is also called a finish cut. A small amount of material is left for the final pass to achieve more accuracy and a finer finish.

Source Code   This program can be written and understood by a person before it is converted to a machine language program that a microprocessor can understand.

Storage   This memory is where programs and data are stored. One of the earlier storage media was paper tape. Now most storage media consist of floppy and hard disks or tapes.

Subroutine   This is a small program that is called by a main program, usually several times. The intent of the subroutine is to reduce the length of the main program. For example, if there were 25 holes to be center-drilled, drilled, and then bored, their positions would be put in a subroutine. The main program would call the subprogram three times for the hole positions. The tool changes and offsets would be located in the main program.

T

Tab Sequential   This very rigid format was one of the first formats used for programming NC machines. This format used a five-word block that included sequence number, preparatory function, X dimension, Y dimension, and miscellaneous function. All five words had to be given for each block even if they were unchanged from the previous block. Because the order of each word was fixed there were no letters to identify each word. There were only numbers for each word separated by a tab code.

Tachometer   This feedback device is used for velocity or speed control. A tachometer puts out a voltage that changes with the speed of rotation. The faster it turns, the more voltage it generates. This signal is used by the motor drive to keep the speed constant. There is one tachometer for each axis the machine has.

Tape   Tape was once very prevalent as a storage device. Programs were developed and punched onto paper tape as a series of holes. The hole pattern determines the character. The tapes were then used to load the program into CNC machines. The earlier NC machines read the tape one "block" at a time.

Tool Offset   See Cutter Compensation.

Tool Path   This is the path that the center of a cutting tool takes as it machines a part. If using a single point tool like a turning tool for a lathe, it is the path that the cutting tip takes.

Tool-Nose Radius Compensation These codes are used on turning centers to adjust the tool path to compensate for the tool tip radius.

Tool Offset These codes can be used to adjust the cutting path for the tool. There are codes for left and right offset and also length compensation codes.

U

U Axis Some machines, such as wire EDM, have additional linear axes of motion beyond the X, Y, and Z axes. The U axis is parallel to the X axis.

Upload This term describes sending a part program or machine parameters from the machine up to a computer.

V

V Axis Some machines, such as wire EDM, have additional linear axes of motion beyond the X, Y, and Z axes. The V axis is parallel to the Y axis.

W

W Axis Some machines, such as wire EDM, have additional linear axes of motion beyond the X, Y, and Z axes. The W axis is parallel to the Z axis.

Word Address This format uses letters to identify each word in a program block. For example, the letter X precedes any X dimension, the letter G precedes all G-codes, and so on.

Work Reference (Zero) Points These are the origins of the part coordinate system. It is possible to have more than one work reference point on the table. This is useful for setting up multiple parts on the table at one time.

X

X Axis This is one of the axes perpendicular to the spindle. The X axis is normally the longest axis of motion on a milling machine. On a vertical mill, the X axis moves to the left and right as you face the machine. On a lathe, the cross feed is the X axis.

Y

Y Axis   This axis is perpendicular to the spindle axis. On a vertical mill the Y axis moves in and out.

Z

Z Axis   This axis of motion is parallel to the spindle. Toward the work is a negative Z move.

Index

....

A

Absolute programming, 27

Arbitrary shape, 176

Arc centerpoints, 254

Arc direction (G02, G03), 254

Arc endpoint, 254

Arc startpoint, 254

ASCII, 413–414

Attribute data, 433

Automated Programmed Tool (APT), 3

Average, 439

Average and range chart, 458

Axes, 22, 217, 219

B

Backlash, 18

Backlash compensation, 19

Ballscrews, 18

Bar feeders, 230

D

E

F

G

H

I

L

M

N

O

P

S